AF581623

Enhancing Mesenchymal Stem Cells (MSCs) for Therapeutic Purposes

Enhancing Mesenchymal Stem Cells (MSCs) for Therapeutic Purposes

Editors

Joni H. Ylostalo
Nisha C. Durand

MDPI • Basel • Beijing • Wuhan • Barcelona • Belgrade • Manchester • Tokyo • Cluj • Tianjin

Editors

Joni H. Ylostalo
Biology
University of Mary
Hardin-Baylor
Belton
United States

Nisha C. Durand
Human Cell Therapy Lab
Center for Regenerative
Medicine
Mayo Clinic Jacksonville
United States

Editorial Office
MDPI
St. Alban-Anlage 66
4052 Basel, Switzerland

This is a reprint of articles from the Special Issue published online in the open access journal *Cells* (ISSN 2073-4409) (available at: www.mdpi.com/journal/cells/special_issues/mesenchymal_stemcell).

For citation purposes, cite each article independently as indicated on the article page online and as indicated below:

LastName, A.A.; LastName, B.B.; LastName, C.C. Article Title. *Journal Name* **Year**, *Volume Number*, Page Range.

ISBN 978-3-0365-4996-5 (Hbk)
ISBN 978-3-0365-4995-8 (PDF)

Contents

About the Editors

Joni H. Ylostalo

Dr. Joni H. Ylostalo is an Associate Professor at the University of Mary Hardin-Baylor (UMHB). Dr. Ylostalo obtained his BSc in Biochemistry and MSc in Biochemistry, with a focus on Biotechnology and Molecular Biology, from the University of Oulu. His MSc research involved biochemical studies of type IX collagen. Dr. Ylostalo obtained his PhD in Biomedical Sciences from Tulane University, where he worked with mesenchymal stem cells (MSCs). His research was focused on the gene expression changes of MSCs during differentiation and culture expansion. Dr. Ylostalo completed his postdoctoral work at Texas A&M, studying the aggregation and 3D culture of MSCs to enhance their therapeutic potential. He continued his MSC research at Texas A&M as a faculty member, focusing on the therapeutic applications of 3D culture-activated MSCs. Since transitioning to UMHB, he has continued his MSC research among other research, teaching, and service opportunities. Dr. Ylostalo has published over 50 articles that span the fields of biochemistry, cell biology, bioinformatics, regenerative medicine, and biology education. He has served as an editor and reviewer for numerous journals and has given plenary talks at various conferences.

Nisha C. Durand

Dr. Nisha Durand is currently the Principal Research Technologist and Operations Manager at the Human Cellular Therapy Lab- Center for Regenerative Medicine, Mayo Clinic Florida. At the Center for Regenerative Medicine, Dr. Durand is responsible for translating research processes into the cGMP environment and developing processes and techniques in support of all phases of cellular product development. In addition, she is also actively involved in the regulatory process, with responsibility for generating pre-clinical safety and efficacy data. Dr. Durand received her PhD in Biochemistry and Molecular Biology from Mayo Clinic Graduate School of Biomedical Sciences in the spring of 2017. For her doctoral thesis, she examined the role of protein and lipid kinases in focal adhesion signaling and cell migration in the context of cancer metastasis. Her current research interests include Mesenchymal Stem/Stromal Cell (MSC) signaling and mechanisms of potentiation. Dr. Durand is an active member of ISCT, and currently serves on both the ISCT's Early-Stage Professional Committee and the Process and Product Development Committees.

Preface to "Enhancing Mesenchymal Stem Cells (MSCs) for Therapeutic Purposes"

The regenerative and immunomodulatory properties of mesenchymal stem cells (MSCs) have made these cells the focus of multiple pre-clinical studies and clinical trials. While the results from these clinical studies have established that MSCs are safe, the efficacy of these cells is not as well-established. In this regard, there have been increased efforts towards generating potentiated/activated MSCs with enhanced therapeutic efficacy. Research on the mechanisms for enhancing MSC potency and efficacy is an area of active study with great potential for translation into clinical settings. The purpose of this book is to bring together recent research from a broad range of topics relating to potentiation strategies for enhancing MSC therapeutic efficacy, including growth factor pre-conditioning, hypoxia, and 3D culture. The research compiled in this book increases the basic understanding of MSC culture techniques and describes some MSC preparations for potential novel therapeutic applications. This book should provide valuable information to scientists and clinicians studying MSCs and update anyone interested in transitioning MSCs to clinics on this area of study. This book was edited by Dr. Joni H. Ylostalo, an Associate Professor of Biology at the University of Mary Hardin-Baylor, and Dr. Nisha Durand, the Principal Research Technologist and Operations Manager at the Human Cellular Therapy Lab—Center for Regenerative Medicine at the Mayo Clinic Florida. We want to acknowledge the journal "*Cells*", MDPI, and specifically the Associate Publisher Dolly Wang for making this book possible.

Joni H. Ylostalo and Nisha C. Durand
Editors

MDPI

Article

Priming with a Combination of FGF2 and HGF Restores the Impaired Osteogenic Differentiation of Adipose-Derived Stem Cells

Jeong Seop Park [1], Doyoung Kim [1] and Hyun Sook Hong [1,2,3,*]

1 Department of Biomedical Science and Technology, Graduate School, Kyung Hee University, Seoul 02447, Korea; godjs@khu.ac.kr (J.S.P.); a0az515@khu.ac.kr (D.K.)
2 East-West Medical Research Institute, Kyung Hee University, Seoul 02447, Korea
3 Kyung Hee Institute of Regenerative Medicine (KIRM), Medical Science Research Institute, Kyung Hee University Medical Center, Seoul 02447, Korea
* Correspondence: hshong@khu.ac.kr; Tel.: +82-2-958-1828

Abstract: Classical aging-associated diseases include osteoporosis, diabetes, hypertension, and arthritis. Osteoporosis causes the bone to become brittle, increasing fracture risk. Among the various treatments for fractures, stem cell transplantation is currently in the spotlight. Poor paracrine/differentiation capacity, owing to donor age or clinical history, limits efficacy. Lower levels of fibroblast growth factor 2 (FGF2) and hepatocyte growth factor (HGF) are involved in cell repopulation, angiogenesis, and bone formation in the elderly ADSCs (ADSC-E) than in the young ADSCs (ADSC-Y). Here, we study the effect of FGF2/HGF priming on the osteogenic potential of ADSC-E, determined by calcium deposition in vitro and ectopic bone formation in vivo. Age-induced FGF2/HGF deficiency was confirmed in ADSCs, and their supplementation enhanced the osteogenic differentiation ability of ADSC-E. Priming with FGF2/HGF caused an early shift of expression of osteogenic markers, including Runt-related transcription factor 2 (Runx-2), osterix, and alkaline phosphatase (ALP) during osteogenic differentiation. FGF2/HGF priming also created an environment favorable to osteogenesis by facilitating the secretion of bone morphogenetic protein 2 (BMP-2) and vascular endothelial growth factor (VEGF). Bone tissue of ADSC-E origin was observed in mice transplanted with FGF/HGF-primed ADSC-E. Collectively, FGF2/HGF priming could enhance the bone-forming capacity in ADSC-E. Therefore, growth factor-mediated cellular priming can enhance ADSC differentiation in bone diseases and thus contributes to the increased efficacy in vivo.

Keywords: adipose-derived stem cell; paracrine potential; osteogenic differentiation; hepatocyte growth factor; fibroblast growth factor 2

Citation: Park, J.S.; Kim, D.; Hong, H.S. Priming with a Combination of FGF2 and HGF Restores the Impaired Osteogenic Differentiation of Adipose-Derived Stem Cells. *Cells* **2022**, *11*, 2042. https://doi.org/10.3390/cells11132042

Academic Editors: Joni H. Ylostalo and Nisha C. Durand

Received: 31 May 2022
Accepted: 26 June 2022
Published: 27 June 2022

1. Introduction

Fractures are one of the leading causes of disability and death, incurring enormous socio-economic costs [1]. In Korea, the 1-year cumulative mortality rate in patients aged ≥50 years after hip fractures was 16.0% (4547/28,426) [2]. Globally, more than nine million people suffer from bone fractures every year, and the incidence tends to increase with age [1–3]. Particularly, underlying conditions such as osteoporosis can aggravate the pathogenesis in skeletal tissue. Osteoporotic fractures were reported to elevate subsequent fracture risk [3–5]. Hence, therapeutic strategies to prevent fatality due to bone loss and fracture are undoubtedly required for an aging society. Clinical treatment of fractures employs surgical immobilization, ultrasound therapy, and bone grafts, but these are limited by their underwhelming regeneration capacity, risk of infection, and side effects [6–9]. Therefore, the development of novel therapies for bone regeneration is the need of the hour.

Currently, stem cell therapy has emerged as an attractive choice for bone regeneration [10]. Stem cells possess the ability of self-renewal with multi-differentiation and paracrine potential [11–13]. Plenty of clinical research has attempted to apply stem cell therapy to incurable diseases and has proved its therapeutic effect in vivo. The application of stem cells for bone disorders has been attempted by the transplantation of stem cells supplemented with scaffold or growth factors [13–15], confirming stem cell-mediated bone-forming capacity.

Adult stem cells available for transplantation reside in various tissues, including bone marrow, adipose tissue, umbilical cord blood, and dental tissues. Bone marrow stem cell (BMSC) is regarded as the primitive stem cell [15–17]. Notably, the osteogenic potential of BMSCs was attested in many studies [18–20], which motivated BMSC utilization to alleviate skeletal defects in the clinic. However, the efficacy of BMSCs is considerably influenced by disease conditions and the donor age [21,22]. BMSCs from aged patients showed low cellular activity and survival during ex vivo culture [23]; therefore, it is challenging to use the patient's BMSCs for autologous transplantation. Nowadays, an adipose-derived stem cell (ADSC) commands attention as an alternative to BMSCs. ADSCs are easily isolated from adipose tissue with low donor-site morbidity and are free from ethical controversy and immunogenic problems [24,25]. Importantly, the cellular activity of ADSCs is reported to be rarely altered by the disease condition compared to BMSCs [26]. These merits place ADSCs in the spotlight of pharmacological and clinical developments.

Various preclinical studies have induced ADSC differentiation into bone tissue [27–29]. Nevertheless, it should be noted that ADSC activity could be affected by age or disease, even though its impact is less than in BMSCs [30–33]. In practice, ADSCs from the elderly show relatively low differentiation potential, slow proliferation, insufficient paracrine ability, and rapid senescence compared to the young [30,33,34]. These data suggest that the autologous application of ADSCs the elderly requires a confident strategy to improve the cellular activity of ADSCs before transplantation. If not, transplanted ADSCs would have a diminished therapeutic impact with undesired effects in vivo. To enhance the therapeutic capability of ADSCs, several methods, such as pre-conditioning by cytokines, genetic manipulation, physical stimuli, extracellular vesicle, and cultures with three-dimensional aggregates have been employed so far [35–39]. However, the problems of biodegradability, anxiety about genetic manipulation, side effects, and unsatisfactory therapeutic impact are still significant challenges [40–42].

To improve the differentiation activity of ADSCs that is impaired by age or disease, reversal of the secretory deficiency is expected to restore its cellular activity fundamentally. Cellular differentiation occurs through the interplay of cytokine/growth factors in an autocrine/paracrine manner. For enhanced differentiation, ADSCs should mount an appropriate cellular response to extracellular stimuli to activate differentiation. However, aging/senescence impairs paracrine potential and receptor occupancy of ADSC, creating a deficiency of essential secretory factors [33,43]. Accordingly, the modulation of the secretory condition of ADSCs is surmised to be decisive for its differentiation potential.

Primary molecular regulators for osteogenesis are fibroblast growth factor 2 (FGF2), transforming growth factor beta-1 (TGF-β1), bone morphogenic protein-2 (BMP-2), and vascular endothelial growth factor (VEGF) [44–46] and they are spontaneously produced in ADSCs and activate specific signaling pathways upon osteogenic stimulation. Of these, FGF2 is a pleiotropic signaling molecule involved in angiogenesis, cell growth, and tissue repair [44,47]. Additionally, FGF2 was found to promote bone formation, accompanied by the enhanced secretion of VEGF [48] and BMP-2 [49]. Importantly, FGF-2 expression is reduced with aging in various tissues [50,51]. TGF-β1 and VEGF have also been reported to enhance bone formation, but their levels are relatively unaffected by aging [33,52,53].

The hepatocyte growth factor (HGF) is a versatile growth factor controlling organogenesis, tissue repair, and bone remodeling via phosphorylation of C-met [54]. HGF is reported to improve bone regeneration via the production of BMP-2 [55,56]. Moreover, HGF induces VEGF expression, aiding bone formation through its angiogenic properties [57]. The secretion of HGF is also reduced by cellular senescence, disease, or donor age [33].

Based on the insufficient secretion induced by age, minimal manipulation of ADSCs with growth factors is estimated to improve osteogenesis. In this study, ADSC-E and ADSC-Y were cultured ex vivo, and their levels of paracrine factors related to osteogenesis were quantitatively compared. Then, a method to restore the osteogenic potential of ADSC-E was established by supplementation of growth factors. The effect of growth factor-priming on ADSCs was determined by early osteogenic marker expression, calcium production in vitro, and bone-forming capacity in vivo.

2. Materials and Methods

2.1. Cell Culture

The elderly's adipose tissues were provided by the Kyung Hee University Medical Center [Seoul, Korea; (IRB# 2016-12-022, donor: 8, 2021-01-011, donor: 20)] with the written agreement of the donors. Adipose tissues immediately harvested from donors aged 50–70 were washed in PBS with 5% Penicillin/streptomycin (Welgene, Daegu, Korea). After washing, the tissues were enzymatically digested using 1% collagenase I for 1 h at 37 °C. The digestion was stopped by adding the same volume of FBS. After centrifugation, the stromal vascular fraction was filtered through a cell strainer (70 μm, Corning, NY, USA) to remove debris and centrifuged at 1500 rpm for 5 min at 4 °C to obtain ADSC pellets. Collected ADSCs were resuspended in α-MEM with 10% FBS and 1% penicillin and streptomycin. ADSCs from healthy young individuals were purchased from ScienCell Research Laboratories (Carlsbad, CA, age: 20–29 years) and cultured. All ADSCs were cultured in a 37 °C, 5% CO2 incubator, and the culture medium was changed every other day. During all experiments, ADSCs between passages 3–5 was used.

2.2. Osteogenic Induction and Growth Factor Treatment

The ADSCs were seeded into 6-well plates (5×10^4 cells/ well). When cell confluency was approximately 80–90%, the culture media was replaced with Stempro osteogenesis differentiation media (Gibco, Grand Island, NY, USA) and cultured for 20 days. During osteogenic induction, the ADSCs were primed with 1 or 5 ng/mL of FGF2 (R&D systems, Minneapolis, MN, USA) or 10 or 50 ng/mL HGF (R&D systems, Minneapolis, MN, USA) or their combination for 1, 3, or 6 days after osteogenic induction. After completion of priming of FGF2 and/or HGF, the ADSCs were maintained in osteogenesis differentiation media for 20 days. On the 20th day after osteogenic induction, cells were fixed with 3.7% formaldehyde (Sigma-Aldrich, ST. Louis, MO, USA) and stained with 2% Alizarin red S solution (Sigma-Aldrich, ST. Louis, MO, USA) for 10 min to visualize calcium deposition. Alizarin red S was eluted using 10% cetylpyridinium chloride solution (Sigma-Aldrich, ST. Louis, MO, USA), and calcium deposition was quantified as the absorbance value at a wavelength of 560 nm (Molecular Devices, Sunnyvale, CA, USA).

2.3. Western Blot

ADSCs at 0, 1, 3, and 6 days after osteogenic induction were washed with PBS and lysed with 1X lysis buffer (Cell Signaling Technology, Danvers, MA, USA). The supernatants were collected by centrifugation at 12,000 rpm for 20 min at 4 °C. The protein concentration was determined by the bicinchoninic acid (BCA) assay (Thermo Fisher, Rockford, IL, USA). The lysates were denatured and electrophoresed using SDS-PAGE and transferred to a nitrocellulose membrane. After blocking with 5% skim milk, the membranes were incubated with the primary antibodies for C-Met and P-Met, FGF2, Runx-2 (Cell Signaling Technology, Danvers, MA, USA), fibroblast growth factor receptor 2 (FGFR2), Osterix, alkaline phosphatase (ALP), and glyceraldehyde 3-phosphate dehydrogenase (GAPDH) (Abcam, Cambridge, UK), followed by anti-IgG horseradish peroxidase-conjugated secondary antibody (Bio-rad, Hercules, CA, USA). The blots were developed by adding ECL (Dogen Bio, Seoul, Korea); chemiluminescence was visualized with an Amersham imager 600 (GE Healthcare, Buckinghamshire, UK). The expression level was quantified using the ImageJ program (Version 1.53e, National Institutes of Health, Bethesda, MD, USA).

2.4. Animals Model

Six-week-old Balb/c nude mice (20–22 g, male) were purchased from Daehan Bio Link (Seoul, Korea). The mice were housed under a 12 h light/dark illumination cycle in an experimental animal room and permitted to adapt for 7 d before the commencement of the experiments. All animals received standard chow and water ad libitum, and this study was approved by the Ethical Committees for Experimental Animals of Kyung Hee University Hospital with the approval number KHMC-IACUC-20-008-01, 02.

2.5. Ectopic Bone Formation and Histological Analysis

The ADSCs were primed for 3 and 6 days under osteogenic induction with FGF2, HGF, and FGF2/HGF, respectively, and then, 2×10^6 ADSCs were mixed with 40 mg of hydroxyapatite/beta-tricalcium phosphate (HA/β-TCP) ceramic powder (Biomatlante, Vigneux-de-Bretagne, France). The ADSC-HA/β-TCP mixture was incubated at 37 °C for 2 h and implanted subcutaneously into the dorsal region of Balb/c nude mice. 12 weeks later, the implants were harvested and fixed in 3.7% formaldehyde. The samples were decalcified in 0.2 M EDTA (PH 7.2~7.4) for 2 weeks and embedded in paraffin.

The paraffin-embedded sample was sectioned to a 5-μm thickness. After deparaffinization and hydration, hematoxylin and eosin (H&E) staining was performed. To detect the transplanted human ADSCs, samples were treated with an antibody for human osteocalcin and incubated with a biotin-conjugated secondary antibody. Enzyme-substrate reaction was carried out with ABC reagent solution. The stained area was visualized with Nova RED (Vector Laboratories, Burlingame, CA, USA), and counterstaining was completed with hematoxylin.

2.6. Statistical Analysis

All data are presented as the mean ± standard deviation. Statistical analyses were performed using GraphPad Prism (GraphPad Software, version 5.01, San Diego, CA, USA). Differences were considered statistically significant at $p < 0.05$ and were interpreted as follows: * $p < 0.05$, ** $p < 0.01$, *** $p < 0.001$. Statistical analysis was performed using an unpaired two-tailed Student's *t*-test.

3. Results

3.1. ADSCs with Weak Osteogenic Potential Is Deficient in Paracrine Factors under Normal Physiological Conditions

For comparative analysis of the cellular activity of young ADSCs (ADSC-Y) and elderly ADSCs (ADSC-E), ADSCs were isolated and cultured from the young and elderly, respectively. Differences in cellular morphology (Figure 1A) and surface maker expression (Table S1) between ADSC-Y and ADSC-E were rarely observed. However, ADSC-Y had a doubling time of approximately 30 h, whereas ADSC-E had doubled in 50 h (Figure 1B). ADSCs were cultured in osteoinductive media for 20 days to evaluate the osteogenic potential, after which alizarin Red S staining was performed to examine calcium deposition (Figure 1C,D). ADSC-Y could differentiate into osteoblasts with high calcium deposition, whereas ADSC-E seldom differentiated into osteoblasts under osteoinductive conditions (Figure 1D,E). This result is consistent with previous studies [33].

Osteogenic differentiation progresses via the activation of Runx-2, a major transcriptional regulator of early osteogenesis. Runx-2 expression was lower in ADSC-E than ADSC-Y during the early phase of osteogenic induction (Figure 1F,G). A higher level of ALP expression was also detected in ADSC-Y (Figure 1F–H). This phenomenon might be related to the loss of osteogenic potential of ADSC-E.

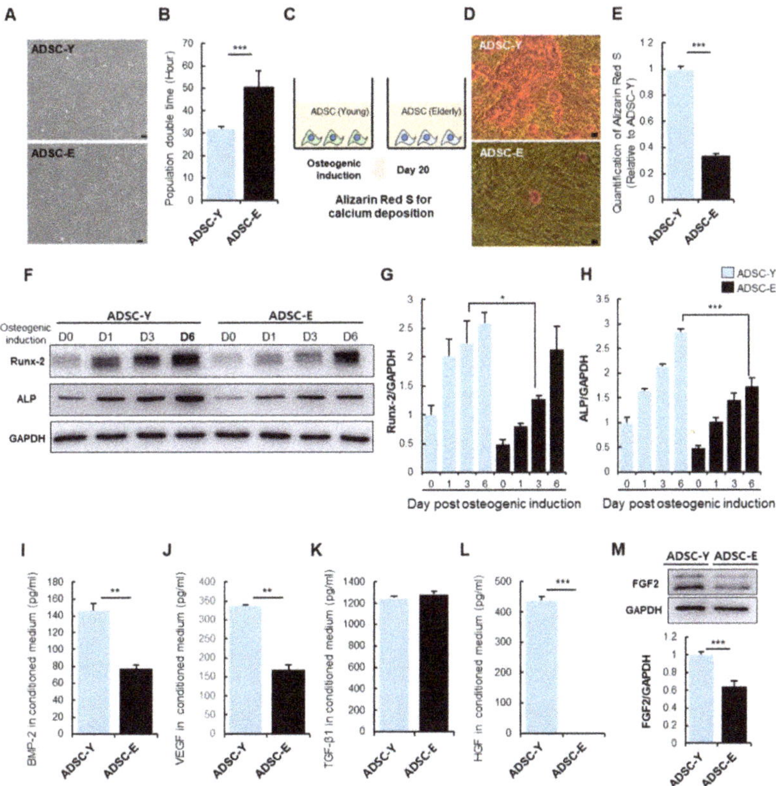

Figure 1. Effect of age on the osteogenic/paracrine potential of ADSCs.(**A**) Comparison of cellular morphology of ADSC-Y and ADSC-E. Scale bar, 100 µm. (**B**) Population doubling time was analyzed. (**C**) Experimental scheme for comparative analysis of osteogenesis in ADSC-Y and ADSC-E. (**D**) Representative images for Alizarin Red S staining of ADSC-Y and ADSC-E after osteogenic induction for 20 days. Scale bar, 100 µm. (**E**) Alizarin Red was quantified with 10% (*w/v*) cetylpyridinium chloride. (**F**) ADSC was lysed at 0, 1, 3, and 6 days after osteogenic induction to analyze osteogenic markers. (**G**,**H**) Expression of Runx-2 and ALP was confirmed by western blot. (**I**–**L**) The level of BMP-2, VEGF, TGF-β1, and HGF in the conditioned medium of ADSC-Y and ADSC-E was quantified by ELISA. (**M**) Protein level of FGF-2 was determined by western blot and quantified by ImageJ. Results are shown as the mean ± SD of three replicate wells for each group. * $p < 0.05$, ** $p < 0.01$, and *** $p < 0.001$.

Runx-2 activation involves signaling molecules such as TGF-β, FGF, and BMP-2 [46,47,49]. Next, to check the paracrine potential of ADSC-Y and ADSC-E, the secretion of osteogenesis-enhancing growth factors, including BMP-2, VEGF, TGF-β1, and HGF, was evaluated. The levels of BMP-2 and VEGF were significantly higher in ADSC-Y than in ADSC-E (Figure 1I,J). Unexpectedly, the production of TGF-β1 in ADSCs was not affected by age and osteogenic capacity (Figure 1K). The difference in HGF secretion was considerable between ADSC-Y and ADSC-E, suggesting that secretion of HGF may be deeply related to the osteogenic potential of ADSCs (Figure 1L). The levels of FGF-2, a representative factor that promotes bone formation, has been reported to decrease in the skin and muscle with age [50,51], which is also observed in ADSCs. ADSC-E had a low expression level of FGF2 compared to ADSC-Y (Figure 1M).

This result suggests that age affects cell repopulation rate and differentiation potential, which may be attributed to the alteration of paracrine factors.

3.2. ADSCs with the Deficient Osteogenic Potential Shows Impaired Paracrine Action in Response to Osteogenic Stimulus

To ascertain the relation between osteogenic potential and paracrine factors, the kinetics of osteogenesis-related paracrine factors during osteogenic induction was examined in ADSC-Y and ADSC-E, respectively.

As shown in Figure 1I, osteogenic stimuli elevated BMP-2 concentration in ADSC-E, similar to that of ADSC-Y (Figure 2A). Therefore, osteogenic stimuli could resolve the deficiency of BMP-2 in ADSC-E. The level of TGF-β1 was sustained in a similar pattern in both ADSC-Y and ADSC-E in osteogenic conditions (Figure 2B). VEGF secretion was gradually elevated in ADSC-Y, while its level almost remained unchanged in ADSC-E for six days (Figure 2C). HGF levels were constantly elevated post osteogenic induction, but HGF in ADSC-E was too low to be detected (Figure 2D). This phenomenon was repeatedly observed by cytokine array (Figure S1). HGF binds to the receptor c-Met and autophosphorylates it to transduce various signaling pathways (5). As predicted, ADSC-Y with active HGF secretion showed higher phosphorylation levels of C-Met, compared to ADSC-E (Figure 2E–G). The comparative analysis revealed that FGF2 and FGF2R expression is maintained higher in ADSC-Y than in ADSC-E, indicating a possible dynamic response of FGF2 signaling in ADSC-Y rather than ADSC-E (Figure 2H–J).

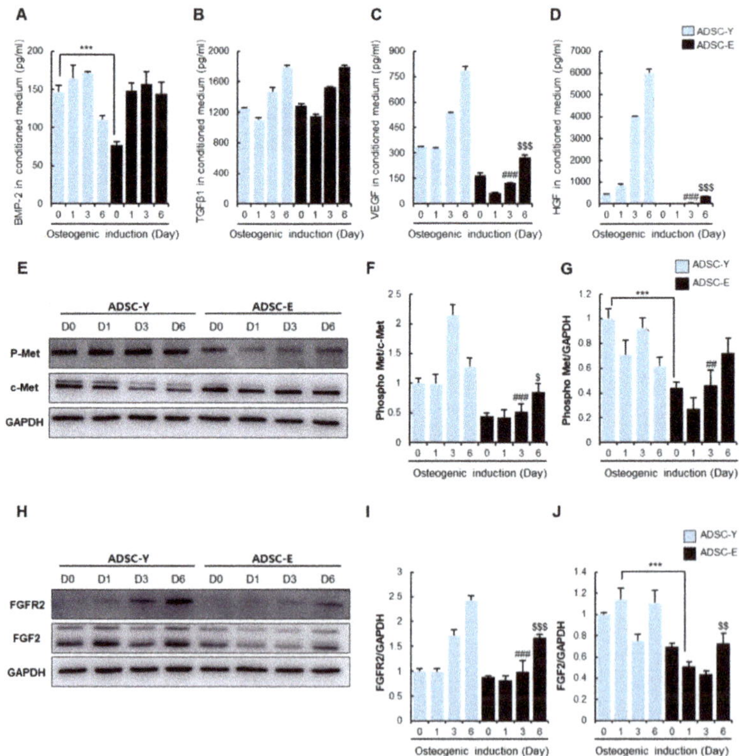

Figure 2. The analysis of the expression pattern of osteogenic factors in ADSC-Y and ADSC-E under osteogenic induction. ADSC-Y and ADSC-E were cultured in osteogenic media, respectively, and its

conditioned medium was collected at 0, 1, 3, and 6 days after osteogenic induction. (**A–D**) Levels of BMP-2, TGF-β1, VEGF, and HGF levels in a conditioned medium were examined by ELISA. (**E–G**) P-Met and C-Met expression levels in ADSCs were analyzed by western blot and quantified relatively to C-Met or GAPDH. (**H–J**) FGFR2 and FGF2 proteins in ADSCs were determined by western blotting and quantified relatively. Results are shown as the mean ± SD of three replicate wells for each group. *** $p < 0.001$, ## $p < 0.01$, ### $p < 0.001$ vs. Day 3 of ADSC-Y and $ $p < 0.05$, $$ $p < 0.01$, $$$ $p < 0.001$ vs. Day 6 of ADSC-Y.

Considering the difference in paracrine potentials from ADSC-Y and ADSC-E during the initial stage of osteoinduction, BMP-2 or TGF-β is not anticipated to be directly related to the loss of osteogenic potential of ADSC-E because of the non-significant difference between ADSC-Y and ADSC-E. It can be surmised that the scarcity of HGF, FGF2, or VEGF directly impinges on the impaired osteogenic potential of ADSC-E.

3.3. FGF2/HGF Priming Promotes Osteogenic Differentiation of ADSCs

This study discovered that VEGF, HGF, and FGF2 are insufficient in ADSCs with low osteogenic activity. To improve the differentiation ability of ADSC-E, the deficient growth factor would need to be restored in ADSC-E. Thus, VEGF, HGF, and FGF2 were considered potential candidates to enhance the osteogenic activity of ADSC-E (Figure 2).

FGF2 and HGF can elevate BMP-2 and VEGF in diverse cell types [48,55–57], and thus, FGF2 or HGF treatment was anticipated to create a VEGF-enriched condition in ADSC-E. It was ultimately examined whether supplementation of FGF2 and/or HGF in ADSC-E can restore osteogenic potential or not.

Modulating the early osteogenic protein expression, including Runx-2 and ALP, is expected to be decisive for osteogenic potential. However, the earliest time window to promote osteogenic protein expression during osteogenic induction is unknown and the development of stem cells to pre-osteoblast with increased Runx-2 was merely known to take 6–7 days in vitro [58].

To clarify the optimal time to stimulate the differentiation of ADSC-E, ADSC-E was treated with FGF2 and/or HGF (FGF2: 1 or 5 ng/mL; HGF: 10 or 50 ng/mL) for 3 or 6 days after osteogenic induction. Thereafter, ADSCs were maintained in osteogenic condition without FGF2 and/or HGF until 20 days (Figure 3A).

FGF2 or HGF priming for six days distinctly enhanced the osteogenic potential of ADSC-E (Figure 3B). Notably, the effect of FGF-2 was rarely observed in ADSC-Y, but a prominent effect was shown in ADSC-E (Figure S2), indicating the significance of FGF2 supplementation for osteogenic potential. FGF2/HGF combination could considerably recover the osteogenic activity of ADSC-E, similar to ADSC-Y (Figures 3 and S3). To determine the precise effect of each condition, Alizarin Red S corresponding to calcium deposition was quantified. A remarkable improvement was observed when primed with a combination of FGF2 and HGF, rather than FGF2 or HGF alone (Figure 3C). The effect of FGF2 or HGF was also determined to be the most effective in 5 ng/mL FGF2 and 50 ng/mL HGF. This dose was used for further experiments.

Treatment with FGF2 and/or HGF for three days post osteogenic induction was not enough to restore ADSC-E's osteogenic activity despite the obvious effect of FGF2 and/or HGF (Figure S4).

These results demonstrated that the initial supplementation of FGF2 and/or HGF for six days could enhance the osteogenic capacity of ADSC-E.

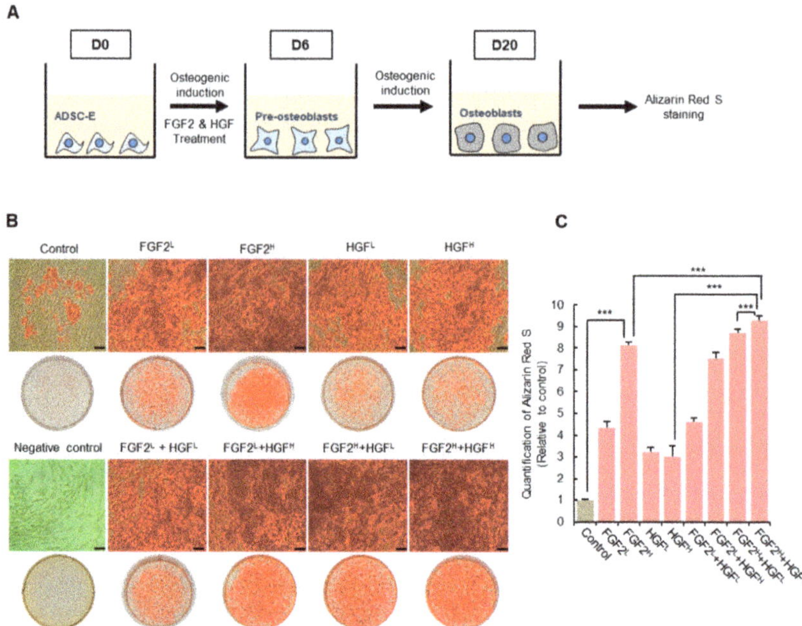

Figure 3. FGF-2/HGF priming enhances the osteogenic potential of ADSC-E. (**A**) Experimental schedule to treat ADSC-E with FGF2 and/or HGF during the osteo-inductive condition. (**B**,**C**) Representative images of Alizarin Red S staining of ADSC-E in each condition and the quantification of calcium deposition. Scale bar, 100 μm. Results are shown as the mean ± SD of four replicate wells for each group. *** $p < 0.001$. $FGF2^L$: 1 ng/mL, $FGF2^H$: 5 ng/mL, HGF^L: 10 ng/mL, HGF^H: 50 ng/mL.

3.4. FGF2/HGF Priming-Mediated Osteogenic Improvement Occurs by Modulation of Early Osteogenic Gene Expression

Stimulation of ADSC-E with FGF2 and/or HGF for six days suitably enhanced osteogenic differentiation of ADSC-E, suggesting the importance of the cellular event that occurred for six days in the presence of FGF2 and/or HGF. Next, osteogenic induction was applied to ADSC-E, and the expression pattern of the representative osteogenic proteins was monitored at 1, 3, and 6 days after osteogenic induction (Figure 4A).

Under the osteogenic condition, the expression of FGFR2 increased in a time-dependent manner, which was pronounced in FGF2 or a combination of FGF2/HGF (Figure 4B,C). Runx-2 showed a time-dependent increase in the non-treated control group, consistent with previous reports [58]. However, FGF2 and/or HGF treatment facilitated Runx-2 expression and shifted its peak to day three. The expression level of Runx-2 was highest when in a combination of FGF2/HGF treatment (Figure 4B,D). Osterix and ALP expressions were slightly changed by FGF2 and/or HGF compared to the control (Figure 4B,E,F). Based on the protein expression profile, FGF2 and/or HGF were predicted to make ADSC-E enter the phase of immature pre-osteoblast from ADSC-E earlier compared to non-treated ADSC-E control.

Among secretory factors with differing basal levels in ADSC-E and ADSC-Y, BMP-2 and VEGF were evaluated. Compared to the control, BMP-2 was sustained at a higher level in the presence of FGF2 and HGF. FGF2 or HGF treatment could not make a significant difference among groups. The combination of FGF2/HGF showed the highest concentration from day one post-treatment (Figure 4G). VEGF secretion was elevated in osteoinduction, and its concentration was exceedingly not affected by FGF2 or HGF priming. However, a combination of FGF2/HGF unambiguously enriched VEGF in ADSC-E (Figure 4H).

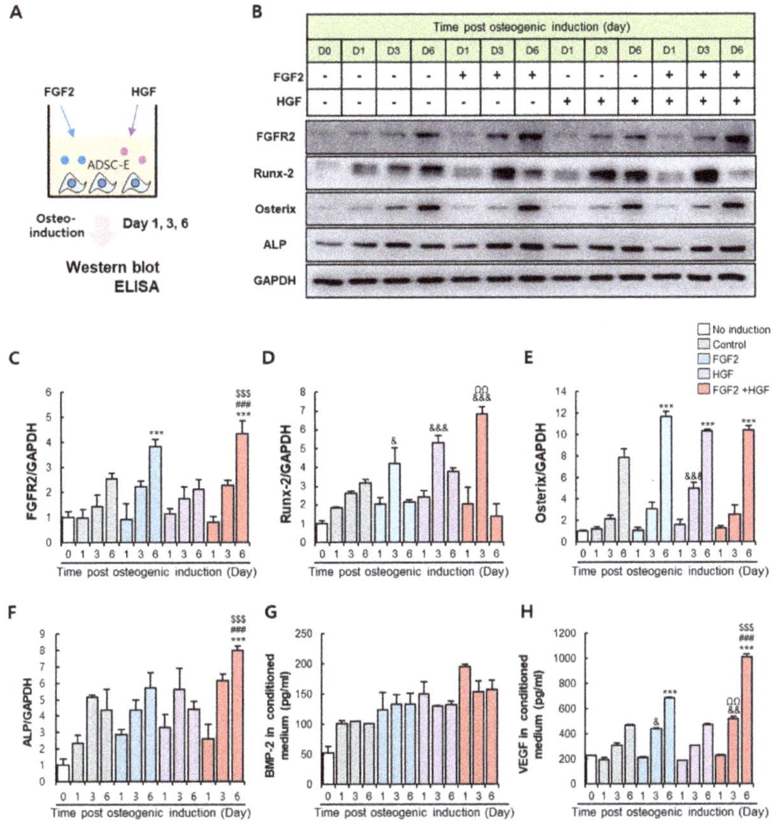

Figure 4. FGF-2 and HGF priming regulate the expression of early osteogenic markers in ADSC-E under osteogenic conditions. (**A**) Experimental scheme for analyzing the effect of FGF-2 and/or HGF priming on osteogenic markers on ADSC-E. (**B–F**) FGFR2, Runx-2, Osterix, and ALP levels in ADSC-E were examined by western blot. (**G**,**H**) Quantification of BMP-2 and VEGF in a conditioned medium was performed by ELISA. Results are shown as the mean ± SD of five replicate wells for each group. & $p < 0.05$, && $p < 0.01$, &&& $p < 0.001$ vs. Day 3 of control, ΩΩ $p < 0.01$ vs. Day 3 of HGF-primed ADSC-E, *** $p < 0.001$ vs. Day 6 of control, ### $p < 0.001$ vs. Day 6 of FGF2-primed ADSC-E, $$$ $p < 0.001$ vs. Day 6 of HGF-primed ADSC-E.

FGF2/HGF priming for six days under osteogenic induction promoted osteogenic protein expression, accompanied by osteogenesis-favored paracrine condition. This environment is expected to provide the cellular environment to improve the osteogenic potential in ADSC-E during an early phase of the osteogenic induction.

3.5. FGF2/HGF Priming Enhances the Bone-Forming Capacity of ADSCs In Vivo

To evaluate the osteogenic capacity of FGF2 and/or HGF-primed ADSCs in vivo, ADSC-E was cultured in the presence of FGF2 and/or HGF in osteogenic media. Three or six days later, ADSC-E was mixed with HA/β-TCP and then transplanted into nude mice to allow in vivo differentiation (Figure 5A). H&E staining showed the formation of osteoid and newly formed bone. FGF2 and/or HGF priming could enhance bone-forming capacity compared to non-treated control, and the combination of FGF2/HGF was superior to FGF2- or HGF-primed cells for both day 3 and day 6. The most osteogenic ability was observed when primed with a combination of FGF2/HGF for 6 days (Figure 5B,C).

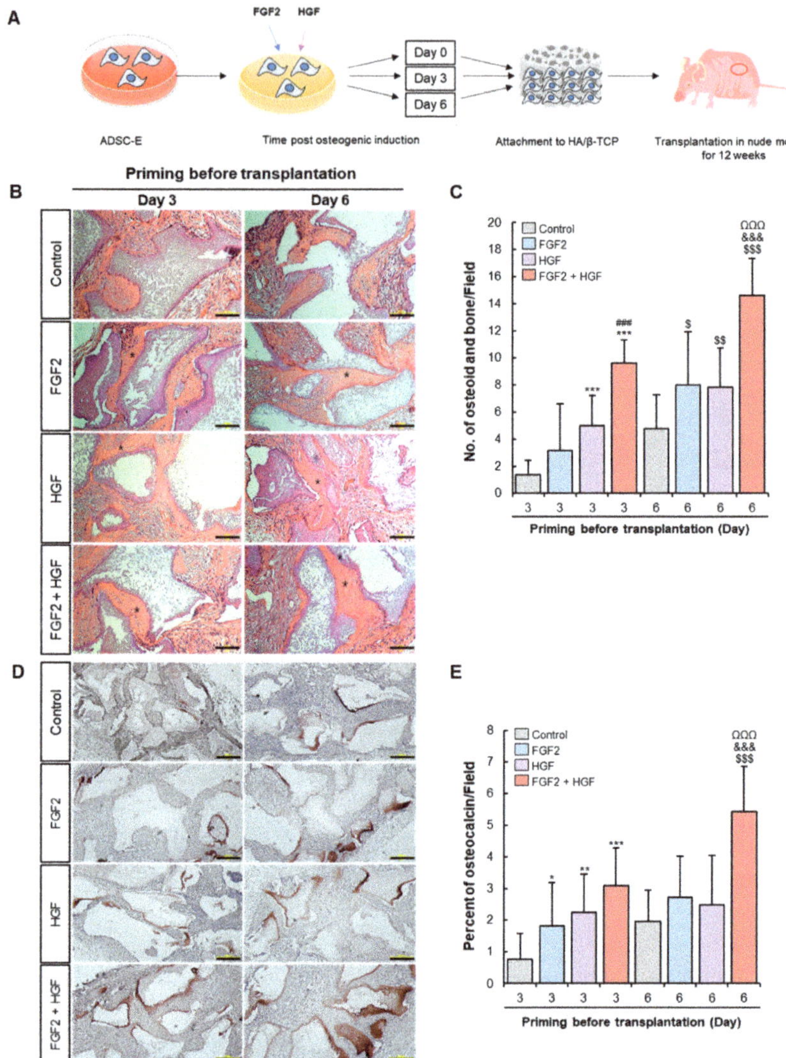

Figure 5. FGF-2 and HGF priming improve the bone forming capacity of ADSC-E in vivo. (**A**) Experimental design for cell transplantation of FGF-2 and/or HGF-primed ADSC-E. (**B**,**C**) H&E staining for the complex of transplanted cells and bone particles. Scale bar, 200 μm. (**D**,**E**) Human osteocalcin stained immunohistochemically was quantified with ImageJ. Scale bar, 200 μm. Results are shown as the mean ± SD of three replicate wells for each group. * $p < 0.05$, ** $p < 0.01$, *** $p < 0.001$ vs. Day 3 of control, ### $p < 0.001$ vs. Day 3 of HGF-primed ADSC-E, $ $p < 0.05$, $$ $p < 0.01$, $$$ $p < 0.001$ vs. Day 6 of control, &&& $p < 0.001$ vs. Day 6 of HGF-primed ADSC-E and ΩΩΩ $p < 0.001$ vs Day 3 of FGF2/HGF-primed ADSC-E.

Osteocalcin is synthesized by osteoblast and is a major bone formation marker. To track bone tissue of human origin, human-specific osteocalcin expression was determined by immunohistochemistry. A small area positive for osteocalcin was observed in the control group; its level increased upon FGF2 and/or HGF priming (Figure 5D). Quantification of the osteocalcin-stained area proves that a considerable expression of osteocalcin is observed upon priming with a combination of FGF2/HGF for 6 days only. That is, the combination of FGF2/HGF works best for ADSC-E and could facilitate the differentiation into osteoblast in vivo.

Collectively, these results corroborate that early priming of combination of FGF2/HGF is required for the improvement of osteogenic potential of ADSC-E in vivo.

4. Discussion

The need for novel treatments for bone regeneration of the elderly has become a major challenge in the clinical field. To address this issue, tissue engineering with stem cells has been attempted, where ADSCs or BMSCs are mainly employed as the reparative cellular source. The therapeutic ability of BMSCs is affected by donor disease or age. BMSCs show a low paracrine/differentiation potential in cases of the elderly with the disease [23,59]. In contrast to BMSCs, the application of ADSCs has advantages, including similar features to BMSCs, easy access to the tissue, and rapid proliferation rate in vitro. Thus, many clinical/non-clinical studies conducted with ADSCs prove the efficacy of ADSCs in critical diseases such as bone, skin, or cartilage defects. However, aged patients primarily need stem cell transplantation: age substantially affects ADSC activity, leading to increased doubling time, insufficient paracrine factors, and decreased differentiation potential. Thus, the functional restoration of ADSCs from the aged was expected to be required before the transplantation, and the optimized strategy to improve ADSC activity would contribute to the better efficacy in vivo.

In this study, the osteogenic potential was comparatively evaluated in ADSC-Y and ADSC-E, showing that ADSC-E has poor osteogenic differentiation potential. Therefore, ADSCs from elderly patients may have low bone-forming capacity when applied autologously. The differentiation process occurs via the binding of soluble growth factors on its receptor, emphasizing the need for sufficient growth factors and their receptors in ADSCs for the differentiation process. In other words, the paracrine potential will have implications on the extent of osteogenic differentiation.

Many growth factors/cytokines that promote osteogenesis have been reported in previous studies. Comparing the generation of soluble factors relating to osteogenesis in ADSC-Y and ADSC-E revealed that FGF2, HGF, and VEGF have a high correlation with age and osteogenic potential, not BMP-2 and TGF-β1. We gleaned that FGF2, HGF, and VEGF could be supplementary factors for ADSC-E. Previous research investigated the interactive action of FGF2, HGF, and VEGF in diverse cell types and ascertain that FGF2 or HGF could promote VEGF secretion [48,57]. Thus, this study has employed FGF2 and HGF as supplementary factors to improve the osteogenic activity of ADSC-E.

Under an osteogenic stimulus, the expression of FGF2 and HGF was extremely low in ADSC-E compared to ADSC-Y. The FGF2 receptor expression was rarely detected in ADSC-E during the osteogenic process. ADSC-Y could have active cellular signaling for FGF2 or HGF, whereas ADSC-E had an inadequate signaling response due to the lack of ligands and receptor proteins. Indeed, the levels of Runx-2, activated by FGF2 or HGF, were rapidly elevated upon osteogenic stimulus in ADSC-Y, but osteogenic growth was slowed down in ADSC-E. This difference in Runx-2 expression is related to the osteogenic potential. This result hints that the supplementation of FGF2 and HGF could be a clue to the weak differentiation potential of ADSC-E, possibly by creating osteogenesis-favored intracellular conditions.

To examine the enhanced osteogenic potential of ADSC-E by FGF-2 and/or HGF priming, ADSC-E was stimulated with FGF2 and/or HGF under osteogenic conditions for 6 days and then cultured for 14 days in osteogenic media. Early enrichment of FGF2 and/or HGF in ADSC-E enhanced osteogenic potential with sufficient calcium deposition. Notably, the effect of a combination of FGF2 and HGF was much better than FGF2 or HGF treatment. Priming ADSC-E with FGF2 and/or HGF for three days facilitated osteogenesis, compared to non-priming control, but its effect was less than for six days. FGF2 and/or HGF priming provoked the early shift of Runx-2 expression in ADSC-E, which might promote the osteogenic process. Additionally, the combination of FGF2 and HGF could increase the protein level of the FGF2 receptor, causing cellular status with dynamic signaling for exogenously added FGF2 and accelerating the secretion of VEGF in ADSC-E. These multiplicative effects of FGF2 and HGF are assumed to contribute to improving the osteogenic potential of ADSC-E.

Despite the apparent effect of FGF2 and/or HGF priming on differentiation of ADSC-E in vitro, transplantation time for osteogenesis in vivo was not determined because the success of bone formation in vivo depends on the commitment of transplanted cells. Additionally, in vitro and in vivo conditions are different. The criteria to determine the ideal cellular condition for transplantation are lubricous, and its result varies depending on the research. ADSC-E was surmised to fail bone formation due to the lack of osteogenic commitment signal, and mature osteoblast is challenging to incorporate into host tissue due to its fully differentiated state. Immature osteoblast is estimated to be appropriate for transplantation, but it is difficult to decide the status of immature osteoblast in osteoinductive conditions.

Runx-2 expression has been used as the initial factor to modulate osteogenesis. Under osteogenic induction in vitro, Runx-2 expression peaks within several days post osteogenic induction and is maintained for the proliferation of osteoblast later.

ADSC-E with FGF2 and/or HGF priming for 3 or 6 days that may correspond to immature osteoblast was transplanted, and ectopic bone-forming capacity was examined. Histological analysis corroborates that the effect of a combination of FGF2 and HGF for 6 days are unequal in aspects of new bone structure and human osteocalcin expression. ADSCs with FGF2 or HGF treatment also could form ectopic bone, but its impact was weaker than the combination of FGF2 and HGF. The result from in vivo ectopic bone formation is consistent with the in vitro data in this study.

This study uncovers the distinct difference in paracrine potential of ADSC-Y and ADSC-E. The alteration of secretory factors is closely correlated with impaired differentiation potential of stem cells. The early priming of stem cells with FGF2 and HGF for supplementation of paracrine potential was enough to recover osteogenic potential, which occurred via the promotion of Runx-2 expression and BMP-2/VEGF secretion. Thus, ex vivo culture of ADSCs with poor cellular activity for transplantation might need extra growth factors, through which a dramatic efficacy would be achieved post-transplantation.

This study primarily focused on ADSCs from the young and elderly. ADSCs with low osteogenic activity can be shown in various pathological situations, including acute/chronic inflammatory disease and aging. Thus, growth factor priming is expected to be broadly used to enhance the differentiation of ADSCs with low osteogenic potential and paracrine activity. Next, the efficacy of FGF2/HGF priming on the osteogenic potential of ADSCs from donors with chronic diseases will be widely evaluated, and its effect on bone defects will be confirmed using a non-clinical osteoporosis model.

Supplementary Materials: The following supporting information can be downloaded at: https://www.mdpi.com/article/10.3390/cells11132042/s1, Figure S1: Effect of age on growth factor expression in ADSC, Figure S2: The effect of FGF2 priming on osteogenic potential of ADSC-E, Figure S3: The effect of FGF2 priming time on osteogenesis of ADSC-E, Figure S4: The osteogenic effect of FGF-2/HGF priming for 3 days in ADSC-E, Table S1: Characterization of human adipose-derived stem cells Figure S5: The effect of combination of FGF-2 and HGF priming on osteogenesis of ADSC-E.

Author Contributions: Conceptualization, H.S.H.; methodology, J.S.P., D.K. and H.S.H. formal analysis, J.S.P. and H.S.H.; investigation, J.S.P.; writing—original draft preparation, J.S.P. and H.S.H.; writing—review and editing, H.S.H.; supervision, H.S.H.; project administration, H.S.H.; funding acquisition, H.S.H. All authors have read and agreed to the published version of the manuscript.

Funding: This study was supported by the Korean Health Technology R&D Project grant [HI18C1492] from the Ministry of Health and Welfare (Sejong, Republic of Korea), and the Technology Innovation Program (or Industrial Strategic Technology Development Program-1415179737) (20018551) funded By the Ministry of Trade, Industry & Energy (MOTIE, Korea).

Institutional Review Board Statement: The study was conducted in accordance with the Declaration of Helsinki, and approved by the Institutional Review Board of Kyung Hee university hospital (Protocol Code 2016-12-022, 2021-01-011) for studies involving humans. The animal study protocol was approved by the Institutional Review Board of Kyung Hee university hospital (protocol code: KHMC-IACUC -20-008-01) for studies involving animals.

Informed Consent Statement: Written informed consent has been obtained from the patient(s) to publish this paper.

Data Availability Statement: The datasets used and/or analyzed during the present study are available from the corresponding author upon reasonable request.

Conflicts of Interest: The authors declare no conflict of interest.

References

1. Majidinia, M.; Sadeghpour, A.; Yousefi, B. The roles of signaling pathways in bone repair and regeneration. *J. Cell. Physiol.* **2018**, *233*, 2937–2948. [CrossRef] [PubMed]
2. Kim, B.S.; Lim, J.Y.; Ha, Y.C. Recent Epidemiology of Hip Fractures in South Korea. *Hip Pelvis* **2020**, *32*, 119–124. [CrossRef] [PubMed]
3. Tanha, K.; Fahimfar, N.; Nematollahi, S.; Sajjadi-Jazi, S.M.; Gharibzadeh, S.; Sanjari, M.; Khalagi, K.; Hajivalizedeh, F.; Raeisi, A.; Larijani, B.; et al. Annual incidence of osteoporotic hip fractures in Iran: A systematic review and meta-analysis. *BMC Geriatr.* **2021**, *21*, 668. [CrossRef] [PubMed]
4. Ström, O.; Borgström, F.; Kanis, J.; Compston, J.; Cooper, C.; McCloskey, E.; Jönsson, B. Osteoporosis: Burden, health care provision and opportunities in the EU. *Arch. Osteoporos.* **2011**, *6*, 59–155. [CrossRef]
5. Charkos, T.G.; Liu, Y.; Oumer, K.S.; Vuoung, A.M.; Yang, S. Effects of β-carotene intake on the risk of fracture: A Bayesian meta-analysis. *BMC Musculoskelet. Disord.* **2020**, *21*, 711. [CrossRef]
6. Gómez-Barrena, E.; Rosset, P.; Lozano, D.; Stanovici, J.; Ermthaller, C.; Gerbhard, F. Bone fracture healing: Cell therapy in delayed unions and nonunions. *Bone* **2015**, *70*, 93–101. [CrossRef]
7. Einhorn, T.A.; Gerstenfeld, L.C. Fracture healing: Mechanisms and interventions. *Nat. Rev. Rheumatol.* **2015**, *11*, 45–54. [CrossRef]
8. Poolman, R.W.; Agoritsas, T.; Siemieniuk, R.A.; Harris, I.A.; Schipper, I.B.; Mollon, B.; Smith, M.; Albin, A.; Nador, S.; Sasges, W.; et al. Low intensity pulsed ultrasound (LIPUS) for bone healing: A clinical practice guideline. *BMJ* **2017**, *356*, j576. [CrossRef]
9. Schmidt, A.H. Autologous bone graft: Is it still the gold standard? *Injury* **2021**, *52* (Suppl. S2), S18–S22. [CrossRef]
10. Iaquinta, M.R.; Mazzoni, E.; Bononi, I.; Rotondo, J.C.; Mazziotta, C.; Montesi, M.; Sprio, S.; Tampieri, A.; Tognon, M.; Martini, F. Adult Stem Cells for Bone Regeneration and Repair. *Front. Cell. Dev. Biol.* **2019**, *7*, 268. [CrossRef]
11. Zakrzewski, W.; Dobrzyński, M.; Szymonowicz, M.; Rybak, Z. Stem cells: Past, present, and future. *Stem Cell Res. Ther.* **2019**, *10*, 68. [CrossRef] [PubMed]
12. Seong, J.M.; Kim, B.-C.; Park, J.-H.; Kwon, I.K.; Mantalaris, A.; Hwang, Y.-S. Stem cells in bone tissue engineering. *Biomed. Mater.* **2010**, *5*, 062001. [CrossRef] [PubMed]
13. Clines, G.A. Prospects for osteoprogenitor stem cells in fracture repair and osteoporosis. *Curr. Opin. Organ. Transplant.* **2010**, *15*, 73–78. [CrossRef] [PubMed]
14. Yun, Y.-R.; Jang, J.H.; Jeon, E.; Kang, W.; Lee, S.; Won, J.-E.; Kim, H.W.; Wall, I. Administration of growth factors for bone regeneration. *Regen. Med.* **2012**, *7*, 369–385. [CrossRef]
15. Stamnitz, S.; Klimczak, A. Mesenchymal Stem Cells, Bioactive Factors, and Scaffolds in Bone Repair: From Research Perspectives to Clinical Practice. *Cells* **2021**, *10*, 1925. [CrossRef]
16. Sagaradze, G.D.; Basalova, N.A.; Efimenko, A.Y.; Tkachuk, V.A. Mesenchymal Stromal Cells as Critical Contributors to Tissue Regeneration. *Front. Cell. Dev. Biol.* **2020**, *8*, 576176. [CrossRef]
17. Emara, K.M.; Diab, R.A.; Emara, A.K. Recent biological trends in management of fracture non-union. *World J. Orthop.* **2015**, *6*, 623–628. [CrossRef]
18. Gamblin, A.-L.; Brennan, M.A.; Renaud, A.; Yagita, H.; Lézot, F.; Heymann, D.; Trichet, V.; Layrolle, P. Bone tissue formation with human mesenchymal stem cells and biphasic calcium phosphate ceramics: The local implication of osteoclasts and macrophages. *Biomaterials* **2014**, *35*, 9660–9667. [CrossRef]
19. Watson, L.; Elliman, S.J.; Coleman, C.M. From isolation to implantation: A concise review of mesenchymal stem cell therapy in bone fracture repair. *Stem Cell Res. Ther.* **2014**, *5*, 51. [CrossRef]
20. Brennan, M.A.; Renaud, A.; Amiaud, J.; Rojewski, M.T.; Schrezenmeier, H.; Heymann, D.; Trichet, V.; Layrolle, P. Pre-clinical studies of bone regeneration with human bone marrow stromal cells and biphasic calcium phosphate. *Stem Cell Res. Ther.* **2014**, *5*, 114. [CrossRef]
21. Chiba, Y.; Kuroda, S.; Osanai, T.; Shichinohe, H.; Houkin, K.; Iwasaki, Y. Impact of ageing on biological features of bone marrow stromal cells (BMSC) in cell transplantation therapy for CNS disorders: Functional enhancement by granulocyte-colony stimulating factor (G-CSF). *Neuropathology* **2012**, *32*, 139–148. [CrossRef] [PubMed]

22. Filion, T.M.; Skelly, J.D.; Huang, H.; Greiner, D.L.; Ayers, D.C.; Song, J. Impaired osteogenesis of T1DM bone marrow-derived stromal cells and periosteum-derived cells and their differential in-vitro responses to growth factor rescue. *Stem Cell Res. Ther.* **2017**, *8*, 65. [CrossRef] [PubMed]
23. Liang, M.; Liu, W.; Peng, Z.; Lv, S.; Guan, Y.; An, G.; Zhang, Y.; Huang, T.; Wang, Y. The therapeutic effect of secretome from human umbilical cord-derived mesenchymal stem cells in age-related osteoporosis. *Artif. Cells Nanomed. Biotechnol.* **2019**, *47*, 1357–1366. [CrossRef] [PubMed]
24. Qomi, R.T.; Sheykhhasan, M. Adipose-derived stromal cell in regenerative medicine: A review. *World J. Stem Cells* **2017**, *9*, 107–117. [CrossRef] [PubMed]
25. Kim, E.H.; Heo, C.Y. Current applications of adipose-derived stem cells and their future perspectives. *World J. Stem Cells* **2014**, *6*, 65–68. [CrossRef] [PubMed]
26. Park, J.S.; Piao, J.; Park, G.; Yoo, K.S.; Hong, H.S. Osteoporotic Conditions Influence the Activity of Adipose-Derived Stem Cells. *Tissue Eng. Regen. Med.* **2020**, *17*, 875–885. [CrossRef]
27. Zhang, Y.; Madhu, V.; Dighe, A.S.; Irvine, J.N.; Cui, Q. Osteogenic response of human adipose-derived stem cells to BMP-6, VEGF, and combined VEGF plus BMP-6 in vitro. *Growth Factors* **2012**, *30*, 333–343. [CrossRef]
28. Zhang, Z.; Ma, Y.; Guo, S.; He, Y.; Bai, G.; Zhang, W. Low-intensity pulsed ultrasound stimulation facilitates in vitro osteogenic differentiation of human adipose-derived stem cells via up-regulation of heat shock protein (HSP)70, HSP90, and bone morphogenetic protein (BMP) signaling pathway. *Biosci. Rep.* **2018**, *38*, BSR20180087. [CrossRef]
29. Chen, S.; Zheng, Y.; Zhang, S.; Jia, L.; Zhou, Y. Promotion Effects of miR-375 on the Osteogenic Differentiation of Human Adipose-Derived Mesenchymal Stem Cells. *Stem Cell Rep.* **2017**, *8*, 773–786. [CrossRef]
30. Jin, Y.; Yang, L.; Zhang, Y.; Gao, W.; Yao, Z.; Song, Y.; Wang, Y. Effects of age on biological and functional characterization of adipose-derived stem cells from patients with end-stage liver disease. *Mol. Med. Rep.* **2017**, *16*, 3510–3518. [CrossRef]
31. Griffin, M.; Ryan, C.M.; Pathan, O.; Abraham, D.; Denton, C.P.; Butler, P.E. Characteristics of human adipose derived stem cells in scleroderma in comparison to sex and age matched normal controls: Implications for regenerative medicine. *Stem Cell Res. Ther.* **2017**, *8*, 23. [CrossRef]
32. Efimenko, A.; Dzhoyashvili, N.; Kalinina, N.; Kochegura, T.; Akchurin, R.; Tkachuk, V.; Parfyonova, Y. Adipose-Derived Mesenchymal Stromal Cells from Aged Patients With Coronary Artery Disease Keep Mesenchymal Stromal Cell Properties but Exhibit Characteristics of Aging and Have Impaired Angiogenic Potential. *Stem Cell Transl. Med.* **2014**, *3*, 32–41. [CrossRef] [PubMed]
33. Park, J.S.; Park, G.; Hong, H.S. Age affects the paracrine activity and differentiation potential of human adipose-derived stem cells. *Mol. Med. Rep.* **2021**, *23*, 160. [CrossRef] [PubMed]
34. Liu, M.; Lei, H.; Dong, P.; Fu, X.; Yang, Z.; Yang, Y.; Ma, J.; Liu, X.; Cao, Y.; Xiao, R. Adipose-Derived Mesenchymal Stem Cells from the Elderly Exhibit Decreased Migration and Differentiation Abilities with Senescent Properties. *Cell Transpl.* **2017**, *26*, 1505–1519. [CrossRef] [PubMed]
35. Zhang, J.; Liu, Y.; Chen, Y.; Yuan, L.; Liu, H.; Wang, J.; Liu, Q.; Zhang, Y. Adipose-Derived Stem Cells: Current Applications and Future Directions in the Regeneration of Multiple Tissues. *Stem Cells Int.* **2020**, *2020*, 8810813. [CrossRef]
36. Sterodimas, A.; de Faria, J.; Nicaretta, B.; Pitanguy, I. Tissue engineering with adipose-derived stem cells (ADSCs): Current and future applications. *J. Plast. Reconstr. Aesthet. Surg.* **2010**, *63*, 1886–1892. [CrossRef]
37. Lee, T.J.; Shim, M.S.; Yu, T.; Choi, K.; Kim, D.I.; Lee, S.H.; Bhang, S.H. Bioreducible polymer micelles based on acid-degradable Poly(ethylene glycol)-poly(amino ketal) enhance the stromal cell-derived factor-1α gene transfection efficacy and therapeutic angiogenesis of human adipose-derived stem cells. *Int. J. Mol. Sci.* **2018**, *19*, 529. [CrossRef]
38. Pan, J.; Alimujiang, M.; Chen, Q.; Shi, H.; Luo, X. Exosomes derived from miR-146a-modified adipose-derived stem cells attenuate acute myocardial infarction−induced myocardial damage via downregulation of early growth response factor 1. *J. Cell. Biochem.* **2019**, *120*, 4433–4443. [CrossRef]
39. Li, M.; Ma, J.; Gao, Y.; Dong, M.; Zheng, Z.; Li, Y.; Tan, R.; She, Z.; Yang, L. Epithelial differentiation of human adipose-derived stem cells (hASCs) undergoing three-dimensional (3D) cultivation with collagen sponge scaffold (CSS) via an indirect co-culture strategy. *Stem Cell Res. Ther.* **2020**, *11*, 141. [CrossRef]
40. Mazini, L.; Rochette, L.; Admou, B.; Amal, S.; Malka, G. Hopes and Limits of Adipose-Derived Stem Cells (ADSCs) and Mesenchymal Stem Cells (MSCs) in Wound Healing. *Int. J. Mol. Sci.* **2020**, *21*, 1306. [CrossRef]
41. Mazini, L.; Ezzoubi, M.; Malka, G. Overview of current adipose-derived stem cell (ADSCs) processing involved in therapeutic advancements: Flow chart and regulation updates before and after COVID-19. *Stem Cell Res. Ther.* **2021**, *12*, 1. [CrossRef] [PubMed]
42. Ma, T.; Sun, J.; Zhao, Z.; Lei, W.; Chen, Y.; Wang, X.; Yang, J.; Shen, Z. A brief review: Adipose-derived stem cells and their therapeutic potential in cardiovascular diseases. *Stem Cell Res. Ther.* **2017**, *8*, 124. [CrossRef] [PubMed]
43. Li, K.; Shi, G.; Lei, X.; Huang, Y.; Li, X.; Bai, L.; Qin, C. Age-related alteration in characteristics, function, and transcription features of ADSCs. *Stem Cell Res. Ther.* **2021**, *12*, 473. [CrossRef] [PubMed]
44. Shafaei, H.; Kalarestaghi, H. Adipose-derived stem cells: An appropriate selection for osteogenic differentiation. *J. Cell Physiol.* **2020**, *235*, 8371–8386. [CrossRef]
45. Schipani, E.; Maes, C.; Carmeliet, G.; Semenza, G.L. Regulation of Osteogenesis-Angiogenesis Coupling by HIFs and VEGF. *J. Bone Miner. Res.* **2009**, *24*, 1347–1353. [CrossRef]

46. Chen, G.; Deng, C.; Li, Y.P. TGF-β and BMP signaling in osteoblast differentiation and bone formation. *Int. J. Biol. Sci.* **2012**, *8*, 272–288. [CrossRef]
47. Qian, X.; Zhang, C.; Chen, G.; Tang, Z.; Liu, Q.; Chen, J.; Tong, X.; Wang, J. Effects of BMP-2 and FGF2 on the Osteogenesis of Bone Marrow-Derived Mesenchymal Stem Cells in Hindlimb-Unloaded Rats. *Cell Biochem. Biophys.* **2014**, *70*, 1127–1136. [CrossRef]
48. Seghezzi, G.; Patel, S.; Ren, C.J.; Gualandris, A.; Pintucci, G.; Robbins, E.S.; Shapiro, R.L.; Galloway, A.C.; Rifkin, D.B.; Mignatti, P. Fibroblast growth factor-2 (FGF-2) induces vascular endothelial growth factor (VEGF) expression in the endothelial cells of forming capillaries: An autocrine mechanism contributing to angiogenesis. *J. Cell Biol.* **1998**, *141*, 1659–1673. [CrossRef]
49. Sun, Z.; Tee, B.C.; Kennedy, K.S.; Kennedy, P.M.; Kim, D.-G.; Mallery, S.R.; Fields, H.W. Scaffold-based delivery of autologous mesenchymal stem cells for mandibular distraction osteogenesis: Preliminary studies in a porcine model. *PLoS ONE* **2013**, *8*, e74672. [CrossRef]
50. Hurley, M.M.; Gronowicz, G.; Zhu, L.; Kuhn, L.T.; Rodner, C.; Xiao, L. Age-related changes in FGF-2, fibroblast growth factor receptors and β-Catenin expression in human mesenchyme-derived progenitor cells. *J. Cell Biochem.* **2016**, *117*, 721–729. [CrossRef]
51. Jung, J.S.; Volk, C.; Marga, C.; Navarrete Santos, A.; Jung, M.; Rujescu, D.; Navarrete, S.A. Adipose-derived stem/stromal cells recapitulate aging biomarkers and show reduced stem cell plasticity affecting their adipogenic differentiation capacity. *Cell Reprogram.* **2019**, *21*, 187–199. [CrossRef] [PubMed]
52. Si, Y.; Kim, S.; Cui, X.; Zheng, L.; Oh, S.J.; Anderson, T.; AlSharabati, M.; Kazamel, M.; Volpicelli-Daley, L.; Bamman, M.M.; et al. Transforming growth gactor beta (TGF-β) is a muscle biomarker of disease progression in ALS and correlates with Smad Expression. *PLoS ONE* **2015**, *10*, e0138425. [CrossRef] [PubMed]
53. Ghorbanzadeh, V.; Pourheydar, B.; Dariushnejad, H.; Ghalibafsabbaghi, A.; Chodari, L. Curcumin improves angiogenesis in the heart of aged rats: Involvement of TSP1/NF-κB/VEGF-A signaling. *Microvasc. Res.* **2022**, *139*, 104258. [CrossRef] [PubMed]
54. Zhang, Y.; Xia, M.; Jin, K.; Wang, S.; Wei, H.; Fan, C.; Wu, Y.; Li, X.; Li, X.; Li, G.; et al. Function of the c-Met receptor tyrosine kinase in carcinogenesis and associated therapeutic opportunities. *Mol. Cancer* **2018**, *17*, 45. [CrossRef] [PubMed]
55. Zhen, R.; Yang, J.; Wang, Y.; Li, Y.; Chen, B.; Song, Y.; Ma, G.; Yang, B. Hepatocyte growth factor improves bone regeneration via the bone morphogenetic protein-2-mediated NF-κB signaling pathway. *Mol. Med. Rep.* **2018**, *17*, 6045–6053. [CrossRef]
56. Tsai, S.Y.; Huang, Y.-L.; Yang, W.-H.; Tang, C.-H. Hepatocyte growth factor-induced BMP-2 expression is mediated by c-Met receptor, FAK, JNK, Runx2, and p300 pathways in human osteoblasts. *Int. Immunopharmacol.* **2012**, *13*, 156–162. [CrossRef]
57. Lin, Y.M.; Huang, Y.L.; Fong, Y.C.; Tsai, C.-H.; Chou, M.C.; Tang, C.H. Hepatocyte growth factor increases vascular endothelial growth factor-A production in human synovial fibroblasts through c-Met receptor pathway. *PLoS ONE* **2012**, *7*, e50924. [CrossRef]
58. Yang, D.; Okamura, H.; Qiu, L. Upregulated osterix expression elicited by Runx2 and Dlx5 is required for the accelerated osteoblast differentiation in PP2A Cα-knockdown cells. *Cell Biol. Int.* **2018**, *42*, 403–410. [CrossRef]
59. Beane, O.S.; Fonseca, V.C.; Cooper, L.L.; Koren, G.; Darling, E.M. Impact of Aging on the regenerative properties of bone marrow-, muscle-, and adipose-derived mesenchymal stem/stromal cells. *PLoS ONE* **2014**, *9*, e115963. [CrossRef]

 cells

MDPI

Article

Combination Therapy of Placenta-Derived Mesenchymal Stem Cells with WKYMVm Promotes Hepatic Function in a Rat Model with Hepatic Disease via Vascular Remodeling

Ji Hye Jun [1], Sohae Park [1], Jae Yeon Kim [1], Ja-Yun Lim [2], Gyu Tae Park [3], Jae Ho Kim [3] and Gi Jin Kim [1,*]

[1] Department of Biomedical Science, CHA University, Seongnam 13488, Korea; jihyejun1015@gmail.com (J.H.J.); sohae11@snu.ac.kr (S.P.); janejaeyeon92@gmail.com (J.Y.K.)
[2] Department of Integrated Biomedical and Life Science, College of Health Science, Korea University, Seoul 02841, Korea; jayun78@korea.ac.kr
[3] Department of Physiology, School of Medicine, Pusan National University, Yangsan 50612, Korea; daramzuy2@naver.com (G.T.P.); jhkimst@pusan.ac.kr (J.H.K.)
* Correspondence: gjkim@cha.ac.kr; Tel.: +81-31-881-7145

Abstract: Changes in the structure and function of blood vessels are important factors that play a primary role in regeneration of injured organs. WKYMVm has been reported as a therapeutic factor that promotes the migration and proliferation of angiogenic cells. Additionally, we previously demonstrated that placenta-derived mesenchymal stem cells (PD-MSCs) induce hepatic regeneration in hepatic failure via antifibrotic effects. Therefore, our objectives were to analyze the combination effect of PD-MSCs and WKYMVm in a rat model with bile duct ligation (BDL) and evaluate their therapeutic mechanism. To analyze the anti-fibrotic and angiogenic effects on liver regeneration, it was analyzed using ELISA, qRT-PCR, Western blot, immunofluorescence, and immunohistochemistry. Collagen accumulation was significantly decreased in PD-MSCs with the WKYMVm combination (Tx+WK) group compared with the nontransplantation (NTx) and PD-MSC-transplanted (Tx) group ($p < 0.05$). Furthermore, the combination of PD-MSCs with WKYMVm significantly promoted hepatic function by increasing hepatocyte proliferation and albumin as well as angiogenesis by activated FPR2 signaling ($p < 0.05$). The combination therapy of PD-MSCs with WKYMVm could be an efficient treatment in hepatic diseases via vascular remodeling. Therefore, the combination therapy of PD-MSCs with WKYMVm could be a new therapeutic strategy in degenerative medicine.

Keywords: liver cirrhosis; placenta-derived mesenchymal stem cells; WKYMVm; combination therapy

Citation: Jun, J.H.; Park, S.; Kim, J.Y.; Lim, J.-Y.; Park, G.T.; Kim, J.H.; Kim, G.J. Combination Therapy of Placenta-Derived Mesenchymal Stem Cells with WKYMVm Promotes Hepatic Function in a Rat Model with Hepatic Disease via Vascular Remodeling. *Cells* **2022**, *11*, 232. https://doi.org/10.3390/cells11020232

Academic Editors: Alexander E. Kalyuzhny

Received: 14 December 2021
Accepted: 7 January 2022
Published: 11 January 2022

1. Introduction

Angiogenesis is essential in many biological processes, including development and wound repair [1]. Progressive hepatic vascular pressure due to repetitive hepatic damage is the major cause of liver cirrhosis [2]. Abnormal hepatocyte–endothelium crosstalk in the injured liver delays regeneration by promoting fibrogenesis and the formation of scar tissues [3]. Additionally, the irregular vascular system not only indicates the fibrotic pathophysiology but also inhibits metabolism in the liver [4]. The hepatic vascular niche, which is mainly composed of liver sinusoidal endothelial cells (LSECs), secretes angiocrine factors such as vascular endothelial growth factor (VEGF), hepatic growth factor (HGF), Wnt, and endoglin (CD105) to promote hepatic regeneration [5,6]. Increased expression of VEGF leads to hepatic regeneration following partial hepatectomy or drug intoxication, and VEGF is a strong modulator of vascular regeneration due to its effects on endothelial cell proliferation and survival [7,8]. Additionally, the secretion of HGF in endothelial cells inhibited the activation of fibroblasts and functioned as a key molecule to prevent fibrosis in a mouse model of liver fibrosis [9]. In endothelial cells, Wnt/β-catenin pathway activation by Wnt ligands can induce cell cycle progression through transcriptional activation of cyclin D1 and regulate endothelial permeability [10,11].

Mesenchymal stem cell (MSC) therapy is a promising strategy for treatment of hepatic diseases through overexpression of HGF or matrix metalloproteinases (MMPs) and a reduction in collagen levels [12,13]. Among MSCs, human placenta-derived MSCs (PD-MSCs), obtained from fetal tissue, have various advantages, such as strong immunosuppressive abilities, multipotent differentiation, and self-renewal properties [14–17]. Previously, we reported that PD-MSCs have therapeutic effects in a carbon tetrachloride (CCl_4)-injured rat model via an autophagic mechanism and IL-6/gp130/signal transducer and activator of transcription 3 (STAT3) pathway [18,19]. Additionally, we demonstrated that PD-MSCs enhance hepatic regeneration via restoration of hepatic lipid metabolism and vascular regeneration by microRNAs (miRNAs) in a rat model of bile duct ligation (BDL) [20,21].

The synthetic peptide WKYMVm, which is composed of Trp-Lys-Tyr-Met-Val-D-Met, is a potent agonist of formyl peptide receptor 2 (FPR2), which belongs to the G-protein coupled receptor family [22,23]. FPR2 promotes vessel growth and cell proliferation [23,24]. The therapeutic effects of WKYMVm have been demonstrated in various degenerative diseases. In ischemic hind limb and myocardial diseases, WKYMVm promoted neovascularization and tissue repair through migration and proliferation of endothelial colony-forming cells (ECFCs) or circulating angiogenic cells (CACs) [25,26]. Topical application of WKYMVm intensified cutaneous wound healing in streptozotocin (STZ)-induced diabetes via the remodeling of von Willebrand factor (vWF)-positive vessels [27]. For treatment of bronchopulmonary dysplasia (BPD), a chronic lung disease, WKYMVm was administered through intraperitoneal injection [28]. Additionally, polylactic-co-glycolic acid (PLGA) could form microspheres encapsulating WKYMVm and consistently released peptides following administration to ischemic hind limbs, extending angiogenic stimulation [29].

However, the effects of combined PD-MSCs and WKYMVm in liver cirrhosis have not been elucidated. Therefore, our objectives were to analyze the effects of PD-MSCs with WKYMVm on hepatic regeneration via vascular regeneration in a BDL-injured rat model of hepatic fibrosis. We investigated the mechanism underlying the therapeutic effects in a liver injury model.

2. Materials and Methods

2.1. Materials

The structure of the WKYMVm peptide was reported by Choi et al. [30]. The synthetic peptide WKYMVm was synthesized at ANYGEN (Kwangju, Korea). The purity of synthesized WKYMVm was >98%.

2.2. Cell Culture

Collection of placental samples from healthy women ($\geq$37 gestational weeks) for research purposes was approved by the Institutional Review Board of CHA Gangnam Medical Center, Seoul, Korea (IRB 07-18). PD-MSCs were isolated and maintained as described previously [31]. A rat hepatocyte-like epithelial cell line, WB-F344, and rat T-HSC/Cl-6 cells, rat hepatic stellate cells (HSCs) transformed with simian virus 40, were maintained in α-MEM supplemented with 10% fetal bovine serum (FBS; Gibco, Langley, OK, USA, 16000-044), and 100 U/mL penicillin–streptomycin (Pen-Strep, P/S; Gibco, 15-140-122). Additionally, human umbilical vein endothelial cells (HUVECs) were cultured in endothelial cell medium (ECM; ScienCell, Carlsbad, CA, USA, 1001) under 5% CO_2 at 37 °C. Furthermore, WB-F344, T-HSC/Cl-6, and HUVECs were treated with lithocholic acid (LCA; 100 μM; Sigma, Burlington, MA, USA, L6250), transforming growth factor-beta (TGF-β; 2 ng/mL; PeproTech, Rocky Hill, NJ, USA, 100-21C), and 5-fluorouracil (5-FU, 1 μg/mL, Sigma, F6627) for 24, 48, and 48 h, respectively. The WKYMVm peptide was administered at a concentration of 1 mM in vitro.

2.3. Animals

Seven-week-old male Sprague Dawley (SD) rats (Orient Bio, Inc., Seongnam, Korea) were maintained in an air-conditioned facility. The common bile duct was ligated. Additionally, detailed protocol is described in Supplemental Figure S1. All animal experimental processes were approved by the protocol consistent with the Institutional Review Board of CHA General Hospital, Seoul, Korea. The experimental protocols were approved by the Institutional Animal Care and Use Committee of CHA University, Seongnam, Korea (IACUC-180023).

2.4. Biochemical Analysis

The harvested serum was examined for aspartate aminotransferase (AST), alanine aminotransferase (ALT), total bilirubin, and albumin (ALB) by Southeast Medi-Chem Institute (Busan, Korea). The experiment was performed in triplicate.

2.5. Cell Proliferation Assay

To confirm the proliferative effects of WKYMVm on PD-MSCs, HUVECs, and WB-F344, we incubated Cell Counting Kit-8 (CCK-8; Dojindo Molecular Technologies, Inc., Rockville, MD, USA, CK04) solution with cells at 37 °C for 2 h. The absorbance at 450 nm for CCK-8 solution was measured with an epoch microplate spectrophotometer (BioTek, Winooski, VT, USA). All reactions were performed in triplicate.

2.6. Senescence-Associated β-Galactosidase (SA-β Gal) Assay

SA-β-gal activity was analyzed using an SA-β-gal staining kit (Cell Signaling, Danvers, MA, USA, 9860S) according to the manufacturer's instructions. The intensity of SA-β-gal was analyzed with a microscope via a high-magnification digital camera (Nikon Instrument, Nikon, Inc., Melville, NY, USA) and the ImageJ program (NIH).

2.7. Differentiation of WKYMVm-Treated PD-MSCs

To evaluate the potential for differentiation, we seeded PD-MSCs at a density of 2×10^4 cells/35 mm plate and maintained them in α-MEM supplemented with 10% FBS, 100 U/mL of Pen/Strep, 25 ng/mL of FGF4 (PeproTech, AF-100-31), and 1 μg/mL of heparin (Sigma, H3149) until they reached 60% confluency. Then, 1 mM of WKYMVm peptide was added to PD-MSCs for 24 h. Adipogenic and osteogenic differentiation was induced by media from the following: StemPro Adipogenesis Differentiation Kit (Gibco) and StemPro Osteogenesis Differentiation Kit (Gibco), respectively. The differentiation media were replaced every 3 days until day 21.

2.8. Tube Formation Assay

For analysis of the angiogenic ability of endothelial cells, HUVECs were stained with Alexa Fluor 488 Ac-LDL (Invitrogen, L23380). HUVECs (5×10^4 cells/well) were precoated with Matrigel (Corning Inc., Corning, New York, NY, USA, 354248) in 24-well culture plates, seeded, and cultured with or without 5-FU and 1 mM of WKYMVm and co-cultivated with PD-MSCs at 37 °C in a 5% CO_2 incubator. The sprouted tube length was measured and quantified using the ImageJ program. All experiments were conducted in triplicate.

2.9. Dextran Permeability Assay

HUVECs were seeded in the upper chamber of a 24-well transwell system and cultured for 24 h to allow growth of a confluent monolayer. Monolayers were treated with 100 μg/mL of 5-FU. After 72 h, monolayers were administered with 1 mM of WKYMVm peptide and co-cultivated with PD-MSCs for 24 h. Dextran permeability was tested by adding 10 μL of 10 mg/mL fluorescein isothiocyanate (FITC)-dextran (Sigma) to the upper chamber for 30 min. After 30 min, 100 μL of conditioned medium in the lower chamber was transferred to a 96-well plate (BD Biosciences, Bedford, MA, USA) and read at excitation and emission wavelengths of 490 and 525 nm, respectively. The experiments were performed in triplicate.

2.10. Histological Analysis

The liver tissues were fixed in 10% (v/v) neutral buffered formalin (NBF; BBC Biochemical, Mount Vernon, WA, USA), embedded in paraffin, cut into 5 μm sections and stained with hematoxylin and eosin (H&E) and Sirius Red. The diameter of the hepatic portal vein and collagen deposition area were measured by ImageJ software through H&E and Sirius Red staining, respectively.

2.11. Immunohistochemistry

To analyze hepatocyte proliferation in tissues following injection with PD-MSCs and WKYMVm or nontransplantation, we examined the expression of proliferating cell nuclear antigen (PCNA) localized in the nucleus. The sectioned slides were incubated in 3% H_2O_2 in methanol to block endogenous peroxidase. After antigen retrieval through microwaving, the slides were reacted with an anti-PCNA antibody (Santa Cruz Biotechnology, Dallas, TX, USA, sc-56) and diluted with antibody diluent (Dako, Santa Clara, CA, USA, S3022) at 4 °C overnight, followed by 30 min with biotinylated secondary anti-rabbit antibody at room temperature (RT). Incubation with horseradish peroxidase-conjugated streptavidin–biotin complex (Dako, K1015) and 3,3-diaminobenzidine (Dako, K1015) was used to induce chromatic signals. The slides were counterstained with Mayer's hematoxylin (Dako, S-3099). Images were detected using a digital slide scanner (3DHISTECH, Ltd., Budapest, Hungary). Finally, the percentage of PCNA-positive hepatocytes was measured in all sections at 400× magnification.

2.12. Immunofluorescence

For analysis of the localization of vWF, α-SMA, active β-catenin, and FPR2 in vivo or in vitro, primary antibodies against vWF (1:200, Abcam, Cambridge, MA, USA, ab6994), α-SMA (1:200; Dako, M0851), active β-catenin (1:100; Cell Signaling Technology, Danvers, MA, USA, 8814S), and FPR2 (1:100; Novus, St. Louis, MO, USA, NLS1878) was added to antibody diluent (Dako, S3022) and reacted with slides at 4 °C overnight. The secondary antibody Alexa Fluor 488 and 594 (1:400; Invitrogen, Camarillo, CA, USA, A21206 and A11012) was added for 1 h. The slides were counterstained with 4,6-diamidino-2-phenylindole (DAPI; Invitrogen, D3571). The images were observed with a confocal microscope (LSM 700; Zeiss, Oberkochen, Germany). The observed images were analyzed with ZEN blue software (Zeiss). The experiments were performed in triplicate.

2.13. Enzyme-Linked Immunosorbent Assay (ELISA)

The concentrations of VEGF, HGF, and TGF-β were analyzed by ELISAs. Their concentrations were measured using human VEGF (Abcam, ab100662), rat VEGF (Abcam, ab100786), human HGF (R&D Systems, DHG00B), rat HGF (R&D Systems, MHG00), and human TGF-β (Abcam, ab100647) ELISA kits in strict accordance with the manufacturer's instructions and detected using a microplate reader (BioTek, Winooski, VT, USA) at 450 nm.

2.14. Western Blot

Protein lysates were subjected to dodecyl sulfate polyacrylamide gel electrophoresis (SDS-PAGE), transferred to polyvinylidene difluoride membranes (PVDF; Bio-Rad Laboratories), and then blocked-in blocking buffer for 1 h. The membranes were subsequently incubated with rabbit anti-vimentin (1:2000; Sigma, v4630), anti-Col I (1:1000; Abcam, ab34710), anti-GAPDH (1:3000; Abfrontier, Seoul, Korea, LF-PA0018), anti-ALB (1:1000; Novus, NBP1-32458), anti-cyclin D1 (1:1000; Abfrontier, LF-MA0325), anti-gp130 (1:1000; Abcam, ab202850) anti-pSTAT3 (1:1000; Cell Signaling Technology, 9134s), mouse anti-E-cadherin (1:1000; Cell Signaling Technology, 14472s), anti-α-SMA (1:1000; Dako, M0851), anti-IL-6 (1:1000; Abcam, ab6672), and anti-tSTAT3 (1:1000; Cell Signaling Technology, 9139s) antibodies at 4 °C overnight. After the reaction, the membranes were treated with anti-rabbit IgG, HRP-linked antibody (1:10,000; Cell Signaling Technology, 7074P2), anti-mouse IgG, or HRP-linked antibody (1:5000; Cell Signaling Technology, 7076s)-conjugated secondary antibody or for 1 h at RT. The bands were detected using a Clarity Western ECL kit (Bio-Rad Laboratories, 1705061). Western blotting was performed in triplicate.

2.15. Quantitative Reverse Transcription-Polymerase Chain Reaction (qRT-PCR)

Total RNA was extracted from rat liver tissues and TRIzol-treated cells (Ambion, Austin, TX, USA, 16696018). Reverse transcription was performed with 250 ng of total RNA and Superscript III reverse transcriptase (Invitrogen). Complementary DNA (cDNA) was synthesized by PCR. Real-time PCR was performed using SYBR Green Master Mix (Roche, Mannheim, Germany) and a CFX Connect Real-Time System (Bio-Rad Laboratories). The PCR conditions were as follows: denaturation at 95 °C for 15 min and 20 s, followed by 40 cycles of 95 °C for 30 s and annealing at 55~60 °C for 40 s. Extension at 70 °C for 15 min and a final extension at 72 °C for 7 min were performed. Gene expression was normalized to that of GAPDH. The primer sequences are shown in Supplemental Table S1. All reactions were performed in at least triplicate.

2.16. Gelatin Zymography

The expression of MMP-2/9 was detected by gelatin zymography. Supernatants from PD-MSCs or HSCs were collected and separated on 12% SDS polyacrylamide gels supplemented with 1 mg/mL of gelatin. The separated gels were washed twice for 40 min with renaturation buffer (Bio-Rad Laboratories, Hercules, CA, USA, AB102-401) and incubated overnight at 37 °C in developing buffer containing 50 mM of Tris-HCl (pH 7.4), 0.2 M of NaCl, 5 mM of CaCl2, and 1% Triton X-100. The next day, the gels were stained with 10% acetic acid 40% methanol containing 0.5% Coomassie Brilliant Blue R-250 (Bio-Rad Laboratories, 1610406) for 3 h and destained with destaining buffer containing 10% acetic acid 40% methanol. The density of unstained bands was used to detect enzyme expression. The intensity of the gel bands was measured using the ImageJ software program. All experiments were performed in triplicate.

2.17. Statistical Analysis

Data are presented as the mean ± standard error of the mean. Differences between different regions of SBEM were analyzed with GraphPad Prism software, and the statistical methods were used for comparisons between pairs. In summary, datasets with more than two groups were analyzed with one-way ANOVA. Datasets with two groups were analyzed with Student's t test. Significance was defined as $p < 0.05$.

3. Results

3.1. Characterization of PD-MSCs Combined with WKYMVm

To confirm the PD-MSC characteristics after WKYMVm treatment, we analyzed the morphology and proliferative capacity of PD-MSCs and the expression of stemness-related markers (e.g., octamer-binding transcription factor 4 (Oct4), sex-determining region Y-box 2, (Sox2), and Nanog), germ lineage markers (e.g., neurofilament-68 (NF-68) and alpha fetoprotein (AFP)) and the human leukocyte antigen g (HLA-G). The WKYMVm-treated PD-MSCs exhibited an elongated, spindle-shaped morphology similar to that of untreated PD-MSCs (Figure 1A). The proliferative capacity of the WKYMVm-treated PD-MSCs was significantly increased compared to that of the untreated PD-MSCs, as shown by CCK-8 assays ($p < 0.05$, Figure 1B). Additionally, there was no difference in the expression levels of stemness markers between the two groups (Figure 1C). To confirm the immunophenotypes of the PD-MSCs treated with WKYMVm, we analyzed the cell surface markers by flow cytometry. The WKYMVm-treated PD-MSCs were positive for the expression of MSC markers such as CD13, CD90, CD105, HLA-ABC, and HLA-G. However, they were negative for hematopoietic lineage markers such as CD34, CD45, and HLA-DR (Figure 1D).

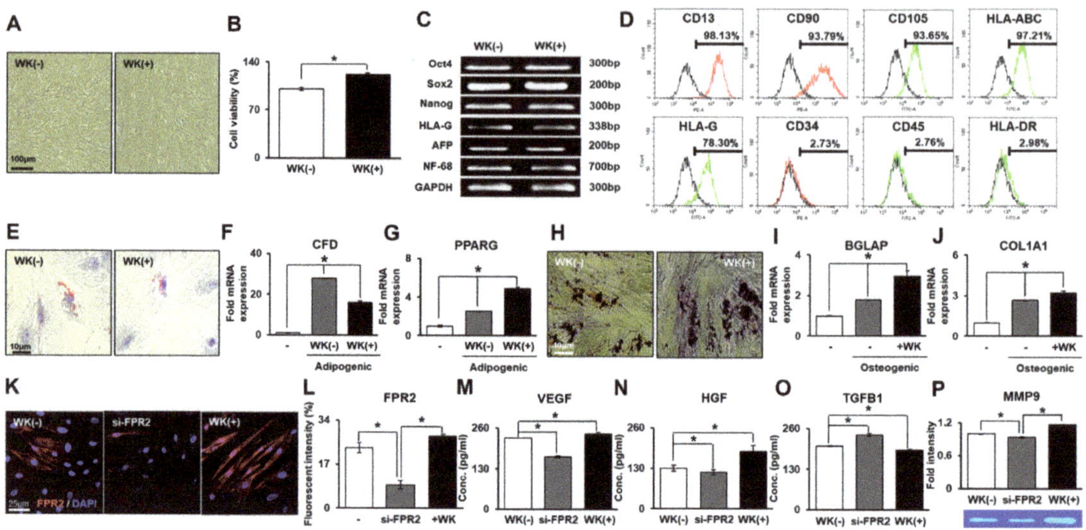

Figure 1. WKYMVm treatment maintains and enhances PD-MSC characteristics and effects. (**A**) Morphology of the PD-MSCs and WKYMVm-treated PD-MSCs. Scale bar = 100 μm. (**B**) Viability of the PD-MSCs and WKYMVm-treated PD-MSCs shown by CCK-8 assays. (**C**) Expression of stemness-related markers in the WKYMVm-treated PD-MSCs shown by RT-PCR. (**D**) MSC surface markers (e.g., hematopoietic, nonhematopoietic, and HLA family) in the WKYMVm-treated PD-MSCs using FACS analysis. (**E**) Lipid droplets in PD-MSCs differentiated into adipocytes shown by Oil Red O staining. Scale bar = 10 μm. Expression of adipogenic-specific markers such as CFD (**F**) and PPARG (**G**) shown by qRT-PCR. (**H**) Calcium deposition of PD-MSCs differentiated into osteocytes shown by von Kossa staining. Scale bar = 40 μm. Expression of osteogenic-specific markers, including BGLAP (**I**) and COL1A1 (**J**), shown through qRT-PCR. (**K**) Expression of FPR2 in PD-MSCs using immunofluorescence. Scale bar = 25 μm. (**L**) Quantification of the fluorescence intensity of FPR2 through immunofluorescence. Expression of VEGF (**M**), HGF (**N**), and TGFB1 (**O**) in the WKYMVm-treated PD-MSC culture supernatants shown by ELISAs. (**P**) The activity of MMP9 in the WKYMVm-treated PD-MSC culture supernatant shown using gelatin zymography. Mean ± SD, * $p < 0.05$ by *t* tests. si-FPR2, siRNA-FPR2-transfected PD-MSCs; +WK, WKYMVm-treated PD-MSCs.

To investigate the differentiation potential of PD-MSCs with WKYMVm, we maintained the cells in adipogenic or osteogenic induction media for 21 days. Lipid droplets stained with Oil Red O and adipogenic markers (e.g., adipsin, CFD; peroxisome proliferator-activated receptor gamma, PPARG) were highly expressed at the mRNA level in the differentiated WKYMVm-treated PD-MSCs ($p < 0.05$, Figure 1E–G). Additionally, calcium deposition stained by von Kossa staining was observed in the WKYMVm-treated PD-MSCs and untreated PD-MSCs (Figure 1H). The levels of osteogenic markers (e.g., osteocalcin, BGLAP; collagen type I, COL1A1) were dramatically increased in the differentiated WKYMVm-treated PD-MSCs versus the undifferentiated PD-MSCs (Figure 1I,J). These data suggest that the WKYMVm-treated PD-MSCs maintain characteristics similar to those of naïve PD-MSCs.

3.2. WKYMVm Enhances the Effects of PD-MSCs

To identify whether PD-MSCs express FPR2, which is the major receptor of the WKYMVm ligand, we analyzed the expression of FPR2 in PD-MSCs by immunofluorescence. As shown in Figure 1L, FPR2 was expressed in naïve PD-MSCs and was decreased by siRNA-FPR2 transfection, while it was significantly increased by WKYMVm treatment of PD-MSCs ($p < 0.05$, Figure 1K,L). To investigate the effect of WKYMVm on PD-MSCs, we performed ELISA and gelatin zymography with PD-MSC culture supernatant. The expression of VEGF and hepatocyte growth factor (HGF) were significantly upregulated in the WKYMVm-treated PD-MSCs compared to the untreated PD-MSCs and the siRNA-FPR2-transfected PD-MSCs ($p < 0.05$, Figure 1M,N). However, transforming growth factor (TGF)-β was significantly decreased, while the activity of MMP-9 was increased in the WKYMVm-treated PD-MSC group ($p < 0.05$, Figure 1O,P). Therefore, these data demonstrate that WKYMVm enhanced the proangiogenic, regenerative, or antifibrotic effects of PD-MSCs via FPR2.

3.3. PD-MSCs Combined with WKYMVm Enhance Hepatic Function in the BDL Rat Model

For analysis of the therapeutic effect of PD-MSCs combined with WKYMVm in a liver-injured rat model, we determined the serum levels of hepatic function markers (e.g., AST, ALT, total bilirubin, and ALB). The BDL rat group (NTx) had significantly increased levels of AST, ALT, and total bilirubin compared to the control. However, PD-MSC transplantation (Tx) slightly reduced the levels of ALT, AST, and total bilirubin. Interestingly, the levels in the PD-MSCs combined with the WKYMVm (Tx+WK) group were evidently lower than those in the NTx and Tx groups ($p < 0.05$, Supplemental Figure S2A–C). Additionally, the serum level of ALB was significantly upregulated in the Tx+WK group compared with the NTx and Tx groups ($p < 0.05$, Supplemental Figure S2D). Therefore, these results suggest that the combined transplantation of PD-MSCs with WKYMVm improves hepatic function in the BDL rat model.

3.4. PD-MSCs Combined with WKYMVm Promote Vascular Regeneration in the BDL Rat Liver

To evaluate the effect of PD-MSCs combined with WKYMVm on hepatic angiogenesis, we analyzed histological changes in portal tracks in liver tissues by H&E staining and the expression levels of angiogenic factors. As shown in Figure 2A, the liver tissues from the NTx group exhibited irregular portal tracts and an increased diameter of the portal vein compared with those of the control group. In contrast, the Tx+WK group showed recovery of the shape of the portal tracts and a decrease in the diameter of the portal vein compared with the NTx and Tx groups ($p < 0.05$, Figure 2A,B). The concentration of VEGF was lower in the NTx group than in the control but significantly increased in the Tx+WK group, which had a much higher level than the Tx group ($p < 0.05$, Figure 2C). The mRNA expression of VEGF in the Tx+WK group was dramatically increased compared to that in the NTx group ($p < 0.05$, Figure 2D). The levels of VEGFR1 and VEGFR2 showed an increasing tendency in the Tx and Tx+WK groups compared to the NTx group ($p < 0.05$, Figure 2E,F).

Additionally, the level of endoglin was significantly upregulated in the Tx+WK group compared to the NTx group ($p < 0.05$, Figure 2G). Furthermore, we examined the expression of the fibrotic marker alpha-smooth muscle actin (α-SMA) and the vascular marker vWF in liver sections using an immunofluorescence assay at 2 weeks after transplantation. Compared with the NTx and Tx groups, the Tx+WK group showed substantially decreased expression of α-SMA and increased vWF expression ($p < 0.05$, Figure 2H–J). These data indicate that PD-MSCs combined with WKYMVm promote vascular regeneration through activation of angiogenic factors in the BDL rat model.

3.5. PD-MSCs Combined with WKYMVm Promote Angiogenic Activity in HUVECs

To further determine the angiogenic effect of PD-MSCs combined with WKYMVm in vitro, we used 5-FU as a functional inhibitor of endothelial cells and cocultured the cells with PD-MSCs and WKYMVm, as shown in Figure 3A. The viability of the 5-FU-treated HUVECs was strongly decreased compared with that of the control, whereas the cells co-cultured with PD-MSCs alone or PD-MSCs with WKYMVm showed a significant increase ($p < 0.05$, Figure 3B). The concentration of VEGF secreted by HUVECs was dramatically increased in the cells cocultured with PD-MSCs and WKYMVm compared to the 5-FU-treated HUVECs and those cocultured with PD-MSCs. Moreover, the VEGF level showed an increase greater than that of the control ($p < 0.05$, Figure 3C). Previous studies have shown that the canonical Wnt/β-catenin signaling pathway plays a regulatory role in vascular regeneration [32]. Thus, we examined whether PD-MSCs and WKYMVm affect the expression of β-catenin in HUVECs. Immunofluorescence showed that reduced active β-catenin expression in the 5-FU-treated HUVECs was significantly enhanced by PD-MSCs and WKYMVm compared with the control ($p < 0.05$, Figure 3D,E).

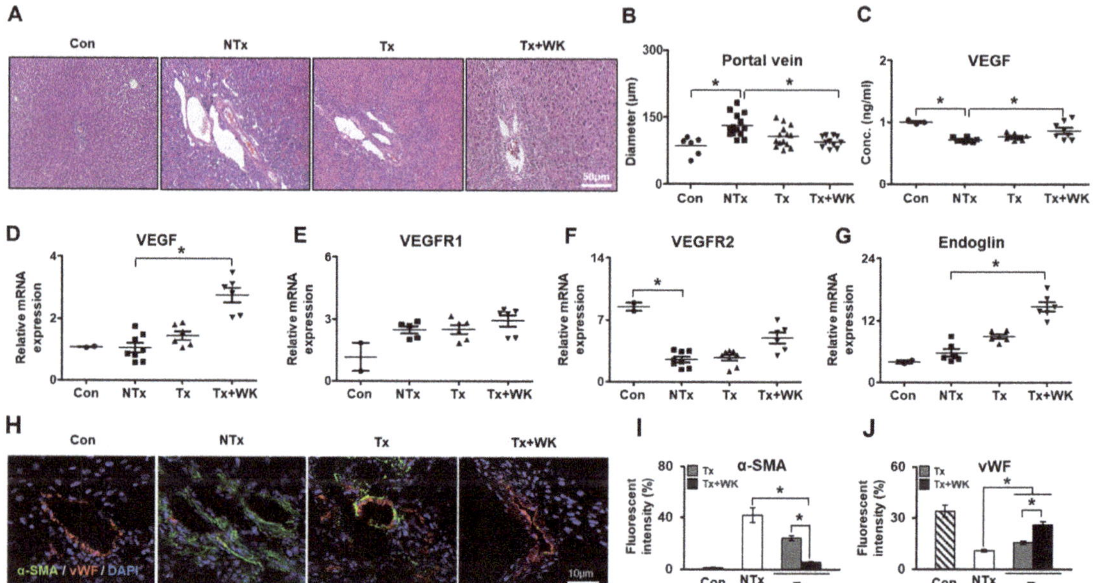

Figure 2. PD-MSCs combined with WKYMVm promoted vascular regeneration in the BDL rat liver. (**A**) Histological phenotype of BDL rat liver by H&E staining. Scale bar = 50 μm. (**B**) Quantification of portal vein diameter in BDL rat liver through H&E staining. (**C**) VEGF level in BDL rat serum shown by ELISAs. mRNA expression of VEGF (**D**), VEGFR1 (**E**), VEGFR2 (**F**), and CD105 (**G**) shown by qRT-PCR. (**H**) Expression of α-SMA and vWF in BDL rat livers by immunofluorescence. Scale bar = 10 μm. Quantified fluorescence intensity of α-SMA (**I**) and vWF (**J**) through immunofluorescence. $n = 3$ rats per group, mean ± SD, * $p < 0.05$ by one-way ANOVA. Con, control; NTx, nontransplantation group; Tx, PD-MSC transplantation group; Tx+WK, PD-MSCs with WKYMVm combined transplantation group; wk, week.

Additionally, the protein level of β-catenin in the nucleus was significantly upregulated in the 5-FU-treated HUVECs cocultured with PD-MSCs and WKYMVm versus the HUVECs with only 5-FU treatment ($p < 0.05$, Figure 3F). Tube formation by HUVECs was significantly increased by PD-MSCs and WKYMVm treatment ($p < 0.05$, Figure 3G,H). However, the expression of TGF-β, which is the key factor of endothelial–mesenchymal transition (End-MT), was significantly decreased by PD-MSCs and WKYMVm compared to that of the HUVECs with 5-FU treatment alone ($p < 0.05$, Figure 3I). To examine the effect of PD-MSCs and WKYMVm on endothelial permeability, we performed a dextran permeability assay, as shown in Figure 3J. PD-MSCs alone and PD-MSCs combined with WKYMVm induced a significant decrease in the dextran permeability of the 5-FU-treated HUVECs ($p < 0.05$, Figure 3K). Taken together, these results suggest that coculture with PD-MSCs and WKYMVm promotes angiogenic activities of ECs by upregulating the expression of VEGF and β-catenin.

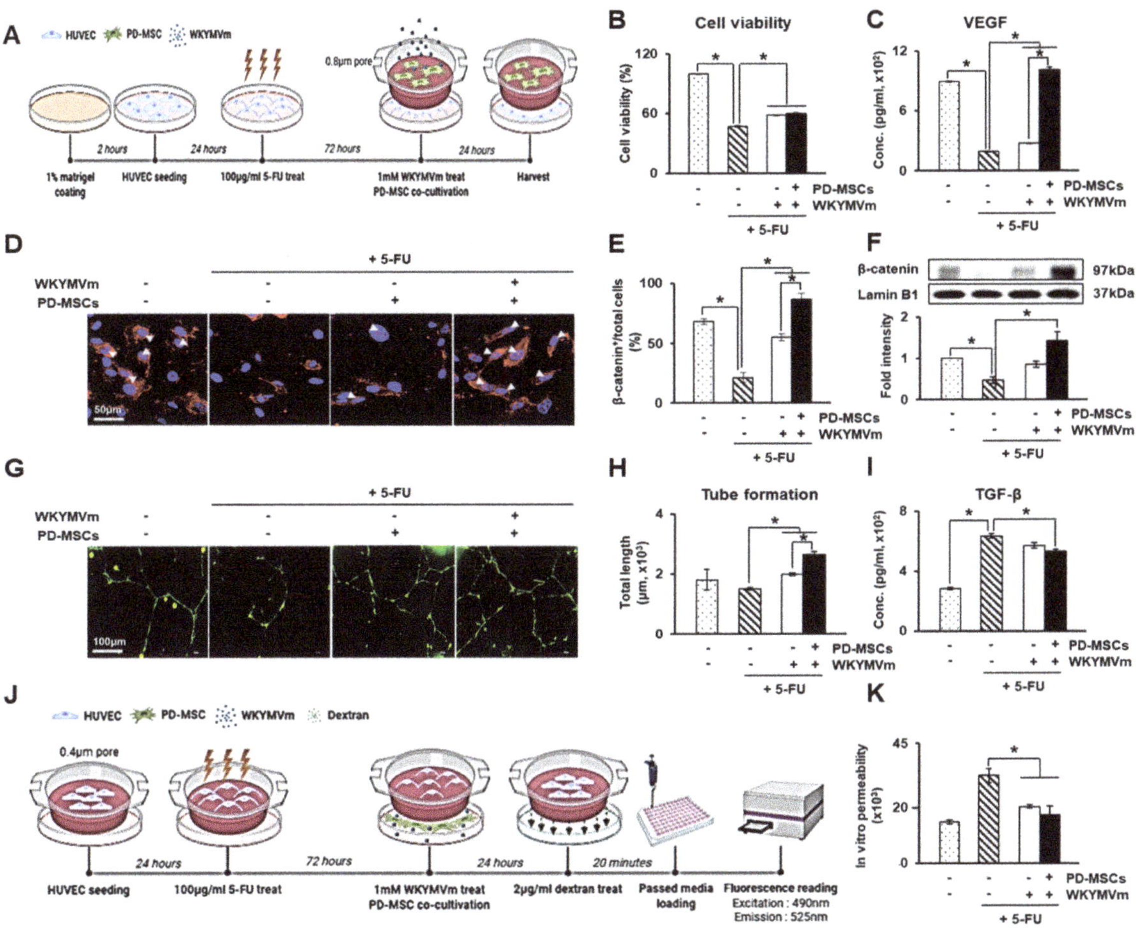

Figure 3. PD-MSCs combined with WKYMVm promote angiogenic activity in HUVECs. (A) In vitro scheme in HUVECs. (B) Cell viability in the 5-FU-treated HUVECs determined by CCK-8 assays. (C) VEGF level in the culture supernatant of HUVECs treated with 5-FU shown by ELISAs. (D) Localization and expression of active β-catenin through immunofluorescence. Scale bar = 50 μm. (E) Quantification of active β-catenin-positive HUVECs versus total HUVECs using immunofluorescence. (F) The protein expression of active β-catenin in the nucleus of HUVECs. (G) Tube formation of HUVECs injured by 5-FU. Scale bar = 100 μm. (H) Quantification of total tube length through tube formation assays. (I) TGF-β levels in the culture supernatant of HUVECs treated with 5-FU shown by ELISAs. (J) Schematic figure of the dextran permeability assay in the HUVECs treated with 5-FU. (K) Dextran permeability of the HUVECs injured by 5-FU. $n = 3$ per group, mean ± SD, * $p < 0.05$ by t tests.

3.6. PD-MSCs Combined with WKYMVm Attenuate Hepatic Fibrosis in the BDL Rat Liver

The effect of PD-MSCs combined with WKYMVm against BDL-induced hepatic fibrosis was evaluated through Sirius Red. As shown in Figure 4A, extensive accumulation of collagen was observed in the liver tissues of the NTx group. In contrast, accumulated collagen was shown to decrease in the Tx group. Additionally, collagen deposition was evidently reduced in the Tx+WK group compared to the NTx group ($p < 0.05$, Figure 4A,B). To verify the antifibrotic effect of the Tx and Tx+WK groups, we examined the expression levels of epithelial markers (e.g., E-cadherin) and mesenchymal and fibrogenic markers (e.g., vimentin, Col I, and α-SMA) at the protein level. As shown by E-cadherin expression, the control maintained basal levels. However, in the Tx and Tx+WK groups, the level of E-cadherin was dramatically increased compared with that in the NTx group ($p < 0.05$, Figure 4C).

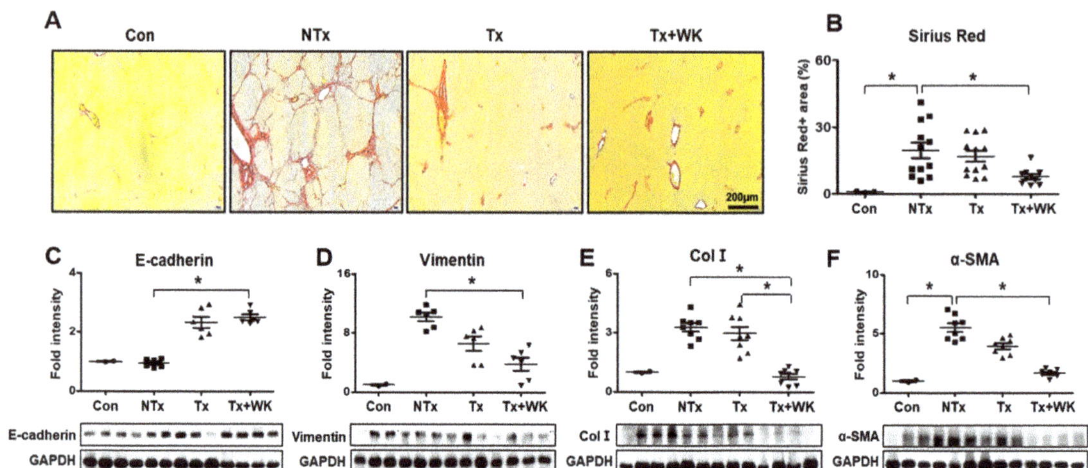

Figure 4. PD-MSCs combined with WKYMVm attenuated hepatic fibrosis in the BDL rat liver. (**A**) Collagen deposition in BDL rat liver shown by Sirius Red staining. Scale bar = 200 μm. (**B**) Quantification of the Sirius Red-positive area in BDL rat livers through Sirius red staining. Protein expression of E-cadherin (**C**), vimentin (**D**), Col I (**E**), and α-SMA (**F**) in BDL rat livers shown by Western blots. Protein expression was normalized to GAPDH expression through Western blot bands. $n = 3$ rats per group, mean ± SD, * $p < 0.05$ by one-way ANOVA. Con, control; NTx, nontransplantation group; Tx, PD-MSC transplantation group; Tx+WK, PD-MSCs with WKYMVm combined transplantation group; wk, week.

In contrast to the expression of E-cadherin, the expression of vimentin and α-SMA showed a decreasing tendency in the Tx group and a significant decrease in the Tx+WK group ($p < 0.05$, Figure 4D,F). In particular, Col I was evidently downregulated in the Tx+WK group compared to the NTx and Tx groups ($p < 0.05$, Figure 4E). These results indicate that administration of PD-MSCs combined with WKYMVm alleviates hepatic fibrosis by upregulating the expression of epithelial markers and inhibiting the expression of mesenchymal markers in a BDL rat model.

3.7. PD-MSCs Combined with WKYMVm Inhibit HSC Activation In Vitro

To further investigate the antifibrotic effect of PD-MSCs and WKYMVm on hepatic stellate cells (HSCs), we cocultured HSCs activated by TGF-β treatment with PD-MSCs and WKYMVm, as shown in a Figure 5A. The mRNA and protein levels of Col I and α-SMA in the TGF-β-treated HSCs were substantially upregulated compared to those of the control, whereas these levels were significantly downregulated by cotreatment with PD-MSCs alone or PD-MSCs and WKYMVm, and PD-MSCs and WKYMVm had greater inhibitory effects ($p < 0.05$, Figure 5B–E). In addition, immunofluorescence data showed that cotreatment with PD-MSCs and WKYMVm significantly reduced α-SMA expression compared to the other treatments ($p < 0.05$, Figure 5F,G). Matrix metalloproteinases (MMPs) play crucial roles in the regulation of hepatic fibrosis and regeneration. MMP2 and MMP9 specialize in degrading extracellular matrices secreted from activated HSCs [33]. Through gelatin zymography, the activities of MMP2 and MMP9 were shown to be reduced in the TGF-β-treated

HSCs compared to the control, but the activities were elevated by cotreatment of PD-MSCs and WKYMVm ($p < 0.05$, Figure 5H,I). These results suggest that treatment with PD-MSCs combined with WKYMVm inhibits fibrogenesis of activated HSCs.

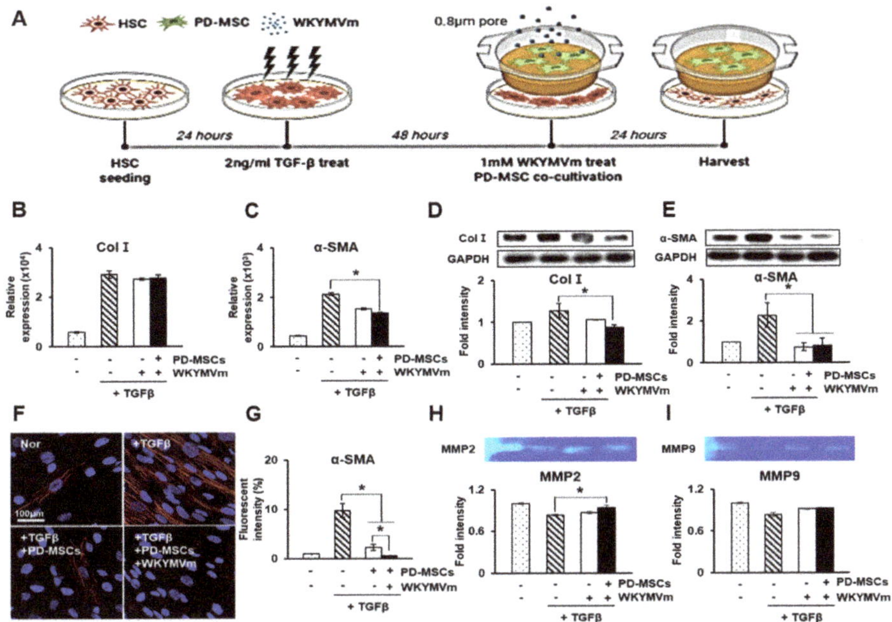

Figure 5. PD-MSCs combined with WKYMVm inhibit HSC activation in vitro. (**A**) Schematic figure of HSC in vitro modeling. mRNA expression of Col I (**B**) and α-SMA (**C**) in HSCs treated with TGF-β shown by qRT-PCR. Protein levels of Col I (**D**) and α-SMA (**E**) shown by Western blots. (**F**) α-SMA expression in HSCs treated with TGF-β, as determined by immunofluorescence. Scale bar = 100 μm. (**G**) Fluorescence intensity of α-SMA, as determined by immunofluorescence. Enzyme activity of MMP-2 (**H**) and MMP-9 (**I**) in the culture supernatant of HSCs treated with TGF-β shown by gelatin zymography. $n = 3$ per group, mean ± SD, * $p < 0.05$ by t tests.

3.8. PD-MSCs Combined with WKYMVm Improve Hepatic Regeneration in the BDL Rat Model

Hepatocyte nuclear factor 1 alpha (HNF1α) is a transcription factor that regulates the expression of several liver-specific genes [34]. The mRNA level of HNF1α was slightly increased in the Tx and Tx+WK groups compared with the NTx group (Figure 6A). The mRNA and protein levels of ALB were significantly increased in the Tx+WK group compared with the NTx and Tx groups ($p < 0.05$, Figure 6B,E). Additionally, the receptors of the WKYMVm peptide are FPR2 and HGF receptors (HGFR; tyrosine-protein kinase Met, c-Met) [35]. The mRNA levels of FPR2 and HGFR in rat liver were maintained at basal levels. In the NTx group, their expression levels showed a decreasing tendency compared with those of the control. However, PD-MSCs with WKYMVm resulted in significantly increased expression compared to that in the NTx group ($p < 0.05$, Figure 6C,D).

Hepatic regeneration is initiated by several growth factors and the IL-6/gp130/STAT3 signaling pathway. To determine whether the combination therapy of PD-MSCs with WKYMVm alleviates hepatic damage in the BDL model, we analyzed the expression of factors related to liver regeneration using Western blotting. We examined the protein levels of IL-6, gp130 and the phosphorylated form of STAT3 in the BDL rat liver. Their expression levels were decreased in the NTx group but gradually increased in the Tx and Tx+WK groups (Figure 6E). The serum level of HGF was also significantly increased in the

Tx+WK group compared to the NTx group ($p < 0.05$, Figure 6H). To further examine the effect of the combination therapy of PD-MSCs with WKYMVm on hepatocyte proliferation, we assessed PCNA expression in liver tissues by immunohistochemistry and cyclin D1 expression by Western blot. Few PCNA-positive hepatocytes were observed in the NTx group, while a significant increase in PCNA-positive cells was found in the Tx+WK group compared to the NTx and Tx groups ($p < 0.05$, Figure 6F,G). Additionally, the level of cyclin D1 was upregulated in the Tx+WK group ($p < 0.05$, Figure 6E). These findings suggest that combination therapy with PD-MSCs and WKYMVm in the rats with BDL promotes hepatic regeneration through induction of hepatocyte-specific proteins and the IL-6/gp130/STAT3 signaling pathway.

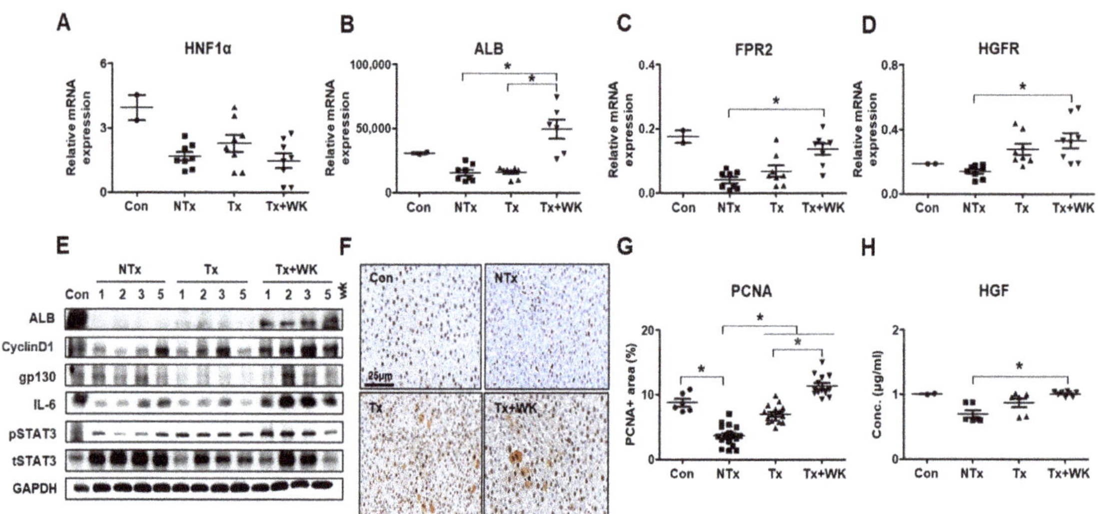

Figure 6. PD-MSCs combined with WKYMVm enhance hepatic regeneration in the BDL rat model. mRNA expression of HNF1α (**A**), ALB (**B**), FPR2 (**C**), and HGFR (**D**) in BDL rat liver shown by qRT-PCR. (**E**) Protein levels of ALB, cyclin D1, gp130, IL-6, p/tSTAT3, and GAPDH in BDL rat livers shown by Western blots. (**F**) Expression of PCNA in BDL rat liver shown by immunohistochemistry. Scale bar = 25 μm. (**G**) Quantification of the PCNA-positive area in BDL rat livers through immunohistochemistry. (**H**) Secreted HGF level in BDL rat serum shown by ELISAs. $n = 3$ rats per group, mean ± SD, * $p < 0.05$ by one-way ANOVA. Con, control; NTx, nontransplantation group; Tx, PD-MSC transplantation group; Tx+WK, PD-MSCs with WKYMVm combined transplantation group; wk, week.

3.9. PD-MSCs Combined with WKYMVm Can Regenerate Damaged Hepatocytes In Vitro

To further analyze their mode of action in vitro, we performed a coculture experiment of LCA-treated WB-F344 rat liver epithelial cells with PD-MSCs and WKYMVm, as shown in Figure 7A. Cell proliferation was measured using a CCK-8 assay. The CCK-8 assay demonstrated that treatment of WB-F344 cells with PD-MSCs and WKYMVm significantly increased proliferation ($p < 0.05$, Figure 7B). Treatment of the LCA-treated WB-F344 cells with PD-MSCs and WKYMVm strongly enhanced the protein expression of ALB and HNF1α, as shown by Western blot, compared to that of the LCA-treated group ($p < 0.05$, Figure 7C,D). Moreover, immunofluorescence data showed downregulated HNF1α expression in the LCA-treated WB-F344. However, HNF1α expression was significantly upregulated by treatment with PD-MSCs and WKYMVm, similar to the control ($p < 0.05$, Figure 7E,F). These data indicate that cotreatment with PD-MSCs and WKYMVm may promote the repair and regeneration of injured hepatocytes.

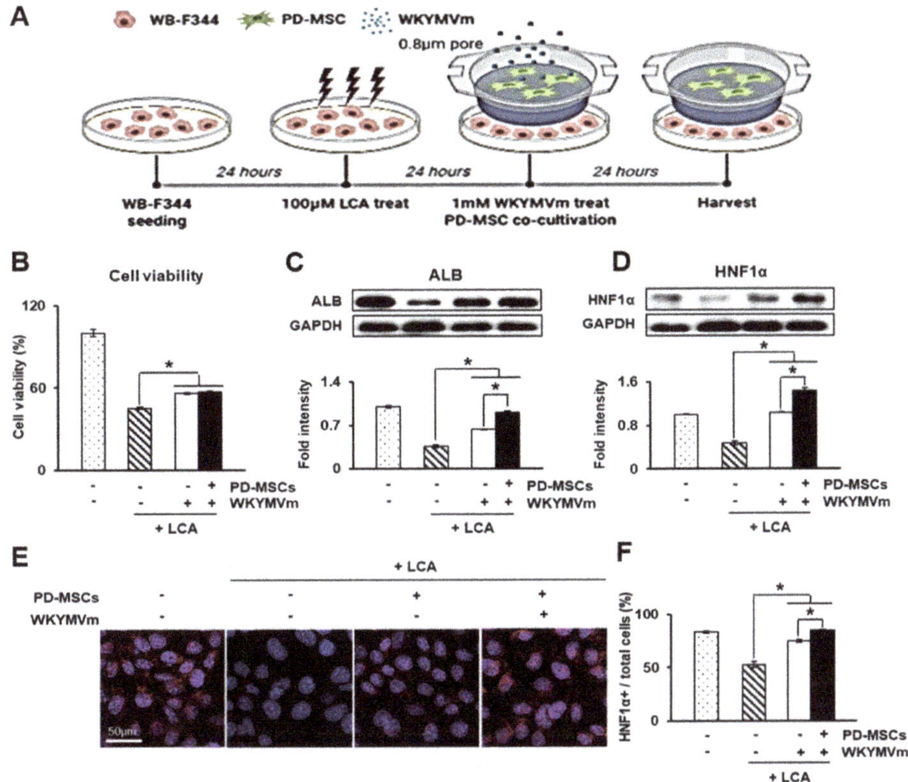

Figure 7. PD-MSCs combined with WKYMVm can regenerate damaged hepatocytes in vitro. (**A**) Schematic figure of WB-F344s cell in vitro modeling. (**B**) Cell viability of the LCA-treated WB-F344s cells shown by CCK-8 assays. Protein expression of ALB (**C**) and HNF1α (**D**) shown using Western blots. Protein expression was normalized to GAPDH expression through Western blot bands. (**E**) Localization and expression of translocated HNF1α in the nucleus shown through immunofluorescence. (**F**) Quantification of translocated HNF1α-positive WB-F344s versus total WB-F344s using immunofluorescence. $n = 3$ per group, mean ± SD, * $p < 0.05$ by t tests.

4. Discussion

A disruption in angiogenesis contributes to numerous ischemic, inflammatory, immune, and malignant disorders [36]. Upon injury to cells or tissues, inflammation or hypoxia results in the generation of angiogenic mediators that regulate the migration of vascular precursor cells from their niche to the site of injury [37]. In our study, we first demonstrated that the combination of PD-MSCs and WKYMVm has proangiogenic and antifibrotic effects and therapeutic effectiveness in a BDL-injured rat model. PD-MSCs and WKYMVm decreased the levels of AST, ALT, and total bilirubin while increasing ALB and hepatokines levels, which are important indices of hepatic function ($p < 0.05$, Supplemental Figure S2A–D). Additionally, PD-MSCs with WKYMVm improved hepatic architectural distortion and collagen deposition in fibrotic liver tissues ($p < 0.05$, Figures 2A and 4A).

During hepatic injury, HSCs activated by TGF-β upregulate the expression of Col I and α-SMA. The NTx group showed dramatically activated HSCs, as demonstrated by the increased expression of vimentin, Col I, and α-SMA in Western blots, while the Tx and Tx+WK groups displayed significantly downregulated expression levels of these molecules ($p < 0.05$, Figure 4D–F). In vitro, PD-MSCs and WKYMVm repressed the expression of α-

SMA and Col I in the rat HSCs, which underwent trans-differentiation by TGF-β ($p < 0.05$, Figure 5B–G). Our in vivo and in vitro studies showed that the combination therapy of PD-MSCs and WKYMVm has an antifibrotic effect in the liver.

In hepatic regeneration, vascular remodeling plays an important role in many physiological and pathological events [38,39]. HSCs and ECs can promote physiological angiogenesis by expressing a variety of angiogenic factors [40]. VEGF is a key regulator of angiogenesis and vascular regeneration and is also known as an essential element for EC viability. In partial hepatectomy or acute hepatic failure, VEGF is evidently increased, and sinusoidal reconstruction is improved during liver regeneration [5,41]. Yang and his colleagues found that VEGF has a dual and opposing role in fibrogenesis and resolution of fibrosis through critical effects of VEGF on vascular permeability [8]. In ischemic heart disease, WKYMVm had proangiogenic effects via VEGF signaling [26]. Additionally, in our study, PD-MSCs with WKYMVm significantly induced vascular regeneration in vivo and in vitro, as demonstrated by the increased expression of angiogenic factors and tube formation and decreased dextran permeability ($p < 0.05$, Figures 2 and 3).

Recently, ECs were shown to differentiate into myofibroblasts through EndMT in the lung, kidney, and heart [42–44]. Ribera and his colleagues showed the mesenchymal phenotype of ECs in CCl_4-induced liver cirrhosis via bone morphogenic protein-7 (BMP-7)/TGF-β signaling [45]. Dufton et al. showed that the ETS-related gene (ERG), a transcription factor in the endothelial lineage, repressed EndMT in a CCl4-injured cirrhotic mouse model. Additionally, ERG attenuated endothelial-dependent hepatic fibrosis by inhibiting SMAD2/3 signaling while maintaining hepatic homeostasis by inducing the SMAD1 pathway [46]. We also demonstrated that the levels of mesenchymal markers, such as vimentin, Col I, and α-SMA, were dramatically decreased in the PD-MSC-transplanted group and in the group treated with PD-MSCs and WKYMVm in cirrhotic rats with BDL ($p < 0.05$, Figure 4).

Park et al. reported that WKYMVm decreases dermal thickness and inhibits scleroderma fibrosis in bleomycin-induced mice by repressing vimentin and phosphorylated SMAD3 expression in myofibroblasts [47]. Additionally, the peptide was demonstrated to inhibit osteoclastogenesis by decreasing the levels of inflammatory cytokines such as IL-1β and tumor necrosis factor-alpha (TNF-α) via the CD9/gp130/STAT3 pathway [48]. However, WKYMVm has a short half-life. Park et al. reported that the anti-inflammatory effect of peptides is constrained since high levels are required for therapeutic efficacy [49]. Therefore, other studies sought to address the limitation of WKYMVm. Additionally, PLGA microspheres were used to overcome the short half-life of WKYMVm [29]. To extend the short elimination half-life of WKYMVm, we developed a combination therapy of PD-MSCs with WKYMVm. For therapy using MSCs, other medications or functional genes have been used to enhance the function of MSCs.

5. Conclusions

The administration of PD-MSCs with WKYMVm peptide improves vascular regeneration by activated FPR2 signaling while attenuating hepatic fibrosis in a rat model of liver cirrhosis. Furthermore, liver regeneration is promoted by PD-MSCs with WKYMVm by increasing the expression of hepatic function markers. Therefore, this study suggests the strong therapeutic efficacy of PD-MSCs combined with WKYMVm as a regulator of hepatic function as well as vascular regeneration in hepatic failure.

Supplementary Materials: The following are available online at https://www.mdpi.com/article/10.3390/cells11020232/s1, Figure S1: Animal experimental scheme, Figure S2: PD-MSCs combined with WKYMVm enhance hepatic function in the BDL rat model, Table S1: Primer sequences using quantitative real time polymerase chain reaction.

Author Contributions: J.H.J.; conceptualization, methodology, validation, formal analysis, investigation, writing—original draft preparation, and visualization, S.P.; writing—review and editing, J.Y.K.; investigation, J.-Y.L.; formal analysis and visualization, G.T.P.; resources, J.H.K.; methodology and resources, G.J.K.; conceptualization, writing—review and editing, project administration and funding acquisition. All authors have read and agreed to the published version of the manuscript.

Funding: This work was supported by the Basic Science Research Program through the National Research Foundation of Korea (NRF) funded by the Ministry of Science, ICT & Future Planning (2017M3A9B4061665).

Institutional Review Board Statement: All participants provided written informed consent prior to the collection of placentas. The collection and use of placentas were approved by the Institutional Review Board of Gangnam CHA General Hospital, Seoul, Korea (IRB 07-18). All animal experimental protocols were approved by the Institutional Animal Care and Use Committee of CHA University, Seongnam, Korea (IACUC-180023).

Informed Consent Statement: Informed consent was obtained from all subjects involved in the study.

Data Availability Statement: Not applicable.

Conflicts of Interest: The authors declare no conflict of interest.

References

1. Otrock, Z.K.; Mahfouz, R.A.R.; Makarem, J.A.; Shamseddine, A.I. Understanding the biology of angiogenesis: Review of the most important molecular mechanisms. *Blood Cells Mol. Dis.* **2007**, *39*, 212–220. [CrossRef] [PubMed]
2. Praktiknjo, M.; Lehmann, J.; Nielsen, M.J.; Schierwagen, R.; Uschner, F.E.; Meyer, C.; Thomas, D.; Strassburg, C.P.; Bendtsen, F.; Møller, S.; et al. Acute decompensation boosts hepatic collagen type III deposition and deteriorates experimental and human cirrhosis. *Hepatol. Commun.* **2018**, *2*, 211–222. [CrossRef] [PubMed]
3. Friedman, S.L.; Sheppard, D.; Duffield, J.S.; Violette, S. Therapy for Fibrotic Diseases: Nearing the Starting Line. *Sci. Transl. Med.* **2013**, *5*, 167sr1. [CrossRef] [PubMed]
4. Liu, L.; Yannam, G.R.; Nishikawa, T.; Yamamoto, T.; Basma, H.; Ito, R.; Nagaya, M.; Dutta-Moscato, J.; Stolz, D.B.; Duan, F.; et al. The microenvironment in hepatocyte regeneration and function in rats with advanced cirrhosis. *Hepatology* **2012**, *55*, 1529–1539. [CrossRef]
5. Poisson, J.; Lemoinne, S.; Boulanger, C.; Durand, F.; Moreau, R.; Valla, D.; Rautou, P.-E. Liver sinusoidal endothelial cells: Physiology and role in liver diseases. *J. Hepatol.* **2017**, *66*, 212–227. [CrossRef] [PubMed]
6. Nolan, D.J.; Ginsberg, M.; Israely, E.; Palikuqi, B.; Poulos, M.G.; James, D.; Ding, B.-S.; Schachterle, W.; Liu, Y.; Rosenwaks, Z.; et al. Molecular Signatures of Tissue-Specific Microvascular Endothelial Cell Heterogeneity in Organ Maintenance and Regeneration. *Dev. Cell* **2013**, *26*, 204–219. [CrossRef] [PubMed]
7. Taniguchi, E.; Sakisaka, S.; Matsuo, K.; Tanikawa, K.; Sata, M. Expression and role of vascular endothelial growth factor in liver regeneration after partial hepatectomy in rats. *J. Histochem. Cytochem.* **2001**, *49*, 121–129. [CrossRef] [PubMed]
8. Yang, L.; Kwon, J.; Popov, Y.; Gajdos, G.B.; Ordog, T.; Brekken, R.A.; Mukhopadhyay, D.; Schuppan, D.; Bi, Y.; Simonetto, D.; et al. Vascular Endothelial Growth Factor Promotes Fibrosis Resolution and Repair in Mice. *Gastroenterology* **2014**, *146*, 1339–1350.e1. [CrossRef]
9. Cao, Z.; Ye, T.; Sun, Y.; Ji, G.; Shido, K.; Chen, Y.; Luo, L.; Na, F.; Li, X.; Huang, Z.; et al. Targeting the vascular and perivascular niches as a regenerative therapy for lung and liver fibrosis. *Sci. Transl. Med.* **2017**, *9*, eaai8710. [CrossRef]
10. Hanai, J.-I.; Dhanabal, M.; Karumanchi, S.A.; Albanese, C.; Waterman, M.; Chan, B.; Ramchandran, R.; Pestell, R.; Sukhatme, V.P. Endostatin Causes G1 Arrest of Endothelial Cells through Inhibition of Cyclin D1. *J. Biol. Chem.* **2002**, *277*, 16464–16469. [CrossRef] [PubMed]
11. Korn, C.; Scholz, B.; Hu, J.; Srivastava, K.; Wojtarowicz, J.; Arnsperger, T.; Adams, R.H.; Boutros, M.; Augustin, H.G.; Augustin, I. Endothelial cell-derived non-canonical Wnt ligands control vascular pruning in angiogenesis. *Development* **2014**, *141*, 1757–1766. [CrossRef] [PubMed]
12. Kim, M.-D.; Kim, S.-S.; Cha, H.-Y.; Jang, S.-H.; Chang, D.-Y.; Kim, W.; Suh-Kim, H.; Lee, J.-H. Therapeutic effect of hepatocyte growth factor-secreting mesenchymal stem cells in a rat model of liver fibrosis. *Exp. Mol. Med.* **2014**, *46*, e110. [CrossRef] [PubMed]
13. Ohnishi, S.; Sumiyoshi, H.; Kitamura, S.; Nagaya, N. Mesenchymal stem cells attenuate cardiac fibroblast proliferation and collagen synthesis through paracrine actions. *FEBS Lett.* **2007**, *581*, 3961–3966. [CrossRef] [PubMed]
14. Kim, S.-H.; Jung, J.; Cho, K.J.; Choi, J.-H.; Lee, H.S.; Kim, G.J.; Lee, S.G. Immunomodulatory Effects of Placenta-derived Mesenchymal Stem Cells on T Cells by Regulation of FoxP3 Expression. *Int. J. Stem Cells* **2018**, *11*, 196–204. [CrossRef] [PubMed]
15. Lee, H.-J.; Jung, J.; Cho, K.J.; Lee, C.K.; Hwang, S.-G.; Kim, G.J. Comparison of in vitro hepatogenic differentiation potential between various placenta-derived stem cells and other adult stem cells as an alternative source of functional hepatocytes. *Differentiation* **2012**, *84*, 223–231. [CrossRef]

16. Kim, M.J.; Shin, K.S.; Jeon, J.H.; Lee, D.R.; Shim, S.H.; Kim, J.K.; Cha, D.-H.; Yoon, T.K.; Kim, G.J. Human chorionic-plate-derived mesenchymal stem cells and Wharton's jelly-derived mesenchymal stem cells: A comparative analysis of their potential as placenta-derived stem cells. *Cell Tissue Res.* **2011**, *346*, 53–64. [CrossRef]
17. Kim, G.D.; Choi, J.H.; Lim, S.M.; Jun, J.H.; Moon, J.W.; Kim, G.J. Alterations in IL-6/STAT3 Signaling by Korean Mistletoe Lectin Regulate the Self-Renewal Activity of Placenta-Derived Mesenchymal Stem Cells. *Nutrients* **2019**, *11*, 2604. [CrossRef] [PubMed]
18. Jung, J.; Moon, J.W.; Choi, J.-H.; Lee, Y.W.; Park, S.H.; Kim, G.J. Epigenetic Alterations of IL-6/STAT3 Signaling by Placental Stem Cells Promote Hepatic Regeneration in a Rat Model with CCl4-induced Liver Injury. *Int. J. Stem Cells* **2015**, *8*, 79–89. [CrossRef] [PubMed]
19. Jung, J.; Choi, J.H.; Lee, Y.; Park, J.-W.; Oh, I.-H.; Hwang, S.-G.; Kim, K.-S.; Kim, G.J. Human Placenta-Derived Mesenchymal Stem Cells Promote Hepatic Regeneration in CCl_4-Injured Rat Liver Model via Increased Autophagic Mechanism. *Stem Cells* **2013**, *31*, 1584–1596. [CrossRef] [PubMed]
20. Bin Lee, Y.; Choi, J.H.; Kim, E.N.; Seok, J.; Lee, H.-J.; Yoon, J.H.; Kim, G.J. Human Chorionic Plate-Derived Mesenchymal Stem Cells Restore Hepatic Lipid Metabolism in a Rat Model of Bile Duct Ligation. *Stem Cells Int.* **2017**, *2017*, 5180579. [CrossRef]
21. Kim, J.Y.; Jun, J.H.; Park, S.Y.; Yang, S.W.; Bae, S.H.; Kim, G.J. Dynamic Regulation of miRNA Expression by Functionally Enhanced Placental Mesenchymal Stem Cells Promotes Hepatic Regeneration in a Rat Model with Bile Duct Ligation. *Int. J. Mol. Sci.* **2019**, *20*, 5299. [CrossRef]
22. Gavins, F.N.E. Are formyl peptide receptors novel targets for therapeutic intervention in ischaemia–reperfusion injury? *Trends Pharmacol. Sci.* **2010**, *31*, 266–276. [CrossRef]
23. Choi, Y.H.; Jang, I.H.; Heo, S.C.; Kim, J.H.; Hwang, N.S. Biomedical therapy using synthetic WKYMVm hexapeptide. *Organogenesis* **2016**, *12*, 53–60. [CrossRef] [PubMed]
24. Jang, I.H.; Heo, S.C.; Kwon, Y.W.; Choi, E.J.; Kim, J.H. Role of formyl peptide receptor 2 in homing of endothelial progenitor cells and therapeutic angiogenesis. *Adv. Biol. Regul.* **2015**, *57*, 162–172. [CrossRef] [PubMed]
25. Heo, S.C.; Kwon, Y.W.; Jang, I.H.; Jeong, G.O.; Yoon, J.W.; Kim, C.D.; Kwon, S.M.; Bae, Y.-S.; Kim, J.H. WKYMVm-induced activation of formyl peptide receptor 2 stimulates ischemic neovasculogenesis by promoting homing of endothelial colony-forming cells. *Stem Cells* **2014**, *32*, 779–790. [CrossRef] [PubMed]
26. Heo, S.C.; Kwon, Y.W.; Jang, I.H.; Jeong, G.O.; Lee, T.W.; Yoon, J.W.; Shin, H.J.; Jeong, H.C.; Ahn, Y.; Ko, T.H.; et al. Formyl Peptide Receptor 2 Is Involved in Cardiac Repair After Myocardial Infarction Through Mobilization of Circulating Angiogenic Cells. *Stem Cells* **2017**, *35*, 654–665. [CrossRef]
27. Kwon, Y.W.; Heo, S.C.; Jang, I.H.; Jeong, G.O.; Yoon, J.W.; Mun, J.-H.; Kim, J.H. Stimulation of cutaneous wound healing by an FPR2-specific peptide agonist WKYMVm. *Wound Repair Regen.* **2015**, *23*, 575–582. [CrossRef]
28. Hao, L.; Lei, X.; Zhou, H.; Marshall, A.J.; Liu, L. Critical role for PI3Kγ-dependent neutrophil reactive oxygen species in WKYMVm-induced microvascular hyperpermeability. *J. Leukoc. Biol.* **2019**, *106*, 1117–1127. [CrossRef]
29. Choi, Y.H.; Heo, S.C.; Kwon, Y.W.; Kim, H.D.; Kim, S.H.L.; Jang, I.H.; Kim, J.H.; Hwang, N.S. Injectable PLGA microspheres encapsulating WKYMVM peptide for neovascularization. *Acta Biomater.* **2015**, *25*, 76–85. [CrossRef]
30. Goodwin, A.M.; DAmore, P.A. Wnt signaling in the vasculature. *Angiogenesis* **2002**, *5*, 1–9. [CrossRef]
31. Roeb, E. Matrix metalloproteinases and liver fibrosis (translational aspects). *Matrix Biol.* **2018**, *68–69*, 463–473. [CrossRef] [PubMed]
32. Costa, R.H.; Kalinichenko, V.V.; Holterman, A.-X.L.; Wang, X. Transcription factors in liver development, differentiation, and regeneration. *Hepatology* **2003**, *38*, 1331–1347. [CrossRef]
33. Cattaneo, F.; Parisi, M.; Ammendola, R. WKYMVm-induced cross-talk between FPR2 and HGF receptor in human prostate epithelial cell line PNT1A. *FEBS Lett.* **2013**, *587*, 1536–1542. [CrossRef]
34. Carmeliet, P. Angiogenesis in life, disease and medicine. *Nature* **2005**, *438*, 932–936. [CrossRef] [PubMed]
35. Shi, X.; Zhang, W.; Yin, L.; Chilian, W.M.; Krieger, J.; Zhang, P. Vascular precursor cells in tissue injury repair. *Transl. Res.* **2017**, *184*, 77–100. [CrossRef] [PubMed]
36. Ober, E.A.; Lemaigre, F.P. Development of the liver: Insights into organ and tissue morphogenesis. *J. Hepatol.* **2018**, *68*, 1049–1062. [CrossRef] [PubMed]
37. Reynaert, H.; Chavez, M.; Geerts, A. Vascular endothelial growth factor and liver regeneration. *J. Hepatol.* **2001**, *34*, 759–761. [CrossRef]
38. Taura, K.; De Minicis, S.; Seki, E.; Hatano, E.; Iwaisako, K.; Osterreicher, C.H.; Kodama, Y.; Miura, K.; Ikai, I.; Uemoto, S.; et al. Hepatic Stellate Cells Secrete Angiopoietin 1 That Induces Angiogenesis in Liver Fibrosis. *Gastroenterology* **2008**, *135*, 1729–1738. [CrossRef]
39. Drixler, T.A.; Vogten, M.J.; Ritchie, E.D.; Van Vroonhoven, T.J.M.V.; Gebbink, M.F.B.G.; Voest, E.E.; Borel Rinkes, I.H.M. Liver Regeneration Is an Angiogenesis- Associated Phenomenon. *Ann. Surg.* **2002**, *236*, 703–712. [CrossRef]
40. Hashimoto, N.; Phan, S.H.; Imaizumi, K.; Matsuo, M.; Nakashima, H.; Kawabe, T.; Shimokata, K.; Hasegawa, Y. Endothelial–Mesenchymal Transition in Bleomycin-Induced Pulmonary Fibrosis. *Am. J. Respir. Cell Mol. Biol.* **2010**, *43*, 161–172. [CrossRef]
41. Potenta, S.; Zeisberg, E.; Kalluri, R. The role of endothelial-to-mesenchymal transition in cancer progression. *Br. J. Cancer* **2008**, *99*, 1375–1379. [CrossRef]

42. Zeisberg, E.M.; Tarnavski, O.; Zeisberg, M.; Dorfman, A.L.; McMullen, J.R.; Gustafsson, E.; Chandraker, A.; Yuan, X.; Pu, W.T.; Roberts, A.B.; et al. Endothelial-to-mesenchymal transition contributes to cardiac fibrosis. *Nat. Med.* **2007**, *13*, 952–961. [CrossRef] [PubMed]
43. Ribera, J.; Pauta, M.; Melgar-Lesmes, P.; Córdoba, B.; Bosch, A.; Calvo, M.; Rodrigo-Torres, D.; Sancho-Bru, P.; Mira, A.; Jiménez, W.; et al. A small population of liver endothelial cells undergoes endothelial-to-mesenchymal transition in response to chronic liver injury. *Am. J. Physiol.-Gastrointest. Liver Physiol.* **2017**, *313*, G492–G504. [CrossRef] [PubMed]
44. Dufton, N.P.; Peghaire, C.R.; Osuna-Almagro, L.; Raimondi, C.; Kalna, V.; Chauhan, A.; Webb, G.; Yang, Y.; Birdsey, G.M.; Lalor, P.; et al. Dynamic regulation of canonical TGFβ signalling by endothelial transcription factor ERG protects from liver fibrogenesis. *Nat. Commun.* **2017**, *8*, 895. [CrossRef] [PubMed]
45. Park, G.T.; Kwon, Y.W.; Lee, T.W.; Kwon, S.G.; Ko, H.-C.; Kim, M.B.; Kim, J.H. Formyl Peptide Receptor 2 Activation Ameliorates Dermal Fibrosis and Inflammation in Bleomycin-Induced Scleroderma. *Front. Immunol.* **2019**, *10*, 2095. [CrossRef] [PubMed]
46. Hu, J.; Li, X.; Chen, Y.; Han, X.; Li, L.; Yang, Z.; Duan, L.; Lu, H.; He, Q. The protective effect of WKYMVm peptide on inflammatory osteolysis through regulating NF-κB and CD9/gp130/STAT3 signalling pathway. *J. Cell. Mol. Med.* **2020**, *24*, 1893–1905. [CrossRef] [PubMed]
47. Kim, S.D.; Kim, Y.-K.; Lee, H.Y.; Kim, Y.-S.; Jeon, S.G.; Baek, S.-H.; Song, D.-K.; Ryu, S.H.; Bae, Y.-S. The Agonists of Formyl Peptide Receptors Prevent Development of Severe Sepsis after Microbial Infection. *J. Immunol.* **2010**, *185*, 4302–4310. [CrossRef]
48. Lee, M.-J.; Jung, J.; Na, K.-H.; Moon, J.S.; Lee, H.-J.; Kim, J.-H.; Kim, G.I.; Kwon, S.-W.; Hwang, S.-G.; Kim, G.J. Anti-fibrotic effect of chorionic plate-derived mesenchymal stem cells isolated from human placenta in a rat model of CCl_4-injured liver: Potential application to the treatment of hepatic diseases. *J. Cell. Biochem.* **2010**, *111*, 1453–1463. [CrossRef]
49. Jun, J.H.; Choi, J.H.; Bae, S.H.; Oh, S.H.; Kim, G.J. Decreased C-reactive protein induces abnormal vascular structure in a rat model of liver dysfunction induced by bile duct ligation. *Clin. Mol. Hepatol.* **2016**, *22*, 372–381. [CrossRef]

Article

Human Mesenchymal Stromal Cell Secretome Promotes the Immunoregulatory Phenotype and Phagocytosis Activity in Human Macrophages

Minna Holopainen [1,2,*], Ulla Impola [1], Petri Lehenkari [3], Saara Laitinen [1,†] and Erja Kerkelä [1,†]

1 Finnish Red Cross Blood Service, FI-00310 Helsinki, Finland; ulla.impola@bloodservice.fi (U.I.); saara.laitinen@bloodservice.fi (S.L.); erja.kerkela@bloodservice.fi (E.K.)

2 Molecular and Integrative Biosciences Research Programme, Faculty of Biological and Environmental Sciences, University of Helsinki, FI-00014 Helsinki, Finland

3 Department of Anatomy and Surgery, Institute of Translational Medicine, University of Oulu and Clinical Research Centre, FI-90014 Oulu, Finland; petri.lehenkari@oulu.fi

* Correspondence: minna.holopainen.fin@gmail.com

† These authors contributed equally to this paper.

Received: 31 August 2020; Accepted: 21 September 2020; Published: 22 September 2020

Abstract: Human mesenchymal stromal/stem cells (hMSCs) show great promise in cell therapy due to their immunomodulatory properties. The overall immunomodulatory response of hMSCs resembles the resolution of inflammation, in which lipid mediators and regulatory macrophages (Mregs) play key roles. We investigated the effect of hMSC cell-cell contact and secretome on macrophages polarized and activated toward Mreg phenotype. Moreover, we studied the effect of supplemented polyunsaturated fatty acids (PUFAs): docosahexaenoic acid (DHA) and arachidonic acid, the precursors of lipid mediators, on hMSC immunomodulation. Our results show that unlike hMSC cell-cell contact, the hMSC secretome markedly increased the CD206 expression in both Mreg-polarized and Mreg-activated macrophages. Moreover, the secretome enhanced the expression of programmed death-ligand 1 on Mreg-polarized macrophages and Mer receptor tyrosine kinase on Mreg-activated macrophages. Remarkably, these changes were translated into improved *Candida albicans* phagocytosis activity of macrophages. Taken together, these results demonstrate that the hMSC secretome promotes the immunoregulatory and proresolving phenotype of Mregs. Intriguingly, DHA supplementation to hMSCs resulted in a more potentiated immunomodulation with increased CD163 expression and decreased gene expression of matrix metalloproteinase 2 in Mreg-polarized macrophages. These findings highlight the potential of PUFA supplementations as an easy and safe method to improve the hMSC therapeutic potential.

Keywords: cell therapy; immunomodulation; polyunsaturated fatty acid; CD206; phagocytosis

1. Introduction

Mesenchymal stromal/stem cells (MSCs) show great promise in cell therapy, such as in the treatment of graft-versus-host disease [1,2] and Crohn's disease [3,4]. MSCs have diverse immunomodulatory effects, which are mediated via cell-cell contact and secreted paracrine factors, such as extracellular vesicles (EVs), tryptophan-degrading enzyme indoleamine-2,3-dioxygenase and lipid mediator prostaglandin E_2 (PGE_2) [5–7]. MSCs are able to, e.g., inhibit the proliferation of T cells and promote the generation of regulatory T cells [5,8]. Moreover, MSCs polarize macrophages toward a more anti-inflammatory phenotype by increasing the expression of multiple cell surface markers, such as CD206, and by enhancing their phagocytosis activity [9–12].

A tight classification of macrophages into different subtypes is redundant due to their plastic and rapidly changing phenotype giving rise to a heterogeneous population [13]. Yet for simplicity, macrophages are typically classified into classically activated, proinflammatory M1 phenotype and to wound healing, anti-inflammatory M2 phenotype. Regulatory macrophages (Mregs) represent an immunoregulatory phenotype that produce anti-inflammatory cytokines, such as interleukin (IL)-10 and transforming growth factor β1 (TGF-β1), potently suppress T-cell function and promote regulatory T-cell phenotype [14,15]. Interestingly, Mregs are also investigated as a potential adjunct therapy in renal transplantations (clinicaltrials.gov: NCT02085629). Although the effects of MSCs on monocytes and type M1 and M2 macrophages have been intensively studied, less is known about the effects of MSCs on Mregs. In our previous study, we observed that human bone marrow-derived MSCs (hBMSCs) and hBMSC-derived EVs (hBMSC-EVs) enhanced the anti-inflammatory phenotype of Mregs [16]. Both hBMSC cell-cell contact and EVs decreased the production of IL-23 and IL-22, which are up-regulated in inflammation and promote T helper 17 cell maintenance and proliferation, respectively. The hBMSCs and EVs also increased the production of PGE_2 [16], which is an essential lipid mediator in MSC function and induces the MSC-mediated skewing of macrophages toward an anti-inflammatory phenotype [11,17].

EVs are small (majority < 300 nm) lipid-bilayered particles secreted by cells through exocytosis and membrane budding. EVs carry intracellular messages by transporting lipids, proteins, nucleic acids, carbohydrates or their metabolites and can mediate immunological effects [18]. Intriguingly, MSC-EVs are able to mediate the therapeutic response of MSCs and have been investigated in various in vivo models, such as acute kidney injury [19], stroke [20,21] and sepsis [22]. Thus, MSC-EVs have emerged as a cell-free therapeutic option for MSCs.

The immunomodulatory response of MSCs resembles the resolution of inflammation, the active dampening phase of inflammation [23]. Lipid mediators, especially the specialized proresolving mediators (SPMs) promote the resolution of inflammation [24] by, e.g., reducing neutrophil trafficking, increasing macrophage polarization toward anti-inflammatory phenotype and macrophage efferocytosis of apoptotic neutrophils. Polyunsaturated fatty acids (PUFAs), such as *n*-3 docosahexaenoic acid (DHA), eicosapentaenoic acid (EPA) and *n*-6 arachidonic acid (AA) are precursors to lipid mediators [25]. DHA is a precursor to resolution-phase SPMs such as D-series resolvins, maresins and protectins and EPA to E-series resolvins. AA is a precursor to proresolving lipoxins, but also for prostaglandins (PGs), thromboxanes and leukotrienes with mainly proinflammatory functions.

We have previously demonstrated that the phospholipid, fatty acid and, importantly, lipid mediator profiles of hBMSCs can be modified with the supplementation of PUFAs [26,27]. hBMSCs cannot efficiently synthetize these long-chained PUFAs from *n*-6 and *n*-3 precursors rendering the PUFA supplementation into the culture medium essential to ensure a sufficient level of precursors for lipid mediator biosynthesis [26]. Interestingly, we also observed that PUFA supplementation to hBMSCs caused the subsequent remodeling of the phospholipid membrane of hBMSC-EVs [27]. Furthermore, EPA and DHA supplementation to murine MSCs increases their immunomodulatory capacity in allergic asthma [28] and sepsis [29,30] models, highlighting the importance of PUFA modifications on MSC immunomodulation.

In this study, we investigated the effect of hBMSC cell-cell contact and secretome on polarized and activated Mregs. Moreover, we investigated the immunomodulatory effect of hBMSCs on Mregs after DHA or AA supplementations, which alter the downstream lipid mediator profile of hBMSCs. Our results demonstrate that the hBMSC secretome skewed macrophages toward a more anti-inflammatory and proresolving phenotype. This effect was even more pronounced by the secretome of DHA-modified hBMSCs. Strikingly, we show for the first time that the hBMSC secretome enhanced the *Candida albicans* phagocytosis activity of macrophages by increasing the CD206 expression.

2. Materials and Methods

2.1. hBMSC Culture and PUFA Supplementation

The patient protocols of the hBMSC isolation were approved by the Ethical Committee of Northern Ostrobothnia Hospital District (ethical approval number: Oulu University hospital EETTMK 21/2011). The hBMSCs were collected from upper femur metaphysis of adult patients after receiving a written informed consent and characterized as described previously [31]. The cells have been characterized according to the guidelines of the International Society of Cell & Gene Therapy [32]. The cells expressed typical MSC markers and lacked the expression of hematopoietic stem cell markers, and the differentiation toward osteoblasts and adipocytes was also tested (data not shown). hBMSCs derived from three donors were inspected in this study.

The cells were thawed, cultured [27] and PUFAs supplemented [26] as described previously. In brief, the hBMSCs at confluence of 80–90% were supplemented with ethanol (purity ≥99.5%, Altia Industrial, Rajamäki, Finland) as a control, DHA or AA (Cayman Chemical, Ann Arbor, MI, USA) as bovine serum albumin (BSA, Merck, Darmstadt, Germany) conjugates. Prior to the PUFA supplementation, the medium was changed to medium containing only 5% fetal bovine serum (FBS, Thermo Fisher Scientific, Waltham, MA, USA) in contrast to 10% FBS in the proliferation medium. The PUFAs dissolved in ethanol were added into 1.5 mM BSA-Dulbecco's phosphate buffered saline (DPBS, Thermo Fisher Scientific) solution, vortexed and immediately added to the cell culture medium. The final PUFA concentration supplemented to the cells was 50 μM. After 24 h, hBMSCs were detached and 50,000 cells/well were added into Mreg polarization assay.

2.2. Mreg Polarization Assay

The use of anonymized peripheral blood mononuclear cells (PBMCs) from blood donors in research is in accordance with the rules of the Finnish Supervisory Authority for Welfare and Health (Valvira, Helsinki, Finland). The layout of the assay is described in Figure 1 and macrophages derived from six different donors were used in the assay. The Mregs were cultured as described in [16] with certain changes. In brief, 2×10^6–4×10^6 PBMCs were plated on 12-well plates (Corning™ Costar™ flat bottom, Thermo Fisher Scientific), incubated for 1–2 h and washed with DPBS. The attached monocytes were incubated in 1.5 mL RPMI-1640 medium (Thermo Fisher Scientific) with 10% FBS (Merck), GlutaMAX™ supplement (Thermo Fisher Scientific) and 5 ng/mL macrophage colony-stimulating factor (M-CSF, PromoCell, Heidelberg, Germany) for 6 days at 37 °C, 5% CO_2. This medium is referred as polarization medium and macrophages obtained in these conditions are referred as Mreg-polarized macrophages from here onwards. At day 6, the medium was changed into the polarization or activation medium [polarization medium with 25 ng/mL interferon (IFN)-γ and 10 ng/mL lipopolysaccharide (LPS, both from Merck)]. Macrophages cultured in the activation medium are referred as Mreg-activated macrophages. The next day, 50,000 control-hBMSCs, DHA-hBMSCs or AA-hBMSCs were added to the bottom of wells (referred as hBMSC cell-cell contact) or to inserts (Corning™ Transwell™, pore size 0.4 μm, Thermo Fisher Scientific) (referred as the hBMSC secretome).

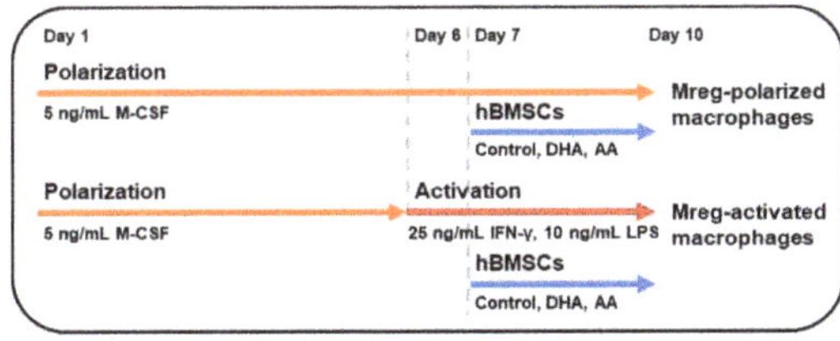

Figure 1. The layout of macrophage polarization assay. Mreg, regulatory macrophage; hBMSC, human bone marrow-derived mesenchymal stromal cell; DHA, docosahexaenoic acid; AA, arachidonic acid; M-CSF, macrophage colony-stimulating factor; IFN, interferon; LPS, lipopolysaccharide.

At day 10, the medium was centrifuged at 300× *g* for 15 min and the supernatant was snap frozen and stored at −80 °C. The cells were washed with DPBS and either detached for flow cytometry with 0.75 µL 4 °C Macrophage Detachment Solution DFX (PromoCell) or scraped into 600 µL RLT lysis buffer (Qiagen, Hilden, Germany) for quantitative polymerase chain reaction (QPCR). The RLT samples were snap frozen and stored at −80 °C.

2.3. Determination of Cytokine Production

The medium samples were thawed on ice and analyzed with human tumor necrosis factor (TNF)-α, IL-10 and IL-23 DuoSet enzyme-linked immunosorbent assays (ELISA) (all from R&D Systems, Minneapolis, MN, USA) according to the manufacturer's protocol. The absorbance was measured with CLARIOstar® microplate reader (BMG Labtech, Ortenberg, Germany).

2.4. Macrophage Phenotyping with Real-Time Quantitative PCR

Only the macrophages with hBMSC secretome samples (hBMSCs cultured on an insert) were analyzed with real-time QPCR, because cell-cell contact samples included both macrophages and hBMSCs. The RNA was extracted using RNeasy Mini Kit (Qiagen) according to the manufacturer's protocol. The concentration and purity of RNA was measured with NanoDrop ND-1000 spectrophotometer (Thermo Fisher Scientific). RNA was converted to complementary DNA with High Capacity cDNA RT Kit (Thermo Fisher Scientific). The gene expression was analyzed with real-time QPCR (CFX96™ Real-Time Systems and C1000™ Thermal Cycler, Bio-Rad, Hercules, CA, USA) using TaqMan® Gene Expression assays and TaqMan® Universal Master Mix II (Thermo Fisher Scientific). The following genes were analyzed: *TGFB1* (ID: Hs00998133_m1), *MMP2* (ID: Hs01548727_m1), *DHRS9* (ID: Hs00608375_m1), *STAT3* (ID: Hs00374280_m1) and *STAT1* (ID: Hs01013996_m1). *HPRT1* was used as the reference gene. Samples were analyzed as duplicates and the results were analyzed with CFX Manager™ 3.0 (Bio-Rad) and with the $2^{-\Delta\Delta Ct}$ method using *HPRT1* as the reference gene [33]. The relative gene expression levels are expressed as log2 fold change relative to the Mreg-polarized or Mreg-activated macrophages cultured without hBMSCs.

2.5. Macrophage Phenotyping with Flow Cytometry

The antibody staining was performed as described in Hyvärinen et al. (2018) using anti-human antibodies PE-CF594-CD86 (clone 2331 FUN-1), APC-CD206 (clone 19.2), BV421-CD163 (clone GHI/61), PerCP-Cy™5.5-CD90 (clone 5E10) (from BD Biosciences, San Diego, CA, USA), FITC-HLA-DR (clone L243), BV 510™-CD274 (PD-L1, B7-H1, clone 29E.2A3), PE/Cy7-CD120b (TNFR2, clone 3G7A02) (from BioLegend, San Diego, CA, USA) and PE-MERTK (clone HMER5DS, Thermo Fisher Scientific). Cell viability was determined with LIVE/DEAD™ Fixable Near-IR Dead Cell Stain Kit (Thermo Fisher Scientific). Data was acquired with BD FACSAria IIU (BD Biosciences) flow cytometer using FACSDiva™ (v8.0.1, BD Biosciences) and analyzed with FlowJo® (v10, BD Biosciences). Gating strategy and doublet discrimination is depicted in Figure S1. Fluorescence positive cells were determined by using isotype controls and CD90 positive cells were excluded from the analysis (Figure S1). The results are presented as median fluorescence intensity (MFI) and frequency of positive cells and as their log2 fold change values relative to the Mreg-polarized or Mreg-activated macrophages cultured without hBMSCs.

2.6. Yeast Heat-Inactivation and CFSE-Staining

Lyophilized *C. albicans* pellets (ATCC® 10231™, Microbiologics, Saint Cloud, MN, USA) were dissolved in 1 mL NaCl Peptone Broth solution (Merck) for 30 min at 37 °C according to the manufacturer's protocol. The yeast solution was incubated for 1 h in 80 °C to kill the cells [34]. Yeast solution (1×10^7 cells/mL) was stained with 2 µM carboxyfluorescein succinimidyl ester (CFSE, Thermo Fisher Scientific) for 15 min at 37 °C. The stained yeast cells were washed with RPMI 1640 and centrifuged at 1000× *g* for 5 min. The pellet was suspended with polarization medium at concentration of 5×10^6 cells/mL and filtered.

The CFSE-stained *C. albicans* were stored o/n at 4 °C. The staining was verified with imaging flow cytometry (Amnis® ImageStream®X Mark II, Luminex Corporation, Austin, TX, USA). The viability of heat-killed yeast was tested by the absence of growth on agar plates after 48 h incubation at 37 °C.

2.7. Phagocytosis Assay with Imaging Flow Cytometry

Polarized macrophages were cultured with control-hBMSCs on inserts as described above. Macrophages derived from five different donors were employed in the phagocytosis assay. On day 10, the hBMSC inserts were removed and the medium was replaced with 1 mL polarization medium containing heat-killed CFSE-stained *C. albicans* at low (5×10^5 cells/well) or high (1.25×10^6 cells/well) concentration. The plates were centrifuged at 30× *g* for 1 min to synchronize the phagocytosis in all wells and incubated for 45 min at 37 °C, 5% CO_2. The cells were washed with DPBS, detached, stained with PE-CF594-CD86 and APC-CD206 and analyzed with imaging flow cytometry.

Amnis® ImageStream®X Mark II 12-channel imaging flow cytometer with the software INSPIRE® (Luminex Corporation) was used for data collection and analysis. Acquisition settings were as follows: Excitation lasers 405 (off), 488 (15 mW), 642 (150 mW) and 785 (6.75 mW) were applied for the excitation of fluorochromes and laser Channels (Ch) 01 and Ch09 (bright field, BF), Ch06 (scattering channel, SSC), plus fluorescence channels Ch02, Ch04 and Ch11 were activated for signal detection.

All acquisition settings were the same in all experiments. Single cells were separated from debris and aggregates in the BF channel using the IDEAS features aspect ratio and area. Samples were acquired at 60× magnification with low flow rate/high sensitivity. At least 2000 events of gated single cells for each sample were collected. Single color controls were used to create a compensation matrix. Unlabeled cells and isotype control samples were used to determine the auto fluorescence and the non-specific background.

Compensated data files were analyzed using algorithms available in the IDEAS® analysis software (v6.2.188.0). Positive events were gated based on cell morphology and the intensity values of each fluorescence channel and cell BF images (Figure S2). Gating and compensation values were used as analysis template for all experimental files. This batch processing of all files assured the comparisons of each experiment with the other.

2.8. Statistical Analysis

Non-parametric tests were performed due to the non-normal distribution of parameters. The results are expressed as medians with interquartile ranges (IQR). Pairwise statistical testing was conducted with Wilcoxon signed rank test. Groupwise statistical testing was conducted with Kruskal-Wallis rank sum test and post hoc pairwise comparisons using Dunn's test for multiple comparisons with Mreg-polarized or Mreg-activated macrophages cultured without hBMSCs. All statistical tests were conducted with R version 3.5.1 and the PMCMR package [35,36] and p-value < 0.05 was considered significant.

3. Results

3.1. Phenotype of Polarized and Activated Macrophages

The phenotypes of macrophages were determined by flow cytometry using markers for T-cell activation costimulatory molecule CD86, major histocompatibility complex class II cell surface receptor human leukocyte antigen (HLA)-DR, mannose receptor CD206, scavenger receptor CD163, programmed death-ligand 1 (PD-L1, also known as CD274), tumor necrosis factor receptor 2 (TNFR2) and Mer receptor tyrosine kinase (MerTK) (Tables S1 and S2). hBMSCs were excluded using CD90 labeling and were detectable only in the samples with hBMSC cell-cell contact (Figure S1). The median frequency of dead cells of Mreg-polarized macrophages was 1.6% (IQR 2.5) and 1.9% (IQR 5.4) for Mreg-activated macrophages.

The phenotype of Mreg-polarized macrophages is shown in Tables S1 and S2. We observed considerable donor-specific variation in response to Mreg polarization conditions and subsequently in the expression of phenotype markers between different buffy coat donors. Approximately, 78% of these macrophages were positive for CD86, 99.7% for HLA-DR, 35% for CD206, 6% for CD163, 59% for PD-L1 and <1% for both TNFR2 and MerTK. The phenotype of Mreg-activated macrophages (mature Mregs) was similar to that of Mreg-polarized macrophages (Tables S1 and S2). Approximately 97% of the activated macrophages were positive for CD86, 99.9% for HLA-DR, 23% for CD206, <1% for CD163, 92% for PD-L1, negative for TNFR2 and <1% for MerTK. In general, the expression levels of phenotype markers were similar between the two macrophage types, but the Mreg-activated macrophages had a higher frequency of CD86+ (pairwise Wilcoxon signed rank test, $p = 0.031$) and PD-L1+ cells ($p = 0.031$) than Mreg-polarized macrophages.

Both macrophage types produced TNF-α and IL-10 (Table 1). The production of TNF-α increased in Mreg-activated macrophages compared with Mreg-polarized macrophages (pairwise Wilcoxon signed rank test, $p = 0.031$) while the production of IL-10 remained at similar levels. The production of IL-23 was negligible. The donor-specific variation was high in the cytokine production.

Both Mreg-polarized and Mreg-activated macrophages expressed TGF-β1, matrix metalloproteinase (MMP)-2, human Mreg marker dehydrogenase/reductase 9 (DHRS9) [15], signal transducer and activator of transcription (STAT)3 and STAT1, determined with QPCR. There were no statistically significant differences in gene expression between the two macrophage types (not shown). Expression was, again, variable depending on the donor.

3.2. hBMSC Secretome Skews Mreg-Polarized and Mreg-Activated Macrophages toward an Anti-Inflammatory and Proresolving Phenotype

The phenotype of both Mreg-polarized and Mreg-activated macrophages was modified especially by the hBMSCs secretome. The effect of hBMSC cell-cell contact and secretome on the cell surface protein expression of Mreg-polarized macrophages is shown in Figure 2 and Tables S1 and S2. Strikingly, the hBMSC secretome increased the log2 fold change of both CD206 MFI (Kruskal-Wallis test with all hBMSC conditions, $p < 0.001$; post hoc test p-values are described in Figure 2) and frequency of CD206+ cells ($p = 0.002$) regardless of the PUFA supplementation. The secretome of control- and DHA-hBMSCs elevated also the log2 fold change of PD-L1 MFI ($p < 0.001$) and DHA-hBMSCs the log2 fold change of CD163 MFI ($p = 0.039$), but the effect was donor-dependent. Interestingly, hBMSC cell-cell contact with all PUFA modifications decreased the log2 fold change of HLA-DR MFI ($p = 0.001$).

The effect of hBMSC cell-cell contact and secretome on the cell surface protein expression of Mreg-activated macrophages is shown in Figure 3 and Tables S1 and S2. Similar to Mreg-polarized macrophages, the log2 fold change of CD206 MFI and the frequency of CD206+ cells increased in the Mreg-activated macrophages by the secretome of hBMSCs with all PUFA modifications (Kruskal-Wallis test of all hBMSC conditions $p < 0.001$; post hoc test p-values are described in Figure 3) and with AA supplementation ($p < 0.001$), respectively. Moreover, the secretome elevated the log2 fold change frequency of MerTK+ cells regardless of the PUFA supplementation ($p = 0.014$). The expression was still low (<15% positive cells) and the increase was variable between donors. The cell-cell contact had an effect on only the log2 fold change frequency of HLA-DR+ cells, which was slightly decreased regardless of the PUFA modification ($p < 0.001$). The most drastic change in both Mreg-polarized and Mreg-activated macrophages was the elevated CD206 expression, which is depicted in Figure 4 (presenting the CD206 data in more detail without the log2 fold change values).

Table 1. Effect of hBMSC cell-cell contact and secretome with Mreg-polarized and Mreg-activated macrophages to cytokine production.

Concentration, pg/mL (IQR)								
		Cell-Cell Contact	Cell-Cell Contact	Cell-Cell Contact	Secretome	Secretome	Secretome	
Cytokine	**Mreg-Polarized**	**Mreg-Polarized +Control-hBMSC**	**Mreg-Polarized +DHA-hBMSC**	**Mreg-Polarized +AA-hBMSC**	**Mreg-Polarized +Control-hBMSC**	**Mreg-Polarized +DHA-hBMSC**	**Mreg-Polarized +AA-hBMSC**	***p*-value [a]**
TNF-α	123.5 (131.5)	247.6 (199.7)	165.4 (278.8)	183.6 (177.2)	181.9 (434.9)	228.0 (273.8)	192.8 (350.3)	0.999
IL-10	39.5 (81.5)	48.0 (31.7)	36.9 (55.4)	36.1 (31.4)	32.5 (40.5)	29.1 (29.6)	53.6 (27.7)	0.975
IL-23	0.0 (0.0)	0.3 (0.9)	0.0 (0.0)	0.0 (0.0)	0.0 (0.0)	0.0 (0.0)	0.0 (0.0)	0.015
	Mreg-activated	**Mreg-activated +control-hBMSC**	**Mreg-activated +DHA-hBMSC**	**Mreg-activated +AA-hBMSC**	**Mreg-activated +control-hBMSC**	**Mreg-activated +DHA-hBMSC**	**Mreg-activated +AA-hBMSC**	***p*-value [a]**
TNF-α	1591.6 (1898.5)	2106.7 (1825.5)	1887.1 (2041.1)	1899.1 (1672.3)	2025.5 (1512.2)	1950.5 (1895.9)	1823.6 (1959.4)	0.998
IL-10	31.0 (69.9)	45.5 (70.4)	39.6 (54.4)	33.9 (76.7)	31.5 (88.2)	45.0 (57.1)	33.3 (69.6)	0.871
IL-23	0.0 (0.1)	0.4 (2.7)	0.4 (1.0)	0.0 (0.2)	0.1 (0.2)	0.0 (0.4)	0.1 (0.4)	0.757

DHA, docosahexaenoic acid; AA, arachidonic acid; TNF, tumor necrosis factor; IL, interleukin; IQR, interquartile range. [a] The statistical significance of variation between groups was determined using the Kruskal-Wallis rank sum test.

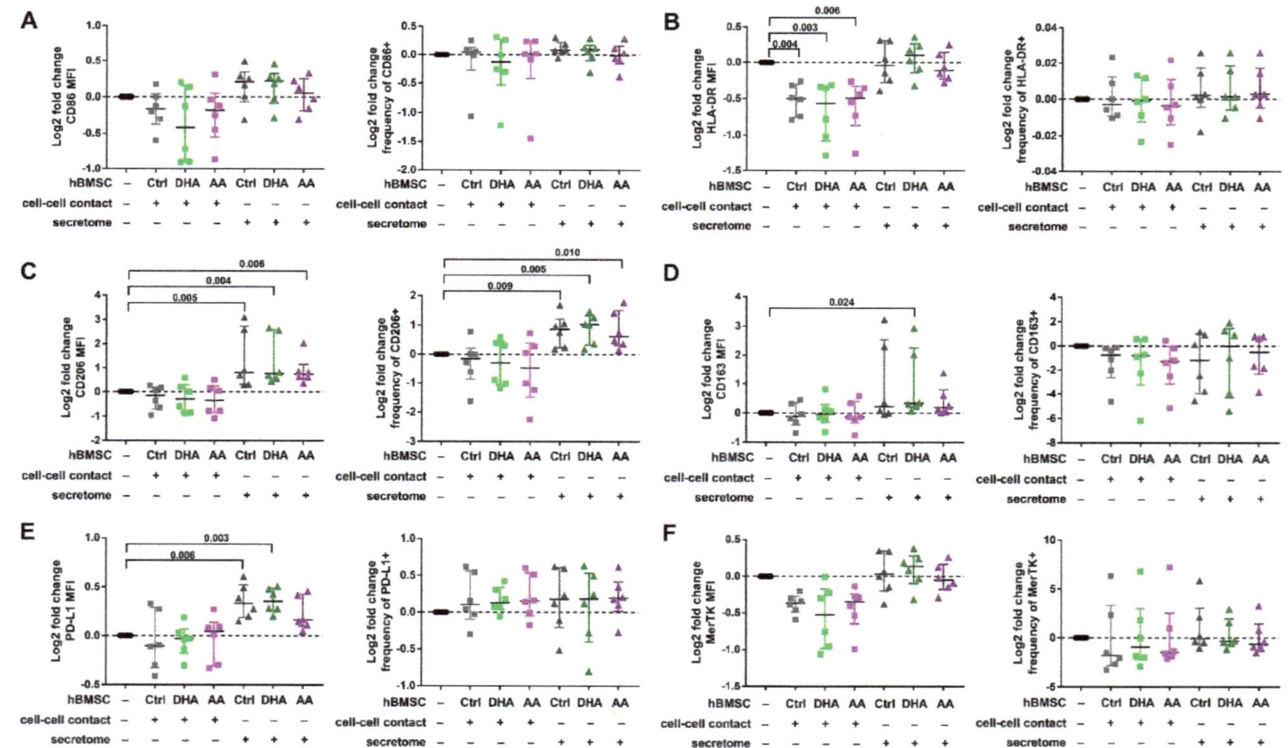

Figure 2. The effect of hBMSC cell-cell contact and secretome on the cell surface the phenotype of Mreg-polarized macrophages. The median fluorescence intensities (MFI, on the left in each panel) and frequencies of positive cells (on the right in each panel) were determined with flow cytometry investigating the expression of (**A**) CD86, (**B**) HLA-DR, (**C**) CD206, (**D**) CD163, (**E**) PD-L1 and (**F**) MerTK. The effect of hBMSCs is visualized with log2 fold changes calculated against macrophages cultured without hBMSCs (represented by the zero line) for each individual donor. The differences among groups were determined with Kruskal-Wallis rank sum test and post hoc using Dunn's test (the latter presented in the figures). The results are expressed as log2 fold changes as medians with interquartile ranges; $n = 6$ biological replicates. Ctrl, control; DHA, docosahexaenoic acid; AA, arachidonic acid.

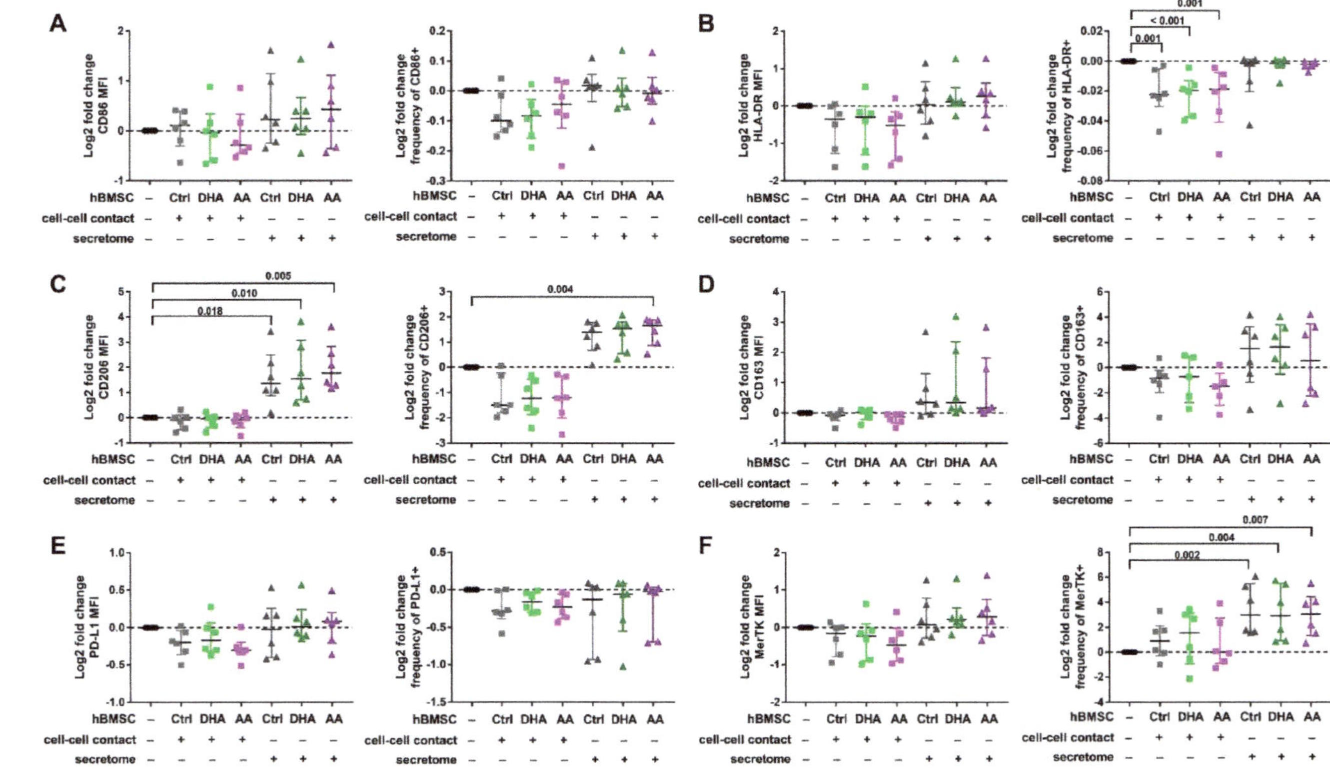

Figure 3. The effect of hBMSC cell-cell contact and secretome on the cell surface the phenotype of Mreg-activated macrophages. The median fluorescence intensities (MFI, on the left in each panel) and frequencies of positive cells (on the right in each panel) were determined with flow cytometry investigating the expression of (**A**) CD86, (**B**) HLA-DR, (**C**) CD206, (**D**) CD163, (**E**) PD-L1 and (**F**) MerTK. The effect of hBMSCs is visualized with log2 fold changes calculated against macrophages cultured without hBMSCs (represented by the zero line) for each individual donor. The differences among groups were determined with Kruskal-Wallis test and post hoc using Dunn's test. The results are expressed as log2 fold changes as medians with interquartile ranges; $n = 6$ biological replicates. Ctrl, control; DHA, docosahexaenoic acid; AA, arachidonic acid.

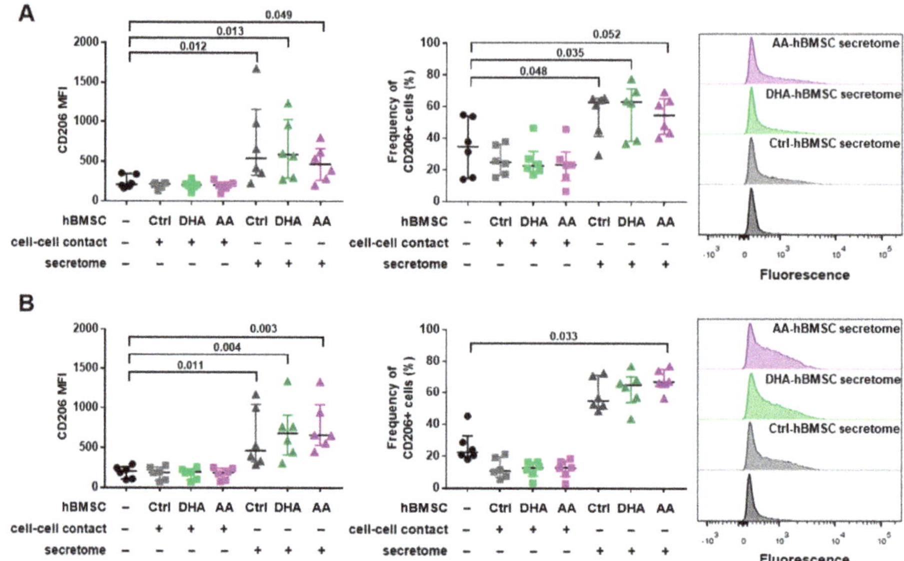

Figure 4. The secretome of hBMSCs increases the expression of CD206. (**A**) The upper panel describes the results of Mreg-polarized macrophages and (**B**) the lower panel the results of Mreg-activated macrophages. The median fluorescence intensities (MFI, left panel) and frequencies of positive cells (middle panel) were determined with flow cytometry. The representative histograms are presented on the right. The differences among groups were determined with Kruskal-Wallis rank sum test and post hoc using Dunn's test. These results are also described as log2 fold changes in Figures 2 and 3. The results are expressed as medians with interquartile ranges; $n = 6$ biological replicates. Ctrl, control; DHA, docosahexaenoic acid; AA, arachidonic acid.

The hBMSCs had no effect on the cytokine production in this experimental setting (Table 1). In the coculture with Mreg-polarized macrophages, the production of IL-23 increased in by the cell-cell contact of control-hBMSCs (Kruskal-Wallis test $p = 0.015$; post hoc Dunn's test, $p = 0.002$); however, the production of IL-23 was at the detection limit and this result should be interpreted with caution. The expression of most genes investigated remained unaltered by the hBMSC secretome (Figure 5). Nevertheless, DHA-hBMSCs resulted in a decreased MMP-2 gene expression in the Mreg-polarized macrophages (Kruskal-Wallis test $p = 0.031$; post hoc Dunn's test, $p = 0.007$) and AA-hBMSCs caused a similar trend. There was also a trend of increased gene expression of the Mreg marker DHRS9 in DHA and AA-hBMSCs coculture.

3.3. Phagocytosis Assay

The *C. albicans* phagocytosis activity of Mreg-polarized macrophages was assessed with/without hBMSCs secretome and the results analyzed with imaging flow cytometry (Figure 6). The macrophages elicited a donor-dependent CD86 and CD206 expression. In contrast to the previous results, hBMSC secretome did not increase the CD206 expression in all of the macrophages derived from different PBMC donors (non-responders $n = 2$, Figure S3). However, in three cases, the CD206 increased by 1.9 to 4.2 fold when compared with the non-responders, which elicited a 0.7 to 0.8-fold decrease in the CD206 expression. The results of the CD206 responders are reported in Figure 6. In the responders, the phagocytosis of low concentration (5×10^5 cells/well) *C. albicans* increased by 1.6 to 5.8 fold in the hBMSC secretome group when compared with Mreg-polarized macrophages alone. In general, the levels of ingested *C. albicans* were higher in the high concentration (1.25×10^6 cells/well) group and the phagocytosis increased by 1.2 to 3.1-fold in the hBMSC secretome group when compared with Mreg-polarized macrophages alone.

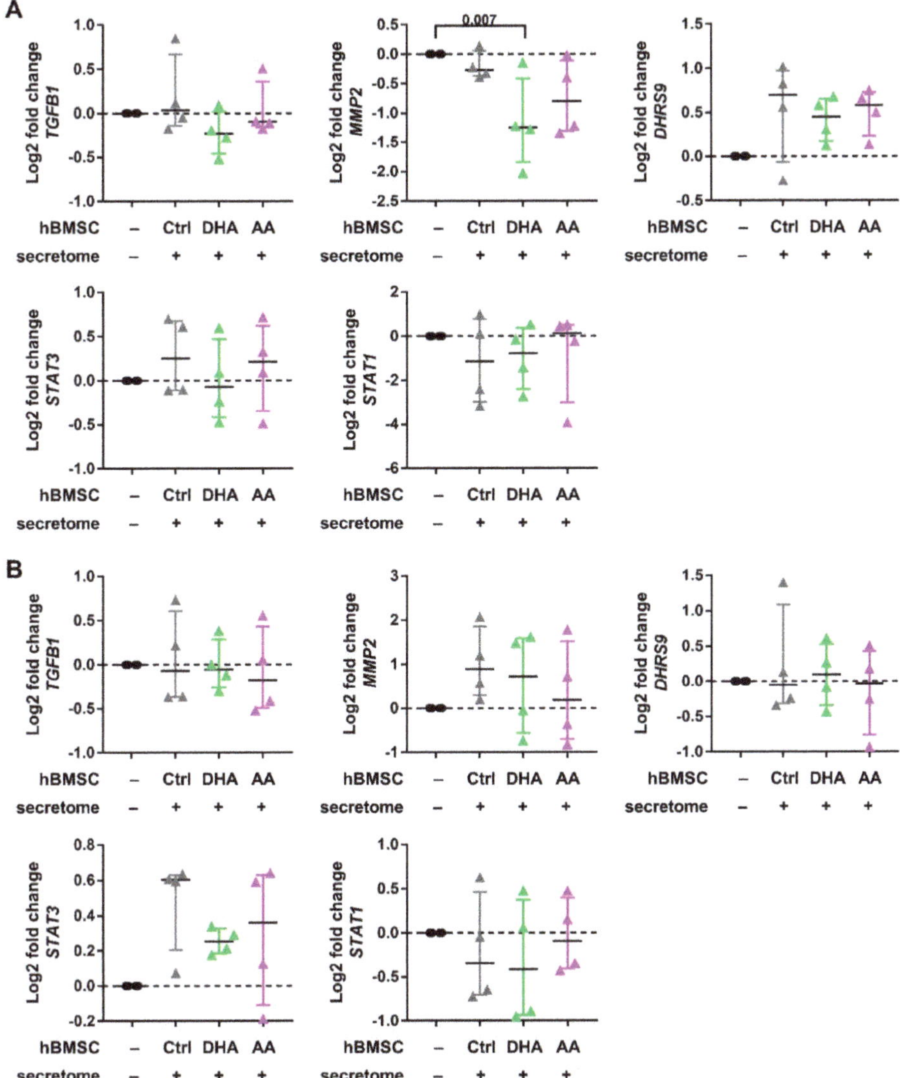

Figure 5. The effect of hBMSC secretome on the gene expression of Mreg-polarized and Mreg-activated macrophages. The gene expression of macrophage phenotype markers was investigated with QPCR. The effect of hBMSCs on (**A**) Mreg-polarized and (**B**) Mreg-activated macrophages is visualized with log2 fold changes calculated against macrophages cultured without hBMSCs (represented by the zero line) for each individual donor. The differences among groups were determined with Kruskal-Wallis rank sum test and post hoc using Dunn's test. The results are expressed as medians with interquartile ranges; $n = 4$ biological replicates. Ctrl, control; DHA, docosahexaenoic acid; AA, arachidonic acid.

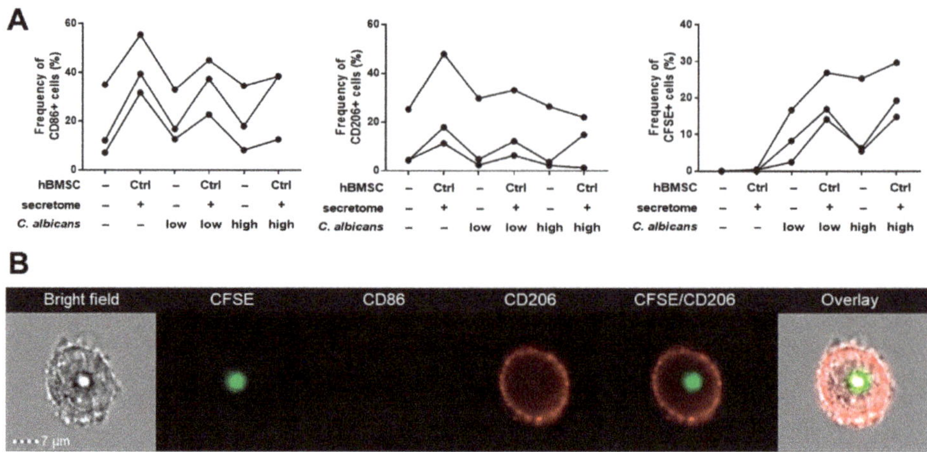

Figure 6. hBMSC secretome improves the *C. albicans* phagocytosis activity of Mreg-polarized macrophages in a CD206-mediated manner. (**A**) The frequency of CD86 and CD206 positive cells and the phagocytosis of CFSE-stained *C. albicans* were determined with imaging flow cytometry, $n = 3$ biological replicates. (**B**) A representative imaging flow cytometry image of a macrophage that has phagocytosed *C. albicans*. Low *C. albicans* concentration, 5×10^5 cells/well; high *C. albicans* concentration, 1.25×10^6 cells/well; CFSE, CFSE-stained *C. albicans*.

4. Discussion

We investigated the effect of hBMSC cell-cell contact and secretome to the phenotype of immunoregulatory macrophages, which were either polarized (Mreg-polarized) or polarized and activated (Mreg-activated) toward Mreg phenotype. Moreover, we supplemented hBMSCs with PUFAs DHA or AA prior to the macrophage coculture to elucidate if PUFAs induced changes in the immunomodulatory capacity of hBMSCs. The hBMSC secretome increased the expression of anti-inflammatory and proresolving phenotype markers in both macrophage types with all hBMSC modifications (control, DHA and AA). In particular, we observed a substantial increase in the CD206 expression. Furthermore, the hBMSC secretome increased donor-dependently the phagocytosis activity of Mreg-polarized macrophages in *C. albicans* phagocytosis assay, a result which was associated with the increased CD206 expression. Intriguingly, the DHA-supplemented hBMSCs induced the most prominent anti-inflammatory changes in Mreg-polarized macrophages while AA supplementation had only a slight effect.

Our aim was to study macrophages both in a more naïve stage (Mreg-polarized macrophages) and as mature Mregs (Mreg-activated macrophages) [37] in order to elucidate the effect of hBMSCs on macrophages at different polarization stages. The phenotypes of these two macrophage types were similar, but the Mreg-activated macrophages manifested a more classically activated phenotype with increased expression of CD86 and PD-L1. Moreover, the Mreg-activated macrophages produced more TNF-α than Mreg-polarized macrophages as expected due to Toll-like receptor engagement by LPS.

Previously, we demonstrated that hBMSC cell-cell contact and hBMSC-EVs enhanced the anti-inflammatory properties of mature Mregs by decreasing the production of IL-23 and IL-22 and increasing the PGE_2 production [16]. In the current study, we did not detect a decrease in the IL-23 production, which was very low when measured with ELISA (previously analyzed with ProcartaPlex Immunoassay), but we observed other anti-inflammatory effects of hBMSCs on these immunoregulatory macrophages. Additionally, the experimental setup and the markers investigated of these two studies were slightly different.

The hBMSC secretome enhanced the anti-inflammatory properties of both Mreg-polarized and Mreg-activated macrophages especially by increasing the CD206 expression. CD206, also known as mannose receptor, is a pattern recognition receptor, which binds to the glycan structures on microbes [38]. Previous studies have shown that human MSC cell-cell contact or secretome can increase the expression of CD206 and overall anti-inflammatory properties of monocytes or M1 macrophages [9,11]. Moreover, human MSC-derived EVs can increase the CD206 and anti-inflammatory properties of murine macrophages [39]. The secretome consists of secreted cytokines, lipid mediators and other molecules and also includes EVs. Thus, our results with macrophages polarized toward Mregs are in agreement with previous studies investigating monocytes or other macrophage subtypes and CD206 expression. In our preceding study with Mregs, we did not observe an increase in the CD206 expression with hBMSC-EV addition [16]; however, the experimental settings of our two studies are not directly comparable. Previously, the EVs added to Mregs were derived from unstimulated hBMSCs and given in two doses. In the current study, the hBMSCs produced the EVs and additional soluble factors constantly in a stimulated environment with macrophage coculture.

Next, we investigated if the increased CD206 expression translated to an increased phagocytosis activity of yeast *C. albicans* by Mreg-polarized macrophages in the presence of hBMSC secretome. The cell surface of *C. albicans* is covered in terminal mannose residues that are recognized by CD206 of macrophages [38,40]. Interestingly, when the hBMSC secretome increased the CD206 expression in the responder macrophages, the phagocytosis of *C. albicans* was also increased indicating an association between these two hBMSC secretome-mediated phenomena. The effect was PBMC donor-dependent, because the donors responded to the polarization and hBMSC secretome differently. Both human and murine MSCs have been shown to induce the phagocytosis of macrophages in various assays [10,12,41] but to our knowledge this is the first study showing that MSCs induce the phagocytosis of *C. albicans*. One of the key aspects in the resolution of inflammation is the clearance of pathogens and apoptotic cells via phagocytosis and efferocytosis [24]. Phagocytosing macrophages become more prevalent during resolution of inflammation when the cell debris and microbes are cleared away in order to achieve homeostasis [24] emphasizing the ability of hBMSC secretome to promote the proresolving phenotype of macrophages.

In addition to CD206, the PD-L1 expression increased by the hBMSC secretome in Mreg-polarized macrophages. When PD-L1 binds to its receptor, a co-inhibitory receptor programmed death 1 (PD-1), T-cell activation and proliferation are inhibited and the immune response is attenuated [42]. PD-L1 is expressed on the surface of proresolving macrophages [43], which indicates that the hBMSC secretome skews the macrophage phenotype in an anti-inflammatory direction. Supporting our results, MSCs have been shown to increase the PD-L1 expression in M2-type macrophages [10].

Strikingly, in Mreg-activated macrophages, the hBMSC secretome increased the expression of MerTK, which was generally negative in the cells. The different PBMC donors; however, responded in a different manner. MerTK is a marker of anti-fibrotic M2c macrophages [44], it is important in the clearance of apoptotic cells [45] and induces SPM production in macrophages [46]. Interestingly, EVs from cardiosphere-derived cells are able to increase the MerTK expression of macrophages via the transfer of microRNA-26a [47], which may indicate that the EVs in the hBMSC secretome are mediating the increased MerTK expression. It has been well established that EVs mediate a large proportion of the immunomodulatory effects of MSCs [48]. Although we did not examine the effect of EVs alone, we can hypothesize that observed anti-inflammatory and proresolving effects of hBMSC secretome were at least in part mediated by the EV fraction.

The secretome of hBMSCs had a larger impact on the phenotype of macrophages than the hBMSC cell-cell contact. Mainly, the cell-cell contact lowered the HLA-DR expression of Mreg-polarized and, to a certain extent, Mreg-activated macrophages. HLA-DR is a proinflammatory marker and is involved in the T-cell activation via antigen presentation. In agreement with our result, human MSCs can diminish the HLA-DR expression of macrophages and monocytes [49,50] indicating that the hBMSC cell-cell contact also rendered the macrophage phenotype more anti-inflammatory.

The PUFA supplementation of hBMSCs prior the coculture with macrophages resulted in a more pronounced anti-inflammatory phenotype of Mregs. In particular, the secretome of DHA-hBMSCs increased the CD163 expression and decreased the gene expression of gelatinase MMP-2 in Mreg-polarized macrophages while control- or AA-hBMSCs had no effect. CD163 is a hemoglobin scavenger receptor inducing an anti-inflammatory response and its increased expression is one of the major changes in the macrophage switch to alternatively activated phenotype [51]. On the other hand, M1 macrophages secrete MMP-2 to induce the degradation of extracellular matrix and recruitment of inflammatory cells to the site of tissue injury [52]. Previous studies have shown that MSCs are able to decrease the gene expression of MMP-2 of macrophages in vitro [53] and in vivo [54] corroborating with our results. The secretome of AA-hBMSCs did not have a significant effect on the PD-L1 expression in Mreg-polarized macrophages while control and DHA-hBMSCs increased this expression. This result hints that AA supplementation could lower the immunosuppressive properties of hBMSCs, because PD-L1 is important in suppressing the T-cell mediated immune response [42]. Contrastingly, the increase in CD206 expression in Mreg-activated macrophages was the most prominent with AA-hBMSC secretome.

Previous in vivo studies have reported that EPA-supplemented murine MSCs had superior therapeutic potential when compared with control MSCs [28,30]. Moreover, DHA-supplemented murine MSCs increased the survival in an in vivo sepsis model when compared with AA-supplemented MSCs [29]. In this study, we observed more prominent anti-inflammatory changes in Mregs with DHA-hBMSCs than with control or AA-hBMSCs; however, the effect was visible in only two phenotype markers. The lack of a substantial effect of PUFA supplementation to hBMSC immunomodulation in this in vitro assay may be due to different reasons. Firstly, the PUFA remodeling of hBMSC membranes takes place relatively quickly, beginning already after 2 h after PUFA supplementation and leading into prominent remodeling after 24 h [27]. In the current in vitro setting, the medium in the 3-day macrophage-hBMSC culture contained 10% FBS. Even though this FBS most likely has modified the cell membranes of hBMSCs and lowered the PUFA content of the membranes, we hypothesized that the initial modifications in the hBMSC membranes would suffice to induce profound changes already at the beginning of the coculture. This hypothesis is supported by the studies, where the EPA- and DHA-supplemented MSCs demonstrated improved therapeutic potential [28–30], even though the cell membranes of these cells would most likely be remodeled in vivo.

Additionally, our in vitro setting focused on changes in polarized and activated Mregs and DHA and AA supplementation. Although limited in this experimental setting, PUFA modifications could still have an impact in hBMSC immunomodulatory properties in other immunological settings. Moreover, EPA supplementation has been shown to be beneficial for MSC immunomodulation [28,30]. EPA is the precursor to PGE_3, which is less proinflammatory than PGE_2 [55] and could be one of the underlying reasons for the pronounced effect of EPA supplementation to MSC immunomodulation. Thus, the effects of PUFA-modified hBMSCs call for further investigations. If PUFA supplementations of MSCs prove out to be beneficial, these supplementations represent an easy and safe way to improve the therapeutic response of MSCs.

The macrophage assays were conducted with primary human PBMCs from individual donors. We acknowledge the challenges of using primary PBMC-derived macrophages as a model due to their high plasticity and variability in responding to cytokines and activation. However, by employing primary human cells, we achieve a more realistic setting than by the use of cell lines, representing a more physiologically accurate situation and acknowledging the varying responses of the patients in the clinic. Indeed, the magnitude of the response to different assay conditions and subsequently, the phenotypes of macrophages derived from individual donors were variable. The hBMSC secretome induced prominent differences in Mreg-polarized and Mreg-activated macrophages. Some of the changes in phenotype markers, as in CD206, were clear in all individual donors by the hBMSC secretome highlighting the significance of this finding. It is also noteworthy that although some of the effects were small, all of the changes skewed macrophages toward an anti-inflammatory and proresolving direction. Moreover, we investigated the effects of hBMSCs derived from three different donors to multiple buffy coat derived PBMCs, which enhances the robustness of our findings.

To conclude, our results demonstrate that the hBMSC secretome can modify macrophages toward immunoregulatory and proresolving phenotype, especially by increasing CD206, PD-L1 and MerTK expression. Moreover, by increasing the CD206 expression, the secretome increased the *C. albicans* phagocytosis activity of Mreg-polarized macrophages. According to our hypothesis, hBMSCs skew macrophages toward a proresolving phenotype that facilitate wound healing and restore homeostasis. Interestingly, DHA-hBMSCs also increased the expression of CD163 and decreased the gene expression of MMP-2 in Mreg-polarized macrophages indicating that the DHA modifications have an impact on the immunomodulatory properties of hBMSCs. These findings highlight the potential of PUFA supplementations as an easy and safe method to improve the hBMSC therapeutic potential.

Supplementary Materials: The following are available online at http://www.mdpi.com/2073-4409/9/9/2142/s1, Figure S1: Gating strategy and CD90 exclusion, Figure S2: Gating strategy in imaging flow cytometry, Figure S3: The phagocytose assay results from the CD206 non-responders, Table S1: Effect of hBMSC cell-cell contact and secretome on the median fluorescence intensity of phenotype markers on Mreg-polarized and Mreg-activated macrophages, Table S2: Effect of hBMSC cell-cell contact and secretome on the frequency of positive cells of phenotype markers on Mreg-polarized and Mreg-activated macrophages.

Author Contributions: Conceptualization, M.H., S.L. and E.K.; Data curation, M.H. and U.I.; Formal analysis, M.H. and U.I.; Funding acquisition, M.H. and S.L.; Investigation, M.H., S.L. and E.L.; Methodology, U.I.; Resources, P.L. and S.L.; Supervision, S.L. and E.K.; Visualization, M.H.; Writing—original draft, M.H.; Writing—review & editing, M.H., U.I., P.L, S.L. and E.K. All authors have read and agreed to the published version of the manuscript.

Funding: This work was supported by Finnish Cultural Foundation (M.H.) and Clinical State Research Funding [EVO/ VTR grant, Finland] (M.H.).

Acknowledgments: We thank Kati Hyvärinen, Lindsay Davies and Kaarina Lähteenmäki for helpful advice in designing the experiments. We also thank Birgitta Rantala, Lotta Andersson and Lotta Sankkila for excellent technical assistance.

Conflicts of Interest: The authors declare no conflict of interest.

References

1. Salmenniemi, U.; Itälä-Remes, M.; Nystedt, J.; Putkonen, M.; Niittyvuopio, R.; Vettenranta, K.; Korhonen, M. Good responses but high TRM in adult patients after MSC therapy for GvHD. *Bone Marrow Transplant.* **2017**, *52*, 606–608. [CrossRef] [PubMed]
2. Kurtzberg, J.; Abdel-Azim, H.; Carpenter, P.; Chaudhury, S.; Horn, B.; Mahadeo, K.; Nemecek, E.; Neudorf, S.; Prasad, V.; Prockop, S.; et al. A Phase 3, single-arm, prospective study of Remestemcel-L, ex vivo culture-expanded adult human mesenchymal stromal cells for the treatment of pediatric patients who failed to respond to steroid treatment for acute graft-versus-host disease. *Biol. Blood Marrow Transplant.* **2020**, *26*, 845–854. [CrossRef]
3. Herreros, M.D.; Garcia-Arranz, M.; Guadalajara, H.; De-La-Quintana, P.; Garcia-Olmo, D. Autologous expanded adipose-derived stem cells for the treatment of complex cryptoglandular perianal fistulas: A phase III randomized clinical trial (FATT 1: Fistula advanced therapy trial 1) and long-term evaluation. *Dis. Colon Rectum* **2012**, *55*, 762–772. [CrossRef] [PubMed]
4. Panés, J.; García-Olmo, D.; Van Assche, G.; Colombel, J.F.; Reinisch, W.; Baumgart, D.C.; Dignass, A.; Nachury, M.; Ferrante, M.; Kazemi-Shirazi, L.; et al. Expanded allogeneic adipose-derived mesenchymal stem cells (Cx601) for complex perianal fistulas in Crohn's disease: A phase 3 randomised, double-blind controlled trial. *Lancet* **2016**, *388*, 1281–1290. [CrossRef]
5. Aggarwal, S.; Pittenger, M.F. Human mesenchymal stem cells modulate allogeneic immune cell responses. *Transplantation* **2005**, *105*, 1815–1822. [CrossRef] [PubMed]
6. Meisel, R.; Zibert, A.; Laryea, M.; Göbel, U.; Däubener, W.; Dilloo, D. Human bone marrow stromal cells inhibit allogeneic T-cell responses by indoleamine 2,3-dioxygenase–mediated tryptophan degradation. *Blood* **2004**, *103*, 4619–4621. [CrossRef]
7. Zhang, B.; Yin, Y.; Lai, R.C.; Tan, S.S.; Choo, A.B.H.; Lim, S.K. Mesenchymal stem cells secrete immunologically active exosomes. *Stem Cells Dev.* **2014**, *23*, 1233–1244. [CrossRef]

8. Di Nicola, M.; Carlo-Stella, C.; Magni, M.; Milanesi, M.; Longoni, P.D.; Grisanti, S.; Gianni, A.M.; Matteucci, P.; Di Nicola, M.; Carlo-Stella, C.; et al. Human bone marrow stromal cells suppress T-lymphocyte proliferation induced by cellular or nonspecific mitogenic stimuli. *Blood* **2002**, *99*, 3838–3843. [CrossRef]
9. Kim, J.; Hematti, P. Mesenchymal stem cell-educated macrophages: A novel type of alternatively activated macrophages. *Exp. Hematol.* **2009**, *37*, 1445–1453. [CrossRef]
10. Abumaree, M.H.; Al Jumah, M.A.; Kalionis, B.; Jawdat, D.; Al Khaldi, A.; Abomaray, F.M.; Fatani, A.S.; Chamley, L.W.; Knawy, B.A. Human placental mesenchymal stem cells (pMSCs) play a role as immune suppressive cells by shifting macrophage differentiation from inflammatory M1 to anti-inflammatory M2 Macrophages. *Stem Cell Rev. Rep.* **2013**, *9*, 620–641. [CrossRef]
11. Chiossone, L.; Conte, R.; Spaggiari, G.M.; Serra, M.; Romei, C.; Bellora, F.; Becchetti, F.; Andaloro, A.; Moretta, L.; Bottino, C. Mesenchymal stromal cells induce peculiar alternatively activated macrophages capable of dampening both innate and adaptive immune responses. *Stem Cells* **2016**, *34*, 1909–1921. [CrossRef] [PubMed]
12. Jackson, M.V.; Morrison, T.J.; Doherty, D.F.; McAuley, D.F.; Matthay, M.A.; Kissenpfennig, A.; O'Kane, C.M.; Krasnodembskaya, A.D. Mitochondrial transfer via tunneling nanotubes is an important mechanism by which mesenchymal stem cells enhance macrophage phagocytosis in the in vitro and in vivo models of ARDS. *Stem Cells* **2016**, *34*, 2210–2223. [CrossRef] [PubMed]
13. Murray, P.J.; Allen, J.E.; Biswas, S.K.; Fisher, E.A.; Gilroy, D.W.; Goerdt, S.; Gordon, S.; Hamilton, J.A.; Ivashkiv, L.B.; Lawrence, T.; et al. Macrophage activation and polarization: Nomenclature and experimental guidelines. *Immunity* **2014**, *41*, 14–20. [CrossRef]
14. Broichhausen, C.; Riquelme, P.; Geissler, E.K.; Hutchinson, J.A. Regulatory macrophages as therapeutic targets and therapeutic agents in solid organ transplantation. *Curr. Opin. Organ Transplant.* **2012**, *17*, 332–342. [CrossRef]
15. Riquelme, P.; Amodio, G.; Macedo, C.; Moreau, A.; Obermajer, N.; Brochhausen, C.; Ahrens, N.; Kekarainen, T.; Fändrich, F.; Cuturi, C.; et al. DHRS9 is a stable marker of human regulatory macrophages. *Transplantation* **2017**, *101*, 2731–2738. [CrossRef]
16. Hyvärinen, K.; Holopainen, M.; Skirdenko, V.; Ruhanen, H.; Lehenkari, P.; Korhonen, M.; Käkelä, R.; Laitinen, S.; Kerkelä, E. Mesenchymal stromal cells and their extracellular vesicles enhance the anti-inflammatory phenotype of regulatory macrophages by downregulating the production of interleukin (IL)-23 and IL-22. *Front. Immunol.* **2018**, *9*, 771. [CrossRef] [PubMed]
17. Ylöstalo, J.H.; Bartosh, T.J.; Coble, K.; Prockop, D.J. Human mesenchymal stem/stromal cells cultured as spheroids are self-activated to produce prostaglandin E2 that directs stimulated macrophages into an anti-inflammatory phenotype. *Stem Cells* **2012**, *30*, 2283–2296. [CrossRef] [PubMed]
18. Van Niel, G.; D'Angelo, G.; Raposo, G. Shedding light on the cell biology of extracellular vesicles. *Nat. Rev. Mol. Cell Biol.* **2018**, *19*, 213–228. [CrossRef] [PubMed]
19. Bruno, S.; Grange, C.; Deregibus, M.C.; Calogero, R.A.; Saviozzi, S.; Collino, F.; Morando, L.; Busca, A.; Falda, M.; Bussolati, B.; et al. Mesenchymal stem cell-derived microvesicles protect against acute tubular injury. *J. Am. Soc. Nephrol.* **2009**, *20*, 1053–1067. [CrossRef] [PubMed]
20. Xin, H.; Li, Y.; Cui, Y.; Yang, J.J.; Zhang, Z.G.; Chopp, M. Systemic administration of exosomes released from mesenchymal stromal cells promote functional recovery and neurovascular plasticity after stroke in rats. *J. Cereb. Blood Flow Metab.* **2013**, *33*, 1711–1715. [CrossRef]
21. Doeppner, T.R.; Herz, J.; Görgens, A.; Schlechter, J.; Ludwig, A.-K.; Radtke, S.; De Miroschedji, K.; Horn, P.A.; Giebel, B.; Hermann, D.M. Extracellular vesicles improve post-stroke neuroregeneration and prevent postischemic immunosuppression. *Stem Cells Transl. Med.* **2015**, *4*, 1131–1143. [CrossRef] [PubMed]
22. Wang, X.; Gu, H.; Qin, D.; Yang, L.; Huang, W.; Essandoh, K.; Wang, Y.; Caldwell, C.C.; Peng, T.; Zingarelli, B.; et al. Exosomal miR-223 contributes to mesenchymal stem cell-elicited cardioprotection in polymicrobial sepsis. *Sci. Rep.* **2015**, *5*, 13721. [CrossRef] [PubMed]
23. English, K. Mechanisms of mesenchymal stromal cell immunomodulation. *Immunol. Cell Biol.* **2013**, *91*, 19–26. [CrossRef] [PubMed]
24. Buckley, C.D.; Gilroy, D.W.; Serhan, C.N. Proresolving lipid mediators and mechanisms in the resolution of acute inflammation. *Immunity* **2014**, *40*, 315–327. [CrossRef] [PubMed]
25. Serhan, C.N.; Petasis, N.A. Resolvins and protectins in inflammation resolution. *Chem. Rev.* **2011**, *111*, 5922–5943. [CrossRef] [PubMed]

26. Tigistu-Sahle, F.; Lampinen, M.; Kilpinen, L.; Holopainen, M.; Lehenkari, P.; Laitinen, S.; Käkelä, R. Metabolism and phospholipid assembly of polyunsaturated fatty acids in human bone marrow mesenchymal stromal cells. *J. Lipid Res.* **2017**, *58*, 92–110. [CrossRef]
27. Holopainen, M.; Colas, R.A.; Valkonen, S.; Tigistu-sahle, F.; Hyvärinen, K.; Mazzacuva, F.; Lehenkari, P.; Käkelä, R.; Dalli, J.; Kerkelä, E.; et al. Polyunsaturated fatty acids modify the extracellular vesicle membranes and increase the production of proresolving lipid mediators of human mesenchymal stromal cells. *BBA Mol. Cell Biol. Lipids* **2019**, *1864*, 1350–1362. [CrossRef]
28. Abreu, S.C.; Lopes-Pacheco, M.; Silva, A.L.; Xisto, D.G.; Oliveira, T.B.; Kitoko, J.Z.; Castro, L.L.; Amorim, N.R.; Martins, V.; Silva, L.H.A.; et al. Eicosapentaenoic acid enhances the effects of mesenchymal stromal cell therapy in experimental allergic asthma. *Front. Immunol.* **2018**, *9*, 1147. [CrossRef]
29. Tsoyi, K.; Hall, S.R.R.; Dalli, J.; Colas, R.A.; Ghanta, S.; Ith, B.; Coronata, A.; Fredenburgh, L.E.; Baron, R.M.; Choi, A.M.K.; et al. Carbon monoxide improves efficacy of mesenchymal stromal cells during sepsis by production of specialized proresolving lipid mediators. *Crit. Care Med.* **2016**, *44*, e1236–e1245. [CrossRef]
30. Silva, J.D.; Lopes-Pacheco, M.; De Castro, L.L.; Kitoko, J.Z.; Trivelin, S.A.; Amorim, N.R.; Capelozzi, V.L.; Morales, M.M.; Gutfilen, B.; De Souza, S.A.L.; et al. Eicosapentaenoic acid potentiates the therapeutic effects of adipose tissue-derived mesenchymal stromal cells on lung and distal organ injury in experimental sepsis. *Stem Cell Res. Ther.* **2019**, *10*, 264. [CrossRef]
31. Leskelä, H.V.; Risteli, J.; Niskanen, S.; Koivunen, J.; Ivaska, K.K.; Lehenkari, P. Osteoblast recruitment from stem cells does not decrease by age at late adulthood. *Biochem. Biophys. Res. Commun.* **2003**, *311*, 1008–1013. [CrossRef] [PubMed]
32. Dominici, M.; Le Blanc, K.; Mueller, I.; Slaper-Cortenbach, I.; Marini, F.C.; Krause, D.S.; Deans, R.J.; Keating, A.; Prockop, D.J.; Horwitz, E.M. Minimal criteria for defining multipotent mesenchymal stromal cells. The International Society for Cellular Therapy position statement. *Cytotherapy* **2006**, *8*, 315–317. [CrossRef] [PubMed]
33. Livak, K.J.; Schmittgen, T.D. Analysis of relative gene expression data using real-time quantitative PCR and the 2-ΔΔCT method. *Methods* **2001**, *25*, 402–408. [CrossRef] [PubMed]
34. Diez-Orejas, R.; Feito, M.J.; Cicuéndez, M.; Rojo, J.M.; Portolés, M.T. Differential effects of graphene oxide nanosheets on Candida albicans phagocytosis by murine peritoneal macrophages. *J. Colloid Interface Sci.* **2018**, *512*, 665–673. [CrossRef] [PubMed]
35. Pohlert, T. The Pairwise Multiple Comparison of Mean Ranks Package (PMCMR). Available online: https://CRAN.R-project.org/package=PMCMR (accessed on 21 September 2020).
36. R Core Team. *R: A Language and Environment for Statistical Computing*; R Foundation for Statistical Computing: Vienna, Austria, 2018.
37. Hutchinson, J.A.; Riquelme, P.; Geissler, E.K.; Fändrich, F. Human regulatory macrophages. *Methods Mol. Biol.* **2011**, *677*, 181–192.
38. Stahl, P.D.; Ezekowitz, R.A.B. The mannose receptor is a pattern recognition receptor involved in host defense. *Curr. Opin. Immunol.* **1998**, *10*, 50–55. [CrossRef]
39. Lo Sicco, C.; Reverberi, D.; Balbi, C.; Ulivi, V.; Principi, E.; Pascucci, L.; Berherini, P.; Bosco, M.C.; Varesio, L.; Franzin, C.; et al. Mesenchymal stem cell-derived extracellular vesicles as mediators of anti-inflammatory effects: Endorsement of macrophage polarization. *Stem Cells Transl. Med.* **2017**, *6*, 1018–1028. [CrossRef]
40. Maródi, L.; Korchak, H.M.; Johnston, R.B., Jr. Mechanisms of host defense against Candida species. Phagocytosis by monocytes and monocyte-derived macrophage. *J. Immunol.* **1991**, *146*, 2783–2789.
41. Deng, W.; Chen, W.; Zhang, Z.; Huang, S.; Kong, W.; Sun, Y.; Tang, X.; Yao, G.; Feng, X.; Chen, W.J.; et al. Mesenchymal stem cells promote CD206 expression and phagocytic activity of macrophages through IL-6 in systemic lupus erythematosus. *Clin. Immunol.* **2015**, *161*, 209–216. [CrossRef]
42. Sun, C.; Mezzadra, R.; Schumacher, T.N. Regulation and function of the PD-L1 checkpoint. *Immunity* **2018**, *48*, 434–452. [CrossRef]
43. Shouval, D.S.; Biswas, A.; Goettel, J.A.; McCann, K.; Conaway, E.; Redhu, N.S.; Mascanfroni, I.D.; AlAdham, Z.; Lavoie, S.; Ibourk, M.; et al. Interleukin-10 receptor signaling in innate immune cells regulates mucosal immune tolerance and anti-inflammatory macrophage function. *Immunity* **2014**, *40*, 706–719. [CrossRef] [PubMed]

44. Zizzo, G.; Hilliard, B.A.; Monestier, M.; Cohen, P.L. Efficient clearance of early apoptotic cells by human macrophages requires "M2c" polarization and MerTK induction. *J. Immunol.* **2012**, *189*, 3508–3520. [CrossRef] [PubMed]
45. Scott, R.S.; McMahon, E.J.; Pop, S.M.; Reap, E.A.; Caricchio, R.; Cohen, P.L.; Shelton Earp, H.; Matsushima, G.K. Phagocytosis and clearance of apoptotic cells is mediated by MER. *Nature* **2001**, *411*, 207–211. [CrossRef]
46. Cai, B.; Thorp, E.B.; Doran, A.C.; Subramanian, M.; Sansbury, B.E.; Lin, C.S.; Spite, M.; Fredman, G.; Tabas, I. MerTK cleavage limits proresolving mediator biosynthesis and exacerbates tissue inflammation. *Proc. Natl. Acad. Sci. USA* **2016**, *113*, 6526–6531. [CrossRef] [PubMed]
47. De Couto, G.; Jaghatspanyan, E.; Deberge, M.; Liu, W.; Luther, K.; Wang, Y.; Tang, J.; Thorp, E.B.; Marbán, E. Mechanism of enhanced MerTK-Dependent macrophage efferocytosis by extracellular vesicles. *Arterioscler. Thromb. Vasc. Biol.* **2019**, *39*, 2082–2096. [CrossRef] [PubMed]
48. Giebel, B.; Kordelas, L.; Börger, V. Clinical potential of mesenchymal stem/stromal cell-derived extracellular vesicles. *Stem Cell Investig.* **2017**, *4*, 84. [CrossRef]
49. Hanson, S.E.; King, S.N.; Kim, J.; Chen, X.; Thibeault, S.L.; Hematti, P. The effect of mesenchymal stromal cell-hyaluronic acid hydrogel constructs on immunophenotype of macrophages. *Tissue Eng. Part A* **2011**, *17*, 2463–2471. [CrossRef]
50. Wise, A.F.; Williams, T.M.; Rudd, S.; Wells, C.A.; Kerr, P.G.; Ricardo, S.D. Human mesenchymal stem cells alter the gene profile of monocytes from patients with Type 2 diabetes and end-stage renal disease. *Regen. Med.* **2016**, *11*, 145–158. [CrossRef]
51. Etzerodt, A.; Moestrup, S.K. CD163 and inflammation: Biological, diagnostic, and therapeutic aspects. *Antioxidants Redox Signal.* **2013**, *18*, 2352–2363. [CrossRef]
52. Parks, W.C.; Wilson, C.L.; López-Boado, Y.S. Matrix metalloproteinases as modulators of inflammation and innate immunity. *Nat. Rev. Immunol.* **2004**, *4*, 617–629. [CrossRef]
53. Hashizume, R.; Yamawaki-Ogata, A.; Ueda, Y.; Wagner, W.R.; Narita, Y. Mesenchymal stem cells attenuate angiotensin II-induced aortic aneurysm growth in apolipoprotein E-deficient mice. *J. Vasc. Surg.* **2011**, *54*, 1743–1752. [CrossRef] [PubMed]
54. Yamawaki-Ogata, A.; Fu, X.; Hashizume, R.; Fujimoto, K.L.; Araki, Y.; Oshima, H.; Narita, Y.; Usui, A. Therapeutic potential of bone marrow-derived mesenchymal stem cells in formed aortic aneurysms of a mouse model. *Eur. J. Cardio Thorac. Surg.* **2014**, *45*, 156–165. [CrossRef] [PubMed]
55. Schmitz, G.; Ecker, J. The opposing effects of n-3 and n-6 fatty acids. *Prog. Lipid Res.* **2008**, *47*, 147–155. [CrossRef] [PubMed]

Article

Vadadustat, a HIF Prolyl Hydroxylase Inhibitor, Improves Immunomodulatory Properties of Human Mesenchymal Stromal Cells

Katarzyna Zielniok [1], Anna Burdzinska [1], Beata Kaleta [2], Radoslaw Zagozdzon [1,2,3] and Leszek Paczek [1,3,*]

1 Department of Immunology, Transplantology and Internal Diseases, Medical University of Warsaw, Nowogrodzka 59, 02-006 Warsaw, Poland; katarzyna.zielniok@wum.edu.pl (K.Z.); anna.burdzinska@wum.edu.pl (A.B.); radoslaw.zagozdzon@wum.edu.pl (R.Z.)

2 Department of Clinical Immunology, Medical University of Warsaw, Nowogrodzka 59, 02-006 Warsaw, Poland; beata.kaleta@wum.edu.pl

3 Department of Bioinformatics, Institute of Biochemistry and Biophysics, Polish Academy of Sciences, Pawińskiego 5A, 02-106 Warsaw, Poland

* Correspondence: leszek.paczek@wum.edu.pl; Tel.: +48-22-502-16-41

Received: 30 September 2020; Accepted: 28 October 2020; Published: 1 November 2020

Abstract: The therapeutic potential of mesenchymal stromal cells (MSCs) is largely attributed to their immunomodulatory properties, which can be further improved by hypoxia priming. In this study, we investigated the immunomodulatory properties of MSCs preconditioned with hypoxia-mimetic Vadadustat (AKB-6548, Akebia). Gene expression analysis of immunomodulatory factors was performed by real-time polymerase chain reaction (real-time PCR) on RNA isolated from six human bone-marrow derived MSCs populations preconditioned for 6 h with 40 μM Vadadustat compared to control MSCs. The effect of Vadadustat preconditioning on MSCs secretome was determined using Proteome Profiler and Luminex, while their immunomodulatory activity was assessed by mixed lymphocyte reaction (MLR) and Culturex transwell migration assays. Real-time PCR revealed that Vadadustat downregulated genes related to immune system: *IL24, IL1B, CXCL8, PDCD1LG1, PDCD1LG2, HIF1A, CCL2* and *IL6*, and upregulated *IL17RD, CCL28* and *LEP*. Vadadustat caused a marked decrease in the secretion of IL6 (by 51%), HGF (by 47%), CCL7 (MCP3) (by 42%) and CXCL8 (by 40%). Vadadustat potentiated the inhibitory effect of MSCs on the proliferation of alloactivated human peripheral blood mononuclear cells (PBMCs), and reduced monocytes-enriched PBMCs chemotaxis towards the MSCs secretome. Preconditioning with Vadadustat may constitute a valuable approach to improve the therapeutic properties of MSCs.

Keywords: mesenchymal stem cells; Vadadustat; AKB-6548; preconditioning; priming; immunomodulation; secretome; chemotaxis

1. Introduction

Human mesenchymal stromal cells (MSCs) therapy has shown a promising potential in the treatment of diseases associated with immune-mediated disorders (reviewed in [1]). Despite the lack of sufficient data explaining the native, physiological function of MSCs in the human body, existing experimental results demonstrate their multipotency [2] and considerable paracrine-mediated immunosuppressive and inflammation resolving activity (reviewed in [3]). As progenitor cells with nonhematopoietic origin, MSCs are isolated from various adult tissues, but the most commonly used source for preclinical and clinical studies is bone marrow, which constitutes the primary niche for this population [4]. Once isolated, MSCs are readily expanded and differentiated in cell culture conditions into several different lineages. Recently, a fundamental shift was made from the initially proposed paradigm of reparative function of MSCs, to the paradigm of MSCs modulating activity, manifested by secretion of numerous bioactive compounds regulating immune response and contributing to tissue regeneration. Encouraging issues favoring the use of MSCs in regenerative therapy are: low immunogenicity, lack of ethical concerns regarding they isolation and use, and an overall minimal risk of their malignant transformation. Moreover, the outstanding potential of MSCs lies in their ability to home to the site of damage and crosstalk with other types of cell in order to limit cell death, diminish an excessive inflammatory response and facilitate the intrinsic tissue regeneration capacity. The therapeutic activity of MSCs seems to be strongly associated with the production of trophic and immunomodulatory factors. A growing number of research indicates, that treatment with MSCs secretome reveals a similar therapeutic effect to MSCs transplantation itself, avoiding the main risks related to allogeneic cell transplantation (reviewed in [5]). Therefore, many attempts have been made to optimize MSCs culture conditions to obtain their most preferred secretory profile. Oxygen tension is one of the major factor closely related to the proliferation, differentiation and stemness of MSCs. However, MSCs are routinely cultured under 21% oxygen pressure conditions that several times exceeds the physiological level available in their natural niches (which ranges from 1% to 7% O_2) [6–8]. There are several reports confirming the beneficial effect of hypoxia and hypoxic preconditioning on migration, regenerative potential, proangiogenic activity and expanded survival of MSCs [9–14]. However, culturing cells under low oxygen conditions is demanding, has some limitations and multiplies the costs of MSCs culture. Therefore, the opportunity of using hypoxia mimetic agents for preconditioning of MSCs seems a highly promising approach. The concept is simple. It assumes the use of a drug targeted at cellular hypoxia sensors, which by switching them off triggers cellular response to hypoxia under normoxic conditions. This response is manifested in a number of transcriptional and translational changes leading to regulation of metabolic, proliferation, transport and survival pathways. Detection of oxygen availability occurs in cells mainly through the prolyl-hydroxylase domain family of enzymes (PHDs), which require molecular oxygen to their biological activity. When enough oxygen is present, PHDs are active and hydroxylate specific proline residues (Pro402 and Pro564) in hypoxia inducible factors alpha (HIF-α)—a three isoforms of transcription factor responsible for the expression of hypoxia adaptation genes, of which HIF-1α and HIF-2α are the most important [15]. Hydroxylation of HIF-α proline residues determines its inactivation, being a signal to its ubiquitination and proteasomal degradation. When the oxygen supplies are low, PHDs are inactivated, which results in stabilization of HIF-α and initiation of mechanisms that adapt cells to hypoxia. Several studies have been made to examine the effects of preconditioning MSCs with PHDs inhibitors (reviewed in [16]). To date, various PHDs inactivation strategies have been used in MSCs research (including gene silencing), but only few studies have been performed using selective PHDs inhibitors. Here, we report for the first time how treatment with Vadadustat—a selective HIF PHDs inhibitor—affects paracrine functions and immunomodulatory properties of MSCs. Our findings reveal new aspects of MSCs preconditioning with pharmacologically induced hypoxia, and we strongly believe that may contribute to the improvement of MSCs-based therapies in the treatment of immune disorders.

2. Materials and Methods

2.1. Isolation and Culture of Human Bone Marrow-Derived Mesenchymal Stromal Cells (BM-MSCs)

BM-MSCs were isolated from bone marrow aspirates of patients without chronic diseases collected during orthopedic surgery (the age and sex profile of donors is provided in Supplementary Table S1). The procedure was performed in accordance with the approval of the Local Bioethics Committee (number KB/115/2016) after receiving informed consent from each patient. Cells were isolated as previously described [17,18]. Briefly, mechanically disassociated bone marrow samples were washed, centrifuged, suspended and seeded on a plastic culture dish (BD Primaria™, BD Biosciences, San Jose, CA, USA) in growth medium composed of low glucose DMEM (Biowest, Riverside, MO, USA) supplemented with 10% FBS (Biowest, Riverside, MO, USA) and antibiotic-antimycotic solution (1% penicillin–streptomycin; 0.5% amphotericin B, Invitrogen, Thermo Fisher Scientific, Waltham, MA, USA). The medium was replaced at day 4, when the first fibroblastic-like colonies of cells were observed do be adhered on a dish. Cells were grown in 5% CO_2/95% humidified air at 37 °C and the medium was replaced every other day. All experiments were performed on at least 6 individual populations (each population isolated from separate donor), between passage 4–6 and fulfilled currently acknowledged criteria for identification of mesenchymal stromal cells (which we previously described in [19]).

2.2. *Human BM-MSCs Identification*

2.2.1. Phenotyping of BM-MSCs by Flow Cytometry

BD Stemflow™ hMSC Analysis Kit (BD Biosciences, San Jose, CA, USA) was used to perform BM-MSCs phenotypic characterization. For the purpose of MSCs characterization, cells at passage 4 were stained with antibodies of surface markers CD105 (PerCP-Cy™ 5.5), CD73 (APC), CD90 (FITC) as well as negative expression markers CD45, CD34, CD11b, CD19, HLA-DR (PE) according to the protocol provided by the manufacturer. Flow cytometry analysis was performed on BD FACS Canto II using BD FACS Diva Software (BD Biosciences, San Jose, CA, USA).

2.2.2. Adipogenic Differentiation

To confirm the ability of isolated BM-MSCs to adipogenic differentiation, cells were grown for 3 weeks in differentiating medium consisting of DMEM-high glucose (Biowest, Riverside, MO, USA) supplemented with 10% FBS (Biowest, Riverside, MO, USA) and 1% penicillin–streptomycin (Invitrogen, Thermo Fisher Scientific, Waltham, MA, USA), 10 μg/mL insulin, 60 μM indomethacin, 1 μM dexamethasone and 500 μM 3-isobutyl-1-methylxanthine (IBMX) (all MilliporeSigma, St. Louis, MO, USA). The differentiation medium was replaced every third day. Accumulation of lipid droplets in cells was visualized under a light microscope after the Oil Red O (MilliporeSigma, St. Louis, MO, USA) staining procedure previously described in [19]).

2.2.3. Osteogenic Differentiation

Confirmation of the osteogenic differentiation ability was achieved by culturing cells for three weeks in an osteogenic medium containing DMEM-low glucose (Biowest, Riverside, MO, USA) supplemented with 10% FBS (Biowest, Riverside, MO, USA) and 1% penicillin–streptomycin (Invitrogen, Thermo Fisher Scientific, Waltham, MA, USA), with 100 nM dexamethasone, 10 mM β-glycerophosphate, 50 μM L-ascorbic acid 2-phosphate (all MilliporeSigma, St. Louis, MO, USA). The medium was replaced every third day. Osteogenic differentiation of cells was evaluated by the visualization of calcium deposits by Alizarin Red staining on fixed with 4% paraformaldehyde cells under light microscope. Additionally, the activity of alkaline phosphatase was evaluated using the colorimetric method as previously described in [19].

2.2.4. Chondrogenic Differentiation

Chondrogenic differentiation was achieved in three-dimensional culture of pelleted cells performed in a 15 mL Falcon tube. Chondrogenic medium was composed of DMEM-high glucose (Biowest, Riverside, MO, USA) supplemented with 0.5% FBS (Biowest, Riverside, MO, USA) and 1% penicillin-streptomycin (Invitrogen, Thermo Fisher Scientific, Waltham, MA, USA), with 100 nM dexamethasone, 1% insulin-transferrin-selenium solution (ITS), 10 ng/mL TGFβ2, 100 μM L-ascorbic acid 2-phosphate and 100 μg/mL sodium pyruvate (all MilliporeSigma, St. Louis, MO, USA). 1×10^6 cells suspended in a chondrogenic medium was pelleted by centrifugation and incubated with differentiation medium for three weeks. Until day two, a spheroid cell structure was observed at the bottom of the tube. The medium was changed every third day. The chondrogenic differentiation was conducted for 3 weeks. Then, the microsphere was fixed with 4% paraformaldehyde and underwent a standard histological procedure of paraffinizing, microtome cutting and hematoxilin and eosin as well as Masson trichrome and toluidine blue staining.

2.3. Preconditioning of Human BM-MSCs with Vadadustat

In the presented study "pharmacological" hypoxia was achieved by culturing cells with the selective PHDs inhibitor, Vadadustat (AKB-6548, Akebia, Cambridge, MA, USA). Based on preliminary data (Western blot analysis of HIF-1α stabilization and the MTT test) we decided to select the Vadadustat concentration of 40 μM for further studies. Vadadustat was dissolved and stored in −80 °C as 5 mM stock solution in DMSO according to the manufacturer instruction. Notably, no more than 0.8% (*v*/*v*) of DMSO was finally present in the culture medium, which did not cause any noticeable cytotoxic effect (MTT analysis presented in Supplementary Figure S1). The control group of MSCs was incubated with the same dose (0.8% *v*/*v*) of DMSO alone.

2.4. BM-MSCs RNA Isolation

For isolation of RNA, cells from 6 donors were cultured in 60 mm dishes until approximately 70% confluency was reached. MSCs were then exposed to experimental conditions. Both, control cells and 40 μM Vadadustat-treated cells were incubated at an atmospheric O_2 concentration. After 6 h treatment all cells were washed and disrupted in 350 μL of RLT buffer from the Qiagen RNeasy Mini Kit (Qiagen, Hilden, Germany). Samples were then stored in −80 °C until further use. RNeasy Mini Kit was used for total RNA extraction from MSCs according to the manufacturer's isolation protocol. The concentration and integrity of collected RNA samples were determined spectrophotometrically using NanoDrop 1000 (NanoDrop Technologies, Thermo Fischer Scientific, Waltham, MA, USA) and Bioanalyzer Chip RNA 7500 series II (Agilent Technologies, Santa Clara, CA, USA).

2.5. Gene Expression Analysis by Real-Time PCR

Reverse transcription was performed using a High Capacity cDNA Reverse Transcription Kit (Applied Biosystems, Thermo Fisher Scientific, Waltham, MA, USA). A quantity of 2 μg of total RNA was converted to cDNA according to producer's instruction. Real-time PCR was conducted using SYBR Select Master Mix (Applied Biosystems, Thermo Fisher Scientific, Waltham, MA, USA) in a 7500 Real-Time PCR System (Applied Biosystems, Thermo Fisher Scientific, Waltham, MA, USA). In each 20 μL reaction 100 ng of cDNA template and 0.5 μM forward and reverse primers was used. PCR reaction was started with two initial steps at 50 °C and 95 °C each for 2 min, followed by 40 cycles of 95 °C for 15 s and 60 °C for 1 min respectively. Standard curves were run on each plate to determine the amplification efficiency. Primer pairs were purchased from the Laboratory of DNA Sequencing and Oligonucleotide Synthesis, Institute of Biochemistry and Biophysics (IBB), Polish Academy of Sciences, Warsaw, Poland (oligo.pl) and -MilliporeSigma, St. Louis, MO, USA). Primer pairs sequences of examined genes are: *IL1B* FRD: 5′-CCACAGACCTTCCAGGAGAATG-3′, REV: 5′-GTGCAGTTCAGTGATCGTACAGG -3′; *IL24* FRD: 5′-CTTCTCTGGAGCCAGGTATCAG-3′, REV: 5′-GGCACTCGTGATGTTATCCTGAG-3′; *CCL28* FRD: 5′-CTGGAAAGAGTGAATATGTGTC-3′, REV: 5′-CTTGACATGAAGGATGACAG-3′; *ICAM1* FRD: 5′-ACCATCTACAGCTTTCCG-3′, REV: 5′-TCACACTTCACTGTCACC-3′; *IL1R1* FRD: 5′-ATTTAAGCAGAAACTACCCG-3′, REV: 5′-TTGCAATCCTTATACCACTG-3′; *LIFR* FRD: 5′-AAGTTTATCCCCATACTCCTAC-3′, REV: 5′-CCTGGTAAATGCCAAGAAAG-3′; *HIF1A* FRD: 5′-GAAACTACTAGTGCCACATC-3′, REV: 5′-GGAACTGTAGTTCTTTGACTC-3′; *IL6* FRD: 5′-GCAGAAAAAGGCAAAGAATC-3′, REV: 5′-CTACATTTGCCGAAGAGC-3′; *CCL2* FRD: 5′-AGACTAACCCAGAAACATCC-3′, REV: 5′-ATTGATTGCATCTGGCTG-3′; *TGFB3* FRD: 5′-TGTTGAGAAGAGAGTCCAAC-3′, REV: 5′-ATCACCTCGTGAATGTTTTC-3′; *IL23A* FRD: 5′-AGATAAATCTACCACCCCAG-3′, REV: 5′-CACATGTCAGTCAGTATTGG-3′; *CXCL8* FRD: 5′-GTTTTTGAAGAGGGCTGAG-3′, REV: 5′-TTTGCTTGAAGTTTCACTGG-3′; *IL17RD* FRD: 5′-AGTAGCTTCAAAAGAACTGG-3′, REV: 5′-CTCGGGTTCTAAAGAAGAAAG-3′; *PDCD1LG1* FRD:

5′-GGCATCCAAGATACAAACTCAA -3′, REV: 5′-CAGAAGTTCCAATGCTGGATTA-3′; *PDCD1LG2* FRD: 5′-GAGCTGTGGCAAGTCCTCAT-3′, REV: 5′-GCAATTCCAGGCTCAACATTA-3′; *B2M* FRD: 5′-TGGAGGCTATCCAGCGTACT-3′, REV: 5′-CGGATGGATGAAACCCAGAC-3′. Primer pairs for *LEP* (Cat. no. qHsaCID0017538) and *TNF* (Cat. no. qHsaCED0037461) were purchased from Bio-Rad Laboratories, Inc (Hercules, CA, USA). The relative quantification of a fold change in gene expression was calculated using the Pffafl method based on ΔCt and amplification efficiency of the transcripts normalized to the B2M (β2-microglobulin) reference gene [20]. The expression of each gene in control samples was appointed as 1. The analysis was performed in triplicate on cell populations from at least 6 BM-MSCs donors.

2.6. Analysis of BM-MSCs Cytokine Secretion by Antibody Array Proteome Profiler

The relative changes in secretory activity of Vadadustat treated BM-MSCs compared to control cells were examined using the Proteome Profiler Human XL Cytokine Array (Cat. no. ARY022B, R&D Systems, Bio-Techne, Minneapolis, MN, USA). The Proteome Profiler membrane-based antibody array enables to simultaneously measure the relative level of 102 human cytokines in a single sample. For the purpose of this assay, BM-MSCs from 6 donors were grown in a standard growth medium on a 6-well plates until approximately 80% confluency was achieved. 24 h before the start of the experiment, all cells were primed with IFNγ (25 ng/mL, MilliporeSigma, St. Louis, MO, USA). Next day, cells were washed and culture medium was replaced with OptiMEM Medium, no phenol red (Gibco, Thermo Fischer Scientific, Waltham, MA, USA) with reduced FBS content to 4% and supplemented with 1.0% penicillin–streptomycin with/without Vadadustat 40 μM. Cells of each population were treated in triplicate. After 24 h treatment cells supernatants were collected in an Eppendorf tube (1.5 mL), centrifuged at 4500 rpm for 5 min, transferred to new tubes, mixed and divided into 200 μL aliquots and frozen in −80 °C. Prior to the analysis, cell supernatants from 6 donors were thawed on ice and pooled. The analysis was performed according to the manufacturer's instruction. Chemiluminescence of membranes was detected with ChemiDoc MP Imaging System (Bio-Rad Laboratories Inc., Hercules, CA, USA) and the integrated optical density of each spot was measured and quantified using Image Lab software (Bio-Rad Laboratories Inc., Hercules, CA, USA).

2.7. Quantitative Analysis of BM-MSCs Cytokine Secretion by Luminex Multiplex Immunoassay

The quantitative analysis of selected cytokines by the Luminex method was performed on cell supernatants, the preparation of which was described above. Samples were not pooled in this analysis, so supernatants from six populations were analyzed separately. The custom Luminex Multiplex kit was purchased in R&D Systems (Bio-Techne, Minneapolis, MN, USA) and contained IL6, CXCL8, IL4, IL10 and HGF analytes. The procedure was performed according to the manufacturer's instructions. The flow based magnetic beads reading was performed on Luminex LX-200 Instrument (Thermo Fisher Scientific, Waltham, MA, USA). All samples were analyzed in duplicate.

2.8. Isolation and Identification of Peripheral Blood Mononuclear Cells (PBMCs)

Human PMBCs used in this study were freshly isolated from buffy coats each time. Buffy coats were purchased at the Regional Blood Donation and Blood Treatment Centre in Warsaw as medical waste from whole blood, which was centrifuged without a density gradient. The isolation of PBMCs was performed within 4 h of collecting whole blood. Buffy coats were first diluted in PBS (without calcium and magnesium) in 50 mL Falcon tubes, and then cells were separated by density gradient centrifugation on Histopaque-1077 (MilliporeSigma, St. Louis, MO, USA). PBMCs were collected from a plasma/Ficoll interface with a Pasteur pipette and transferred to a new 50 mL falcon tube. Isolated PBMCs were then washed four times in PBS to rinse cells pellets and to reduce platelet contamination. Finally, cells were suspended in growth medium composed of RPMI-1640 (Thermo Fischer Scientific, Waltham, MA, USA) with 10% human serum (Biowest, Riverside, MO, USA) and 1% penicillin-streptomycin solution (Invitrogen, Thermo Fisher Scientific, Waltham, MA,

USA) and counted. To obtain the monocyte-enriched population used in the migration assay, isolated PBMCs were seeded on Primaria™ Tissue Culture Dishes at a density of 75×10^4/mL in RPMI-1640 (Biowest, Riverside, MO, USA) without serum. After two hours, the plates were vigorously washed five times with PBS, then the adherent cells remaining on the dishes were scraped off and suspended for further processing. The flow cytometric analyses of the PBMCs and monocyte-enriched were performed by CD3 (PerCP Mouse anti Human Clone SP34-2), CD14 (FITC Mouse Anti-Human Clone M5E2) and CD16 (PE-Cy™7 Mouse Anti-Human Clone 3G8) staining (all BD Biosciences, San Jose, CA, USA) on BD FACS Canto II using BD FACS Diva Software (BD Biosciences, San Jose, CA, USA) and analyzed by FCSExpress7 (De Novo Software, Glendale, CA, USA).

2.9. Mixed Lymphocyte Reaction (MLR) Assay

For the purpose of MLR assay human PBMCs were isolated from buffy coats from 6 healthy blood donors. The assay was performed in three independent sets of experiments on two donors each. Supernatants from 6 populations of IFNγ (25 ng/mL) primed BM-MSCs treated for 24 h with/without Vadadustat 40 μM were used to determine the effect of Vadadustat pretreatment on immunomodulatory activity of MSCs secretome. In this study, half of the isolated PBMCs were inactivated for 90 min with γ-irradiation. Next, 1×10^5 both responder (active) and irradiated (stimulatory) PBMCs were seeded into wells of 96-well plates in a combination of auto- (AA_{ir}, BB_{ir}) and allo- (AB_{ir}, BA_{ir}) stimulation. Cells were maintained in RPMI-1640 (Thermo Fischer Scientific, Waltham, MA, USA) supplemented with 10% FBS (Gibco, Thermo Fisher Scientific, Waltham, MA) and antibiotic–antimycotic solution (1% penicillin-streptomycin; 0.5% amphotericin B, Invitrogen, Thermo Fisher Scientific, Waltham, MA). The MLR assay were performed using 96-well plates. In the part of the wells where the direct effect of Vadadustat on auto- and allostimulated PBMCs as well as its effect on the interaction between MSCs and PBMCs were studied, 40 μM Vadadustat was added to the experimental wells daily as a stock solution. Control wells were treated daily with equivalent volumes of DMSO. In the remaining wells, in which the indirect effect of Vadadustat pre-conditioning on the interaction between MSCs and PBMCs was studied, a 1:1 mixture of RPMI-1640 growth medium and supernatants from 24 h cell culture of control or Vadadustat preconditioned MSCs was added once at the beginning of the experiment. Plates were then cultured for 5 days at 37 °C in a humidified atmosphere with 5% CO_2. After 5 days of cell culture, PBMCs were pulsed with 1 μCi/well of 3H-thymidine (113 Ci/nmol, NEN) for the last 18 h of incubation and then harvested with an automated cell harvester (Skatron). The 3H-thymidine incorporation into cells was measured based on the level of radioactivity reported as 'Corrected Counts per Minute' (CCPM) using a scintillation counter (Wallac, PerkinElmer, Inc., Waltham, MA, USA). All treatments were performed in triplicate.

2.10. Transwell Migration Assay

The effect of 40 μM Vadadustat preconditioning on the chemotactic properties of the BM-MSCs secretome was investigated using a "96 Well Cell Migration Assay" reagent kit from Cultrex® (cat. no. 3465-096-K) (R&D Systems, Bio-Techne, Minneapolis, MN, USA), which utilize a simplified design of a Boyden chamber with polyethylene terephthalate (PET) membrane with pores of 8 μm size. For the migration test, we used monocyte-enriched PBMCs ($n = 4$) suspended in RPMI (Biowest, Riverside, MO, USA) containing 0.5% human serum (Biowest, Riverside, MO, USA) at a density of 4×10^6/mL. 50 μL of cell suspension from each donor were applied to the upper chambers of the plate (2×10^4 cells per well), each in duplicate. Quantities of 150 μL per well of growth medium (RPMI with 0.5% human serum) or freshly thawed, pooled supernatants from cultures of 7 MSCs populations were applied to the bottom chambers of the plate. Each of three treatments: growth medium alone, supernatants from control MSCs and MSCs preconditioned with Vadadustat was applied in duplicate. Plates were then incubated under standard conditions (37 °C, 5% CO_2) for 48 h. After incubation, the upper chambers were carefully aspirated and the cells that migrated to the bottom compartments of the plate were detached using a cell dissociation solution with calcein acetomethylester (calcein-AM). Afterwards, plates were incubated at 37 °C for 30 min. During this time, cells internalized calcein-AM, and cellular esterases then cleaved it into free calcein. Released calcein possess strong fluorescence, that was used to estimate the number of migrated cells. After incubation, plates were disassembled and bottom chambers were fluorescently read at 485 nm excitation and 520 nm emission on Perkin Elmer Victor X4 plate reader (PerkinElmer, Inc., Waltham, MA, USA). The degree of cell migration was assessed by comparing fluorescence in the wells with MSCs culture supernatants to fluorescence in wells with growth medium alone, and expressed as the ratio of migrating cells.

2.11. Statistical Analysis

The results were statistically analyzed using STATISTICA 13.1 software (Tibco, Palo Alto, CA, USA). Shapiro-Wilk test was used to analyze the data distribution within groups. Wilcoxon matched-pairs signed-rank test was used to determine statistical significance between two groups of related data with abnormal distribution. Student's t-test was used to evaluate significance between two groups of related data with confirmed normal distribution. A p-value of < 0.05 (*) was considered statistically significant, and $p < 0.01$ (**), or $p < 0.001$ (***) as highly significant. Graphs are presented as mean ± SEM (standard error of the mean) unless otherwise indicated.

3. Results

3.1. Isolation and Characterization of Human BM-MSCs

MSCs isolated from bone marrow were identified according to the International Society for Cell and Gene Therapy (ISCT) statement established in 2006 [21]. All isolated cell populations were proven to form colonies and adhere to a plastic culture surface (a representative population is shown in Figure 1b). The mean expression of surface markers from the 7 BM-MSCs population was: CD73—99.5%, CD90—98.8%, CD105—99.4%, and no antigens CD45, CD34, CD11b, CD19 and HLA-DR were detected on 97.6% cells (Supplementary Table S2). Figure 1a shows a representative panel of BM-MSCs phenotyping results using flow cytometry (full panel in Supplementary Table S3). Tri-lineage differentiation capability was confirmed as shown in Figure 1c–e.

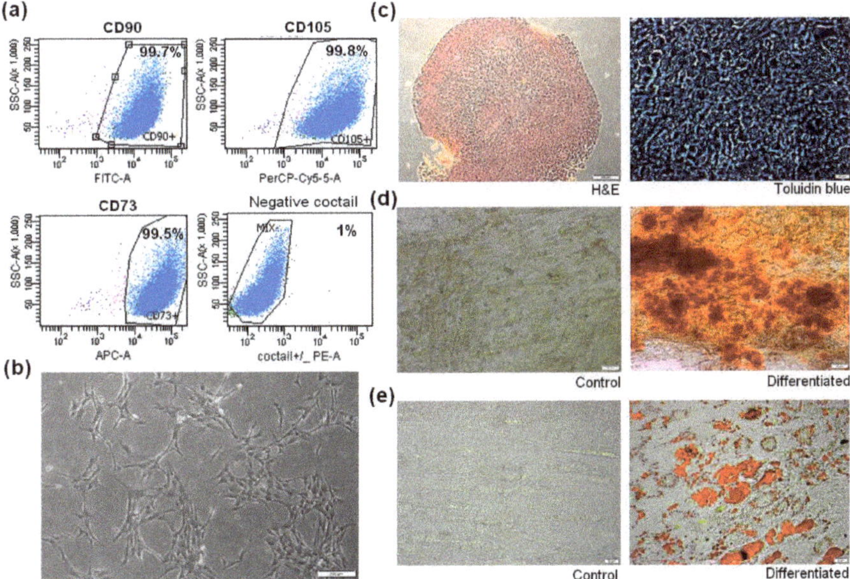

Figure 1. Identification and morphology of human bone marrow-derived mesenchymal stromal cells

(hBM-MSCs). (a) Flow cytometry analysis of representative MSCs population. MSCs were positive for CD73, CD90, CD105 and negative for CD34, CD45, CD11b, CD19, HLA-DR. (b) Representative image of undifferentiated BM-MSCs morphology cultured under standard growth condition (21% O_2 and 5% CO_2). Light microscopy, image scale 200 µm (c) chondrogenic differentiation of BM-MSCs. Hematoxylin-eosin (HE) and Toluidine blue staining of BM-MSCs microsphere section. Light microscopy, HE staining scale 200 µm, Toluidine blue staining scale 20 µm. (d) Osteogenic differentiation of BM-MSCs. Control and differentiated cells stained with Alizarin red. Light microscopy, scale 50 µm. (e) Adipogenic differentiation of BM-MSCs. Oil Red O staining of control and differentiated cells. Light microscopy, scale 20 µm.

3.2. Vadadustat Preconditioning of BM-MSCs Affected the Expression of Genes Associated with the Regulation of Immune Responses

Due to our particular interest in the regulation of immune functions by MSCs, we analyzed the expression of genes related to their immunomodulatory properties. The list of examined genes was based on our previous study evaluating Vadadustat-induced changes in the MSCs transcriptome obtained by RNA sequencing (currently under review). The list included genes encoding factors secreted by MSCs in response to immune stimuli (CCL2, IL6, CXCL8 and TNF), proteins involved in signaling pathways activated by cytokines (e.g., IL17RD, LIFR, IL6R and ICAM1) and other immune regulatory molecules PDCD1L1, PDCD1L2 or LEP (genes listed in Figure 2). We performed real-time PCR analysis of selected genes on the 6 BM-MSCs populations incubated for 6 h under standard grown conditions (control) or with 40 µM Vadadustat (Figure 2).

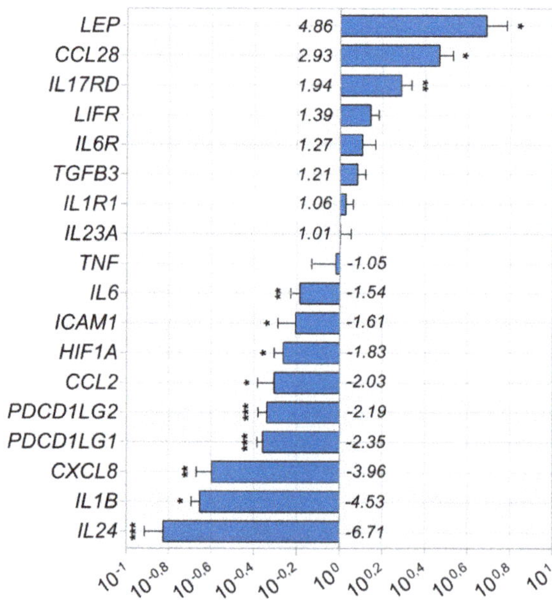

Figure 2. Relative expression of chosen immune system-related genes performed by real-time PCR on 6 bone marrow-derived mesenchymal stromal cells (BM-MSCs) populations incubated for 6 h with Vadadustat or under standard culture conditions (control). Results are presented as fold change in gene expression in MSCs cultured with Vadadustat versus control MSCs ± SEM (standard error of the mean), calculated by the Pfaffl method [20]. * indicates the statistically significant ($p < 0.05$), ** ($p < 0.01$) and *** ($p < 0.001$) highly statistically significant differences obtained by Student's paired t-test (for data with normal distribution) or Wilcoxon matched-pairs signed rank test (for data with abnormal distribution) of ΔCt values of Vadadustat-treated samples in relation to ΔCt of control samples.

The results clearly demonstrate, that Vadadustat strongly down-regulated the expression of *IL24* (−6.71), *IL1B* (−4.53), *CXCL8* (−3.96) and slightly *PDCD1LG1* (−2.35), *PDCD1LG2* (−2.19), *CCL2* (−2.03), *HIF1A* (−1.83), *ICAM1* (−1.61) and *IL6* (−1.54), and up-regulated the expression of *IL17RD* (1.94), *CCL28* (2.93) and *LEP* (4.86) compared with control cells. The results obtain for *TNF* (−1.05), *IL23A* (1.01), *IL1R1* (1.06), *TGFB3* (1.21), *IL6R* (1.27), *LIFR* (1.39) did not differ significantly between control cells and treated with Vadadustat.

3.3. Functional Activity of BM-MSCs Preconditioned with 40 μM Vadadustat

Genes expression analysis of Vadadustat-preconditioned BM-MSCs showed promising results due to its potential influence on MSCs immunomodulatory properties. At the next stage, it was crucial to determine whether these changes are reflected in the functional activity of MSCs. Several analyzes related to MSCs activity were performed to determine whether this method of MSCs preconditioning could enhance their immunosuppressive potential.

3.3.1. Preconditioning with Vadadustat Changed the Secretory Profile of BM-MSCs

The effect of Vadadustat preconditioning on BM-MSCs cytokine and chemokine secretion profile was determined after 24 h cells treatment with 40 μM Vadadustat by the antibody based Proteome profiler array (Figure 3a,b).

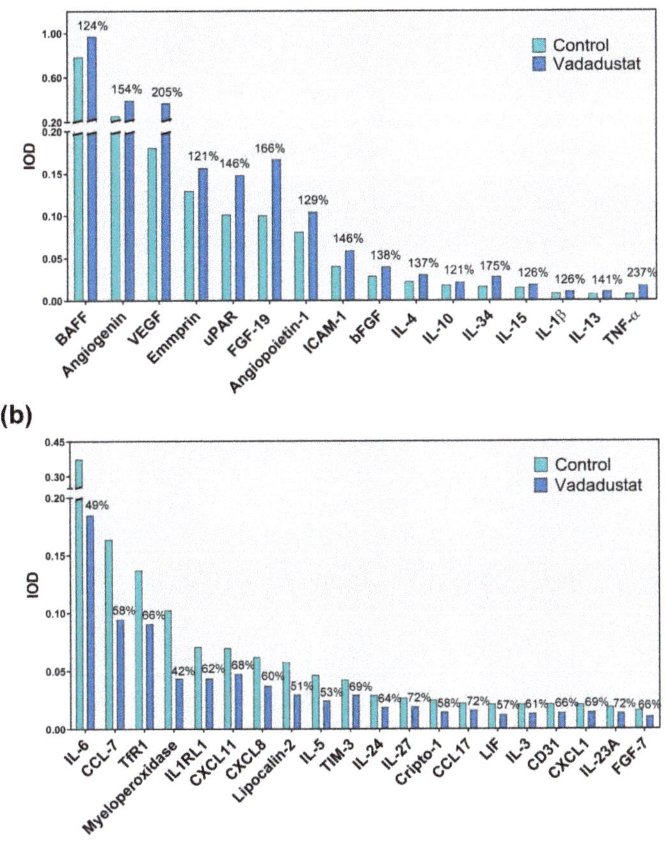

Figure 3. *Cont.*

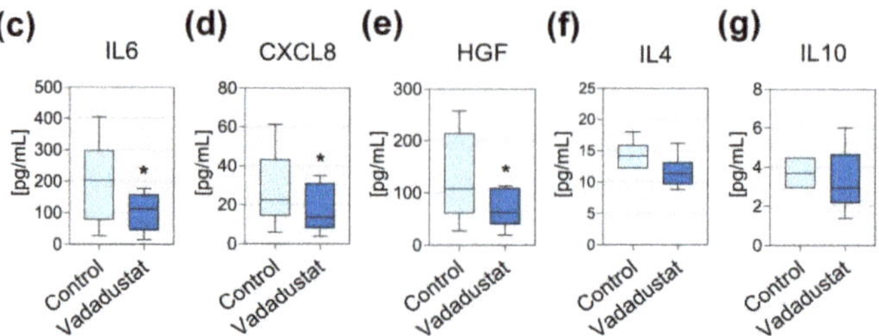

Figure 3. Analysis of the secretome from 6 populations of human bone marrow-derived mesenchymal stromal cells (hBM-MSCs) pretreated with Vadadustat for 24 h. (**a**,**b**) Proteome profiler analysis of cytokines and chemokines whose secretion by MSCs: (**a**) increased or (**b**) decreased by at least 20% as a result of 24 h pretreatment with 40 µM Vadadustat. Bars represent IOD (integrated optical density) of antibodies-dots measured by chemiluminescent detection. The values above the bars show what percentage of control MSCs secretion (indicated as 100%) are values obtained after incubation with Vadadustat. Samples from 6 MSCs populations were pooled for the analysis. (**c**–**g**) Quantitative evaluation of (**c**) IL6, (**d**) CXCL8, (**e**) HGF, (**f**) IL4 and (**g**) IL10 level in the secretome of 6 BM-MSCs populations cultured under control conditions or preconditioned with Vadadustat, conducted using Luminex assay. The samples were not pooled for analysis. Boxes show quartiles of secreted cytokine amount in pg/mL with median, whiskers represent "min to max" values. * indicates the statistically significant difference ($p < 0.05$) by Wilcoxon matched-pairs signed-rank test in the groups of related data with abnormal distribution, and by Student's t-test in the groups of related data with confirmed normal distribution.

The analysis of cytokines and chemokines found in the pooled supernatants of 6 BM-MSCs populations treated with Vadadustat showed a number of factors whose secretion changed when compared to the control cell supernatants. The observed changes concerned both compounds whose secretion increased (Figure 3a) as well as those whose secretion decreased (Figure 3b). Analysis of the results indicated that among compounds whose secretion was downregulated by Vadadustat treatment, the most prominent decrease was observed among the cytokines secreted in large quantities by MSCs. The marked decrease in secretion was noted for myeloperoxidase (to 42% of the control value), IL6 (to 49%) and CCL7 (to 58%). A decrease in secretion was also noted among compounds secreted by MSCs in smaller amounts: lipocalin 2 to 51%, IL5 to 53%, LIF to 57%, Criptol to 59%, CXCL8 to 60%, IL3 to 61% and IL24 to 64%. In addition, the level of five flagship cytokines related to MSC immunomodulation: IL6, CXCL8, HGF, IL4, IL10 was measured in MSCs secretome using the Luminex method. MSCs preconditioning with Vadadustat resulted in a reduction of IL6 secretion from an average of 199 pg/mL to 103 pg/mL (49% decrease, Figure 3c). The CXCL8 secretion decreased on average from 24.59 pg/mL in control cells to 15.46 pg/mL in those treated with Vadadustat (Figure 3d). HGF secretion was reduced from an average of 128 pg/mL to 68 pg/mL (47% decrease, Figure 3e). There were no statistically significant changes in the secretion of IL4 (14 ng/mL vs. 11 pg/mL, Figure 3f) and IL10 (3.7 pg/mL vs. 3.1 pg/mL, Figure 3g) related to Vadadustat pretreatment.

3.3.2. Vadadustat Significantly Increased the Inhibitory Effect of MSCs on Proliferation of Allostimulated PBMCs

A series of MLR assays were performed to evaluate the effect of Vadadustat preconditioning on the inhibitory properties of BM-MSCs on PBMCs. The assays examined both the direct effect of Vadadustat on the PBMCs and MSCs–PBMCs interaction as well as the indirect effect of incubating PBMCs with supernatants from Vadadustat-preconditioned MSCs. PBMCs were obtained from 6 donors and 6 BM-MSC populations were used. The results presented in Figure 4a clearly demonstrate that Vadadustat enhanced an immunosuppressive activity of MSCs on PBMCs.

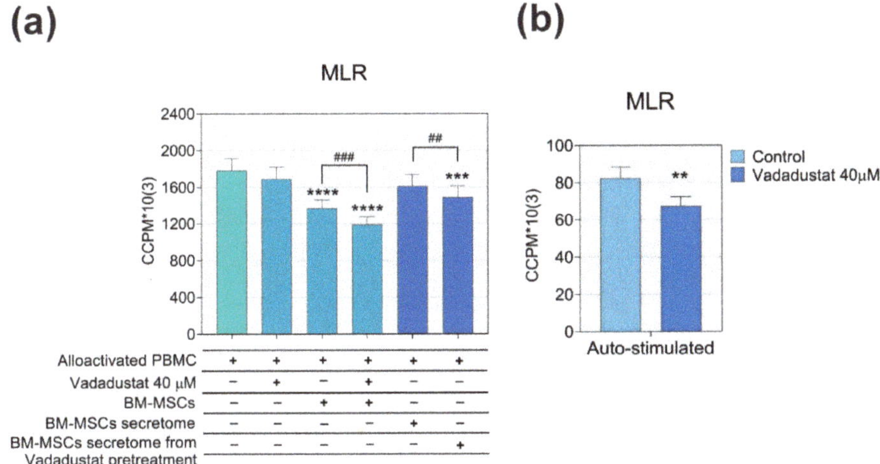

Figure 4. Mixed lymphocyte reaction (MLR) assay. The effect of 40 μM Vadadustat on the allo- and auto-responsiveness of leukocytes. (**a**,**b**) Assessment of peripheral blood mononuclear cells (PBMCs) activation and proliferation capacity in response to direct and indirect effects of Vadadustat by MLR. MLR was performed on freshly isolated PBMCs from 6 donors, 6 populations of bone marrow-derived mesenchymal stromal cells (BM-MSCs) and supernatants pooled from the culture of 6 BM-MSCs populations. The experiment was conducted for 5 days. Results are presented as: (**a**) Effect of Vadadustat on PBMCs alloreactivity and PBMCs–MSCs interaction, and (**b**) effect of Vadadustat on PBMCs autoreactivity. Both assessed by measuring radioactivity of 3H-thymidine-incorporated cells and reported as mean 'Corrected Counts per Minute' (CCPM) ± SEM (standard error of the mean). ** indicates the statistically significant difference for $p < 0.01$, *** for $p < 0.001$, **** for $p < 0.0001$ by Friedman test with Dunn's multiple comparison of mean rank of each group with a mean rank of an alloactivated PBMC (control) in the groups of related data with abnormal distribution. ## indicates statistically significant ($p < 0.01$) and ### ($p < 0.001$) highly significant differences between the two treatment groups analyzed by Wilcoxon matched-pairs signed-rank test in the groups of related data with abnormal distribution, and by Student's t-test in the groups of related data with confirmed normal distribution.

The PBMCs co-culture with control MSCs resulted in a statistically highly significant 22% decrease in PBMCs proliferation compared to the value of allostimulated PBMCs (control). Furthermore, five-day treatment of PBMCs-MSCs culture with Vadadustat resulted in a greater, 28% decrease in allostimulated PBMCs proliferation compared to control cells. When the same PBMCs were cultured for 5 days with the MSCs secretome (1:1 mixture of MSCs supernatants and RPMI growth medium), a significant inhibition of PBMCs proliferation was also noted. The secretome of control MSCs caused a decrease in PBMCs alloreactivity by 9%, and of MSCs preconditioned with Vadadustat by 16% compared to allostimulated PBMCs. However, in both direct and indirect Vadadustat treatments, a similar percentage decrease in PBMCs proliferation between the MSCs alone and preconditioned with Vadadustat was determined (6–7%). In both treatments, the effect of Vadadustat was statistically highly significant. Due to the very short half-life of Vadadustat (4.5 h according to the manufacturer [22]) in the collected secretome of Vadadustat preconditioned MSCs (24 h incubation) there should no longer be an active inhibitor. We noted lack of HIF-1a stabilization after 24 h MSCs incubation with Vadadustat, confirmed by Western blot analysis (data not shown). Thus, we argue that the observed suppressive effect of secretome from Vadadustat preconditioned MSCs was related only to the secretory activity of cells and not to Vadadustat itself. However, in some part Vadadustat possess the direct effect on PBMCs proliferation as well. Treatment of allostimulated PBMCs with Vadadustat caused a 5% decrease in their proliferation. A much larger decrease in proliferation resulting from Vadadustat was noted in autostimulated PBMCs (Figure 4b), which reached 18%. However, it should be noted that PBMCs autoresponsiveness remained at a much lower level. To summarize, the observed suppressive effect of Vadadustat on the reactivity of PBMCs was both, associated with the changes in MSCs activity as well as the direct effect on PBMCs.

3.3.3. Secretome from Vadadustat Preconditioned MSCs Significantly Reduced PBMCs Migration

96 Well Cell Migration Assay was performed to assess the effect of Vadadustat preconditioning on the chemotactic properties of BM-MSCs secretome. We quantified the degree of 5 donors' PBMCs migration through a 8 micron PET membrane in response to stimulating and/or inhibiting compounds contained in pooled supernatants collected from 7 MSCs populations incubated for 24 h with Vadadustat, under standard growth conditions or with growth medium alone. For the purpose of migration analysis, we used PBMCs fraction with enriched monocyte content. We obtained the monocyte-enriched PBMCs by pre-culturing the cells on plates for 2 h, and applying a 5× PBS wash to leave only adherent cells. Phenotypic analysis of precultured PBMCs by flow cytometry showed that nearly 76% of the cells used in the assay were monocytes ($CD14^{+}$), of which 66.5% were activated monocytes ($CD14^{+}CD16^{+}$) (Figure 5a).

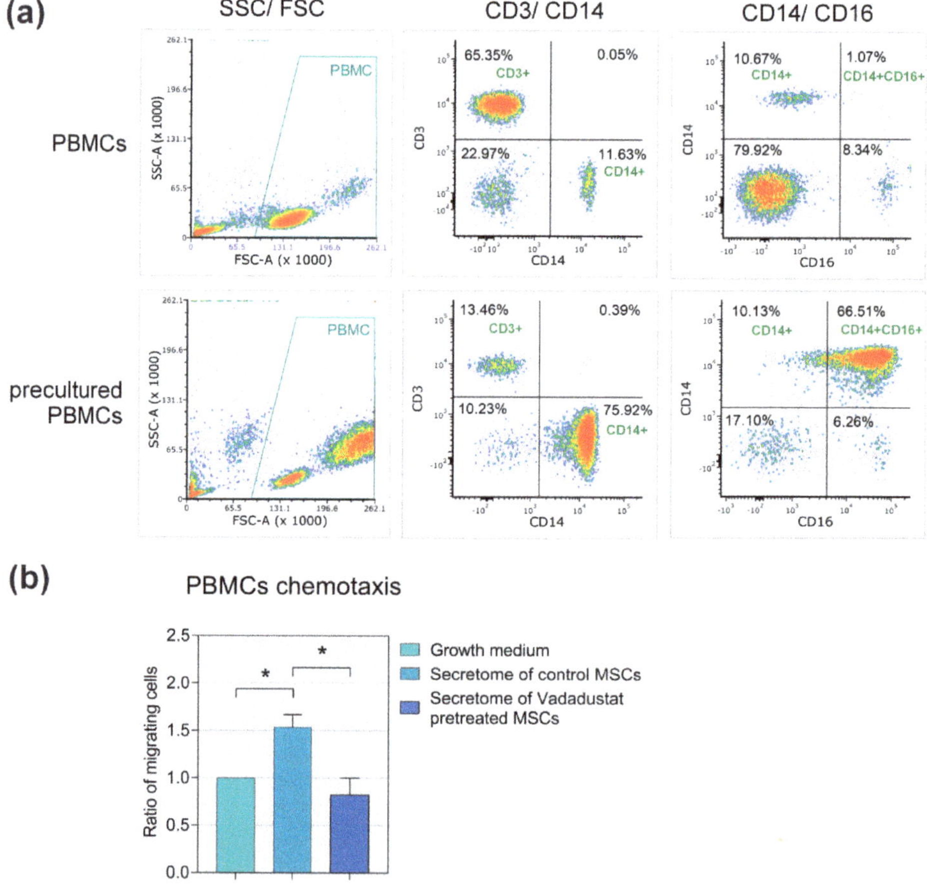

Figure 5. Assessment of the chemotactic properties of the secretome of Vadadustat preconditioned bone marrow-derived mesenchymal stromal cells (BM-MSCs). (**a**) Flow cytometric analysis of lymphocyte

populations in peripheral blood mononuclear cells (PBMCs) isolated or precultured in plates based on CD3 (PerCP), CD14 (FITC) and CD16 (PE-Cy7) staining. PBMCs were gated based on FSC and SSC, percentages of monocytes ($CD14^+$) and activated monocytes ($CD14^+CD16^+$) in PBMCs are shown in quadrants. (**b**) The analysis of chemotactic properties of secretome from 6 BM-MSCs populations cultured for 24 h under standard conditions (control) and with Vadadustat on the migration of monocyte-enriched peripheral blood mononuclear cells (PBMCs) (precultured PBMCs). The rate of cell migration is presented as the ratio of migrating cells ± SEM (standard error of the mean), obtained by comparing the fluorescence of PBMCs migrated to the growth medium (denoted as 1) with the fluorescence of PBMCs migrated into the secretome of control and Vadadustat pretreated MSCs. * Indicates the statistically significant ($p < 0.05$) differences obtain by Wilcoxon matched-pairs signed-rank test in the groups of related data with abnormal distribution, and by Student's t-test in the groups of related data with confirmed normal distribution.

Cell migration from the inserts to the basal compartment of plate wells was assessed after 48 h of monocyte-enriched PBMCs incubation. The results presented in Figure 5b show that within 48 h of incubation there was a significant, 53% increase in PBMCs/monocytes migration in wells containing secretome of control MSCs compared to wells with growth medium alone. Conversely, there was a statistically significant 46% decrease in cell migration in wells with secretome of Vadadustat-preconditioned MSCs compared to migration in wells with secretome of control MSCs. Moreover, there was a slight (17%) decrease in cell migration by comparing Vadadustat supernatants with the growth medium alone, although this effect was not statistically significant.

4. Discussion

In recent years, great efforts have been made to develop methods for obtaining more effective and safer MSCs for use in cell therapy. Many studies were carried out to determine the role of MSCs in the regulation of the immune system, showing that their immunomodulatory capacity is a very plastic feature [23]. The plasticity of MSCs immunomodulation is associated with the ability to elicit markedly different modulatory responses, which results from the current state of inflammatory mediators in their microenvironment. Development of a chronic inflammatory microenvironment, resulted from loss of peripheral immune tolerance and excessive stimulation of innate and adaptive immune responses, is associated with the course of autoimmune diseases. MSCs can target such an inflammatory microenvironment by paracrine actions, demonstrating broad immunosuppressive, anti-fibrogenic, anti-apoptotic and pro-angiogenic effects [24]. Immunomodulation attributed to the therapeutic activity of MSCs is related to their function to modulate the proliferation, differentiation, adhesion, and migration of immune cells under disease conditions. Since the immunosuppressive nature of MSCs activity is generally therapeutically desirable, many approaches have been developed to modulate the culture conditions of MSCs in order to obtain their inflammatory-resolving phenotype. While a number of MSCs preconditioning strategies are currently being investigated, cytokine priming and recently hypoxic pretreatment appear to be the major approaches used to increase MSCs immunomodulatory properties [25,26]. Moreover, recent findings indicate that hypoxia inducible factor-1α (HIF-1α) is a major regulator of the immunomodulatory functions of MSCs [27,28]. Although the effect of HIF-1α stabilization by hypoxia mimetic agents on MSCs properties has already been studied (cobalt chloride, deferoxamine, ciclopirox olamine, N-acetylcysteine, FG-4497, AKB-4924 [29–33], we used for the first time Vadadustat (AKB-6548)—a novel oral PHD2 inhibitor tested in phase III clinical trials that works through the mechanism of active site iron chelation in the submicromolar range [15]. Our research has shown that Vadadustat pretreatment enhances the immunosuppressive potential of MSCs. Vadadustat significantly enhanced the suppressive effect of MSCs on PBMCs proliferation (MLR test), and this effect was partially associated with the modulation of MSCs secretome. However, the suppressive capacity of MSCs was higher in direct contact with PBMCs. This may indicate that changes in both, compounds secreted by MSCs and presented on their surface are responsible for enhancing the immunosuppressive effect of MSCs pretreated with Vadadustat. Moreover, Vadadustat significantly diminished the chemotactic properties of the MSCs secretome, as assessed by the monocyte-enriched PBMCs migration assay. It is difficult to discuss all factors whose regulation may have an effect on the immunosuppressive capacity of Vadadustat-preconditioned MSCs, but some of them are of particular importance. First of all, Vadadustat significantly decreased the expression of *IL6* and the level of secreted IL6 in relation to control MSCs. Considering that the level of IL6 can raise many thousand-fold in the course of inflammation and autoimmune diseases, we believe that MSCs preconditioning with Vadadustat may appear to be a very promising approach for the use in therapy of autoimmune diseases. Another immune-related factor that is highly regulated by Vadadustat is CXCL8 (IL8). Preconditioning with Vadadustat significantly reduces expression and secretion of CXCL8 by MSCs, as demonstrated by real-time -PCR, proteome profiler and Luminex analyzes. CXCL8 is a chemokine considered to be proinflammatory and chemotactic, especially to neutrophils. They are the most abundant group of leukocytes, constituting the indispensable line of innate immune defense against infectious diseases and their role in regulating the immune response is recently increasingly emphasized. However, neutrophils infiltration and released neutrophil extracellular traps (NETs) are also mentioned as contributing to the development of autoimmune diseases [34], especially rheumatoid arthritis (RA) [35], ANCA-associated vasculitis (AAV) and systemic lupus erythematosus (SLE) [36]. Other chemotactic factors negatively regulated in MSCs by Vadadustat were CCL7 (MCP-3) and CCL2 (MCP-1), that are both a potent monocyte-attracting chemokines. The decrease in their secretion may constitute one of the factors responsible for inhibiting monocyte-enriched PBMCs migration in the chemotaxis assay, especially when considered together with the HGF, CCL11 and CCL17. It appears that the inhibitory effect of Vadadustat preconditioning on the chemotactic properties of MSCs secretome may be therapeutically positive, as abnormal infiltration and activation of monocytes and macrophages are observed in many autoimmune diseases (reviewed in [37]). Conversely, real-time -PCR analysis showed a significant increase in the expression of another chemokine—*CCL28* after Vadadustat pretreatment. While CCL28 is responsible for the recruitment of various immune cells (which express CCR10 and CCR3) for mucosal tissue and inflammatory sites, some data indicate that it is responsible for recruiting Treg, maintaining tolerance of self antigens and preventing autoimmune diseases [38]. While there is a study demonstrating that MSCs do not express *TNF* [39], we have received its expression and increase of TNF secretion level by 237% in the secretome of Vadadustat-preconditioned MSCs (however, signal intensity indicates that this level is extremely low). It seems that such small amounts may be responsible for maintaining the MSCs immunosuppressive phenotype rather than providing wider pro-inflammatory signaling. It should also be noted that detecting such small amounts of secreted factors might give false results. Due to differences in the amount of secreted factors between MSCs populations, analyzes of pooled samples may not reflect the true trend, especially when it comes to factors secreted in small quantities. As demonstrated by Luminex, IL4 and IL10 are secreted by MSCs in very small amounts. What is more, their level decreases after Vadadustat treatment—rather than increases as the proteome profiler analysis showed—however, not statistically significantly. Therefore, the regulation of IL4 and IL10 secretion by Vadadustat pretreated MSCs, described by many authors as one of the mechanisms of MSCs immunomodulation, in this case does not seems to play a major role. It should be mentioned, however, that Vadadustat was shown to significantly reduce the secretion of another member of the IL10 family—IL24. The results obtained by real-time PCR as well as the proteome profiler showed a significant decrease in its expression and secretion. Although this cytokine is involved in the process of wound healing, the overproduction of IL24 underlies pro-inflammatory autoimmune diseases such as psoriasis, allergic contact dermatitis, atopic dermatitis, rheumatoid arthritis and inflammatory bowel disease [40,41]. Therefore, a decrease in the secretion of IL24 by MSCs as a result of Vadadustat treatment seems to be a beneficial effect when considering the use of Vadadustat preconditioned MSCs in the treatment of patients with autoimmune diseases.

In addition to soluble factors, contact-dependent signals are also responsible for MSCs immunosuppressive activity. Our results showed that Vadadustat pretreatment caused a decrease in ICAM1 expression, while after 24 h increase in secretion of its soluble form was noted. ICAM1 is an adhesion molecule ligand for LFA-1, leukocyte integrin crucial for T cell trafficking, activation and proliferation [42]. Binding of ICAM1 with LFA-1 is involved in leukocyte endothelial transmigration. Soluble ICAM1 binding to LFA-1 was shown to inhibit lymphocyte attachment to endothelial cells (Rieckmann et al., 1995), and anti-ICAM1 or LFA-1 antibodies inhibit autoreactive T cell proliferation [43]. Therefore, the downregulation of ICAM1 expression together with increase in secretion of soluble ICAM1 after Vadadustat may constitute another mechanism for MSCs immunosuppressive activity, particularly promising in the context of inhibiting T cell autoreactivity. A more equivocal result was a Vadadustat-mediated decrease in gene expression of another cell surface molecules: PD-L1 (*PDCD1LG1*) and PD-L2 (*PDCD1L2*). Both PD-L1 and PD-L2 represent cell surface ligands for the PD-1 receptor (programmed cell death protein-1) expressed on T and B cells as an immunological checkpoint molecule. PD-1 is critical for modulating adaptive immunity by negatively regulating T-cell activation and preventing excessive or self-oriented immune responses. It is known that licensing of MSCs with proinflammatory cytokines (IFNγ, TNFα) increases the expression of PD-L1 and PD-L2 on their cell surfaces [44,45]. MSCs inhibition of T cell proliferation was reported to function through the contact dependent interaction of PD-1/PD-L1 [46–48]. The role of PD1 pathway in the immunomodulatory activity of MSCs is even more complex, since secretion of soluble PD-L1 and PD-L2 by BM-MSCs has also been reported [45]. We observed that the secretome from Vadadustat preconditioned MSCs inhibited the alloreactivity of PBMCs more than the secretome from control cells. Therefore, we suppose that despite the decrease in gene expression for PD-L1 and PD-L2 after pretreatment with Vadadustat, MSCs may increase the secretion of their soluble forms. However, further research is needed to define the role of Vadadustat in the by PD-L1 and PD-L2-mediated immunomodulatory function of MSCs.

In this study, we demonstrated that HIF-1 prolyl hydroxylase inhibition by Vadadustat positively affects the immunomodulatory properties of hMSCs. Preconditioning with Vadadustat has several particularly valuable features when considering the use of MSCs or MSCs secretome in the treatment of autoimmune diseases. Vadadustat, which is currently being tested for the maintenance treatment of patients with anemia secondary to chronic kidney disease in Phase III clinical studies (NCT02648347, NCT02680574, NCT04313153), aspires to become an effective tool enhancing the therapeutic activity of MSCs in the field of cell therapies.

Supplementary Materials: The following are available online at http://www.mdpi.com/2073-4409/9/11/2396/s1, Table S1: BM-MSCs donor profile, Table S2: Flow cytometry analysis of BM-MSCs populations, Table S3: Flow cytometry results of representative BM-MSCs population, Figure S1: MTT assay.

Author Contributions: Conceptualization, funding acquisition and supervision, A.B., R.Z. and L.P.; methodology, K.Z. and A.B.; investigation, K.Z., A.B. and B.K.; formal analysis, K.Z. and A.B.; visualization, K.Z.; writing—original draft, K.Z.; writing—review and editing, A.B., B.K., R.Z. and L.P. All authors have read and agreed to the published version of the manuscript.

Funding: This research was funded by the National Science Centre, Poland, grant number 2017/25/B/NZ6/01380).

Acknowledgments: We would like to thank Aleksander Roszczyk, Michał Zych and Monika Kniotek for their support in flow cytometry and Agnieszka Graczyk-Jarzynka for her support in acquisition the result of the migration assay.

Conflicts of Interest: The authors declare no conflict of interest.

References

1. Shammaa, R.; El-Kadiry, A.E.; Abusarah, J.; Rafei, M. Mesenchymal Stem Cells Beyond Regenerative Medicine. *Front. Cell Dev. Biol.* **2020**, *8*, 72. [CrossRef] [PubMed]
2. Pittenger, M.F.; Mackay, A.M.; Beck, S.C.; Jaiswal, R.K.; Douglas, R.; Mosca, J.D.; Moorman, M.A.; Simonetti, D.W.; Craig, S.; Marshak, D.R. Multilineage potential of adult human mesenchymal stem cells. *Science* **1999**, *284*, 143–147. [CrossRef] [PubMed]
3. Najar, M.; Raicevic, G.; Fayyad-Kazan, H.; Bron, D.; Toungouz, M.; Lagneaux, L. Mesenchymal stromal cells and immunomodulation: A gathering of regulatory immune cells. *Cytotherapy* **2016**, *18*, 160–171. [CrossRef] [PubMed]

4. Wang, S.; Zhu, R.; Li, H.; Li, J.; Han, Q.; Zhao, R.C. Mesenchymal stem cells and immune disorders: From basic science to clinical transition. *Front. Med.* **2019**, *13*, 138–151. [CrossRef]
5. Harrell, C.R.; Fellabaum, C.; Jovicic, N.; Djonov, V.; Arsenijevic, N.; Volarevic, V. Molecular Mechanisms Responsible for Therapeutic Potential of Mesenchymal Stem Cell-Derived Secretome. *Cells* **2019**, *8*, 467. [CrossRef] [PubMed]
6. Chow, D.C.; Wenning, L.A.; Miller, W.M.; Papoutsakis, E.T. Modeling pO(2) distributions in the bone marrow hematopoietic compartment. II. Modified Kroghian models. *Biophys. J.* **2001**, *81*, 685–696. [CrossRef]
7. Eliasson, P.; Jonsson, J.I. The hematopoietic stem cell niche: Low in oxygen but a nice place to be. *J. Cell Physiol.* **2010**, *222*, 17–22. [CrossRef]
8. Antoniou, E.S.; Sund, S.; Homsi, E.N.; Challenger, L.F.; Rameshwar, P. A theoretical simulation of hematopoietic stem cells during oxygen fluctuations: Prediction of bone marrow responses during hemorrhagic shock. *Shock* **2004**, *22*, 415–422. [CrossRef]
9. Kim, D.S.; Ko, Y.J.; Lee, M.W.; Park, H.J.; Park, Y.J.; Kim, D.I.; Sung, K.W.; Koo, H.H.; Yoo, K.H. Effect of low oxygen tension on the biological characteristics of human bone marrow mesenchymal stem cells. *Cell Stress Chaperones* **2016**, *21*, 1089–1099. [CrossRef]
10. Hao, D.; He, C.; Ma, B.; Lankford, L.; Reynaga, L.; Farmer, D.L.; Guo, F.; Wang, A. Hypoxic Preconditioning Enhances Survival and Proangiogenic Capacity of Human First Trimester Chorionic Villus-Derived Mesenchymal Stem Cells for Fetal Tissue Engineering. *Stem Cells Int.* **2019**, *2019*, 9695239. [CrossRef]
11. Lambertini, E.; Penolazzi, L.; Angelozzi, M.; Bergamin, L.S.; Manferdini, C.; Vieceli Dalla Sega, F.; Paolella, F.; Lisignoli, G.; Piva, R. Hypoxia Preconditioning of Human MSCs: A Direct Evidence of HIF-1alpha and Collagen Type XV Correlation. *Cell Physiol. Biochem.* **2018**, *51*, 2237–2249. [CrossRef] [PubMed]
12. Rosova, I.; Dao, M.; Capoccia, B.; Link, D.; Nolta, J.A. Hypoxic preconditioning results in increased motility and improved therapeutic potential of human mesenchymal stem cells. *Stem Cells* **2008**, *26*, 2173–2182. [CrossRef]
13. Bader, A.M.; Klose, K.; Bieback, K.; Korinth, D.; Schneider, M.; Seifert, M.; Choi, Y.H.; Kurtz, A.; Falk, V.; Stamm, C. Hypoxic Preconditioning Increases Survival and Pro-Angiogenic Capacity of Human Cord Blood Mesenchymal Stromal Cells In Vitro. *PLoS ONE* **2015**, *10*, e0138477. [CrossRef]
14. Beegle, J.; Lakatos, K.; Kalomoiris, S.; Stewart, H.; Isseroff, R.R.; Nolta, J.A.; Fierro, F.A. Hypoxic preconditioning of mesenchymal stromal cells induces metabolic changes, enhances survival, and promotes cell retention in vivo. *Stem Cells* **2015**, *33*, 1818–1828. [CrossRef] [PubMed]
15. Yeh, T.L.; Leissing, T.M.; Abboud, M.I.; Thinnes, C.C.; Atasoylu, O.; Holt-Martyn, J.P.; Zhang, D.; Tumber, A.; Lippl, K.; Lohans, C.T.; et al. Molecular and cellular mechanisms of HIF prolyl hydroxylase inhibitors in clinical trials. *Chem. Sci.* **2017**, *8*, 7651–7668. [CrossRef] [PubMed]
16. Esfahani, M.; Karimi, F.; Afshar, S.; Niknazar, S.; Sohrabi, S.; Najafi, R. Prolyl hydroxylase inhibitors act as agents to enhance the efficiency of cell therapy. *Expert Opin. Biol. Ther.* **2015**, *15*, 1739–1755. [CrossRef]
17. Burdzinska, A.; Dybowski, B.; Zarychta-Wisniewska, W.; Kulesza, A.; Zagozdzon, R.; Gajewski, Z.; Paczek, L. The Anatomy of Caprine Female Urethra and Characteristics of Muscle and Bone Marrow Derived Caprine Cells for Autologous Cell Therapy Testing. *Anat. Rec. (Hoboken)* **2017**, *300*, 577–588. [CrossRef]
18. Burdzinska, A.; Dybowski, B.; Zarychta-Wisniewska, W.; Kulesza, A.; Butrym, M.; Zagozdzon, R.; Graczyk-Jarzynka, A.; Radziszewski, P.; Gajewski, Z.; Paczek, L. Intraurethral co-transplantation of bone marrow mesenchymal stem cells and muscle-derived cells improves the urethral closure. *Stem Cell Res. Ther.* **2018**, *9*, 239. [CrossRef]
19. Zarychta-Wisniewska, W.; Burdzinska, A.; Kulesza, A.; Gala, K.; Kaleta, B.; Zielniok, K.; Siennicka, K.; Sabat, M.; Paczek, L. Bmp-12 activates tenogenic pathway in human adipose stem cells and affects their immunomodulatory and secretory properties. *BMC Cell Biol.* **2017**, *18*, 13. [CrossRef]
20. Pfaffl, M.W. A new mathematical model for relative quantification in real-time RT-PCR. *Nucleic Acids Res.* **2001**, *29*, e45. [CrossRef]
21. Dominici, M.; Le Blanc, K.; Mueller, I.; Slaper-Cortenbach, I.; Marini, F.; Krause, D.; Deans, R.; Keating, A.; Prockop, D.; Horwitz, E. Minimal criteria for defining multipotent mesenchymal stromal cells. The International Society for Cellular Therapy position statement. *Cytotherapy* **2006**, *8*, 315–317. [CrossRef] [PubMed]
22. Vadadadustat Manufacturer's Website. Available online: https://www.medchemexpress.com/Vadadustat.html (accessed on 20 August 2020).

23. Wang, Y.; Chen, X.; Cao, W.; Shi, Y. Plasticity of mesenchymal stem cells in immunomodulation: Pathological and therapeutic implications. *Nat. Immunol.* **2014**, *15*, 1009–1016. [CrossRef]
24. Cheng, R.-J.; Xiong, A.-J.; Li, Y.-H.; Pan, S.-Y.; Zhang, Q.-P.; Zhao, Y.; Liu, Y.; Marion, T.N. Mesenchymal Stem Cells: Allogeneic MSC May Be Immunosuppressive but Autologous MSC Are Dysfunctional in Lupus Patients. *Front. Cell Dev. Biol.* **2019**, *7*. [CrossRef] [PubMed]
25. Saparov, A.; Ogay, V.; Nurgozhin, T.; Jumabay, M.; Chen, W.C.W. Preconditioning of Human Mesenchymal Stem Cells to Enhance Their Regulation of the Immune Response. *Stem Cells Int.* **2016**, *2016*, 3924858. [CrossRef]
26. Kadle, R.L.; Abdou, S.A.; Villarreal-Ponce, A.P.; Soares, M.A.; Sultan, D.L.; David, J.A.; Massie, J.; Rifkin, W.J.; Rabbani, P.; Ceradini, D.J. Microenvironmental cues enhance mesenchymal stem cell-mediated immunomodulation and regulatory T-cell expansion. *PLoS ONE* **2018**, *13*, e0193178. [CrossRef] [PubMed]
27. Contreras-Lopez, R.; Elizondo-Vega, R.; Paredes, M.J.; Luque-Campos, N.; Torres, M.J.; Tejedor, G.; Vega-Letter, A.M.; Figueroa-Valdes, A.; Pradenas, C.; Oyarce, K.; et al. HIF1alpha-dependent metabolic reprogramming governs mesenchymal stem/stromal cell immunoregulatory functions. *FASEB J.* **2020**. [CrossRef]
28. Martinez, V.G.; Ontoria-Oviedo, I.; Ricardo, C.P.; Harding, S.E.; Sacedon, R.; Varas, A.; Zapata, A.; Sepulveda, P.; Vicente, A. Overexpression of hypoxia-inducible factor 1 alpha improves immunomodulation by dental mesenchymal stem cells. *Stem Cell Res. Ther.* **2017**, *8*, 208. [CrossRef]
29. Gharibi, B.; Farzadi, S.; Ghuman, M.; Hughes, F.J. Inhibition of Akt/mTOR attenuates age-related changes in mesenchymal stem cells. *Stem Cells* **2014**, *32*, 2256–2266. [CrossRef]
30. Liu, X.B.; Wang, J.A.; Ogle, M.E.; Wei, L. Prolyl hydroxylase inhibitor dimethyloxalylglycine enhances mesenchymal stem cell survival. *J. Cell Biochem.* **2009**, *106*, 903–911. [CrossRef]
31. Najafi, R.; Sharifi, A.M. Deferoxamine preconditioning potentiates mesenchymal stem cell homing in vitro and in streptozotocin-diabetic rats. *Expert Opin. Biol. Ther.* **2013**, *13*, 959–972. [CrossRef]
32. Gupta, N.; Nizet, V. Stabilization of Hypoxia-Inducible Factor-1 Alpha Augments the Therapeutic Capacity of Bone Marrow-Derived Mesenchymal Stem Cells in Experimental Pneumonia. *Front. Med. (Lausanne)* **2018**, *5*, 131. [CrossRef] [PubMed]
33. Wang, Q.; Zhu, H.; Zhou, W.G.; Guo, X.C.; Wu, M.J.; Xu, Z.Y.; Jiang, J.F.; Shen, C.; Liu, H.Q. N-acetylcysteine-pretreated human embryonic mesenchymal stem cell administration protects against bleomycin-induced lung injury. *Am. J. Med. Sci.* **2013**, *346*, 113–122. [CrossRef]
34. Thieblemont, N.; Wright, H.L.; Edwards, S.W.; Witko-Sarsat, V. Human neutrophils in auto-immunity. *Semin. Immunol.* **2016**, *28*, 159–173. [CrossRef] [PubMed]
35. O'Neil, L.J.; Kaplan, M.J. Neutrophils in Rheumatoid Arthritis: Breaking Immune Tolerance and Fueling Disease. *Trends Mol. Med.* **2019**, *25*, 215–227. [CrossRef]
36. Frangou, E.; Vassilopoulos, D.; Boletis, J.; Boumpas, D.T. An emerging role of neutrophils and NETosis in chronic inflammation and fibrosis in systemic lupus erythematosus (SLE) and ANCA-associated vasculitides (AAV): Implications for the pathogenesis and treatment. *Autoimmun. Rev.* **2019**, *18*, 751–760. [CrossRef] [PubMed]
37. Ma, W.T.; Gao, F.; Gu, K.; Chen, D.K. The Role of Monocytes and Macrophages in Autoimmune Diseases: A Comprehensive Review. *Front. Immunol.* **2019**, *10*, 1140. [CrossRef]
38. Eksteen, B.; Miles, A.; Curbishley, S.M.; Tselepis, C.; Grant, A.J.; Walker, L.S.; Adams, D.H. Epithelial inflammation is associated with CCL28 production and the recruitment of regulatory T cells expressing CCR10. *J. Immunol.* **2006**, *177*, 593–603. [CrossRef]
39. van den Berk, L.C.; Jansen, B.J.; Siebers-Vermeulen, K.G.; Roelofs, H.; Figdor, C.G.; Adema, G.J.; Torensma, R. Mesenchymal stem cells respond to TNF but do not produce TNF. *J. Leukoc. Biol.* **2010**, *87*, 283–289. [CrossRef]
40. Chen, J.; Caspi, R.R.; Chong, W.P. IL-20 receptor cytokines in autoimmune diseases. *J. Leukoc. Biol.* **2018**, *104*, 953–959. [CrossRef]
41. Mitamura, Y.; Nunomura, S.; Furue, M.; Izuhara, K. IL-24: A new player in the pathogenesis of pro-inflammatory and allergic skin diseases. *Allergol. Int.* **2020**. [CrossRef]
42. Kandula, S.; Abraham, C. LFA-1 on CD4+ T cells is required for optimal antigen-dependent activation in vivo. *J. Immunol.* **2004**, *173*, 4443–4451. [CrossRef] [PubMed]
43. Roep, B.O.; Heidenthal, E.; de Vries, R.R.; Kolb, H.; Martin, S. Soluble forms of intercellular adhesion molecule-1 in insulin-dependent diabetes mellitus. *Lancet* **1994**, *343*, 1590–1593. [CrossRef]

44. Chinnadurai, R.; Copland, I.B.; Patel, S.R.; Galipeau, J. IDO-independent suppression of T cell effector function by IFN-gamma-licensed human mesenchymal stromal cells. *J. Immunol.* **2014**, *192*, 1491–1501. [CrossRef] [PubMed]
45. Davies, L.C.; Heldring, N.; Kadri, N.; Le Blanc, K. Mesenchymal Stromal Cell Secretion of Programmed Death-1 Ligands Regulates T Cell Mediated Immunosuppression. *Stem Cells* **2017**, *35*, 766–776. [CrossRef]
46. Augello, A.; Tasso, R.; Negrini, S.M.; Amateis, A.; Indiveri, F.; Cancedda, R.; Pennesi, G. Bone marrow mesenchymal progenitor cells inhibit lymphocyte proliferation by activation of the programmed death 1 pathway. *Eur. J. Immunol.* **2005**, *35*, 1482–1490. [CrossRef]
47. Sheng, H.; Wang, Y.; Jin, Y.; Zhang, Q.; Zhang, Y.; Wang, L.; Shen, B.; Yin, S.; Liu, W.; Cui, L.; et al. A critical role of IFNgamma in priming MSC-mediated suppression of T cell proliferation through up-regulation of B7-H1. *Cell Res.* **2008**, *18*, 846–857. [CrossRef] [PubMed]
48. Luz-Crawford, P.; Noël, D.; Fernandez, X.; Khoury, M.; Figueroa, F.; Carrión, F.; Jorgensen, C.; Djouad, F. Mesenchymal stem cells repress Th17 molecular program through the PD-1 pathway. *PLoS ONE* **2012**, *7*, e45272. [CrossRef] [PubMed]

Publisher's Note: MDPI stays neutral with regard to jurisdictional claims in published maps and institutional affiliations.

 cells

Article

Pre-Conditioning with IFN-γ and Hypoxia Enhances the Angiogenic Potential of iPSC-Derived MSC Secretome

Suya Wang [1], Felix Umrath [1], Wanjing Cen [1], António José Salgado [2,3], Siegmar Reinert [1] and Dorothea Alexander [1,*]

1 Department of Oral and Maxillofacial Surgery, University Hospital Tübingen, 72076 Tübingen, Germany; suya.wang@student.uni-tuebingen.de (S.W.); felix.umrath@med.uni-tuebingen.de (F.U.); cenwanjingwj@gmail.com (W.C.); siegmar.reinert@med.uni-tuebingen.de (S.R.)
2 Life and Health Sciences Research Institute (ICVS), School of Medicine, University of Minho, Campus de Gualtar, 4710-057 Braga, Portugal; asalgado@med.uminho.pt
3 ICVS/3B's–PT Government Associate Laboratory, University of Minho, 4710-057 Braga, Portugal
* Correspondence: dorothea.alexander@med.uni-tuebingen.de

Abstract: Induced pluripotent stem cell (iPSC) derived mesenchymal stem cells (iMSCs) represent a promising source of progenitor cells for approaches in the field of bone regeneration. Bone formation is a multi-step process in which osteogenesis and angiogenesis are both involved. Many reports show that the secretome of mesenchymal stromal stem cells (MSCs) influences the microenvironment upon injury, promoting cytoprotection, angiogenesis, and tissue repair of the damaged area. However, the effects of iPSC-derived MSCs secretome on angiogenesis have seldom been investigated. In the present study, the angiogenic properties of IFN-γ pre-conditioned iMSC secretomes were analyzed. We detected a higher expression of the pro-angiogenic genes and proteins of iMSCs and their secretome under IFN-γ and hypoxic stimulation (IFN-H). Tube formation and wound healing assays revealed a higher angiogenic potential of HUVECs in the presence of IFN-γ conditioned iMSC secretome. Sprouting assays demonstrated that within Coll/HA scaffolds, HUVECs spheroids formed significantly more and longer sprouts in the presence of IFN-γ conditioned iMSC secretome. Through gene expression analyses, pro-angiogenic genes (FLT-1, KDR, MET, TIMP-1, HIF-1α, IL-8, and VCAM-1) in HUVECs showed a significant up-regulation and down-regulation of two anti-angiogenic genes (TIMP-4 and IGFBP-1) compared to the data obtained in the other groups. Our results demonstrate that the iMSC secretome, pre-conditioned under inflammatory and hypoxic conditions, induced the highest angiogenic properties of HUVECs. We conclude that pre-activated iMSCs enhance their efficacy and represent a suitable cell source for collagen/hydroxyapatite with angiogenic properties.

Keywords: iPSC-derived MSCs; iMSC secretome; pre-conditioning; angiogenesis; IFN-γ; hypoxia; potentiation of iMSC efficacy

Citation: Wang, S.; Umrath, F.; Cen, W.; Salgado, A.J.; Reinert, S.; Alexander, D. Pre-Conditioning with IFN-γ and Hypoxia Enhances the Angiogenic Potential of iPSC-Derived MSC Secretome. *Cells* **2022**, *11*, 988. https://doi.org10.3390/cells11060988

Academic Editors: Joni H. Ylostalo and Nisha C. Durand

Received: 11 February 2022
Accepted: 10 March 2022
Published: 14 March 2022

1. Introduction

The reconstruction of large tissue defects is one of the main challenges in the field of oral and maxillofacial surgery. Despite some significant limitations, including donor site morbidity, restricted availability, and poor bone quality [1], autologous grafting has remained the gold standard for bone over the years [2]. In order to overcome these drawbacks, tissue engineering with stem cells, signal molecules, and scaffolds has attracted attention in the area of regenerative medicine [3,4]. This therapeutic procedure benefits from the regenerative capacity of the human body through the application of adult stem cells in combination with optimized synthetic materials [5]. As oxygen and nutrient supply is essential for survival within the graft, the use of suitable cells to promote angiogenesis and to recruit endothelial cells has gained in importance.

Mesenchymal stem cells (MSCs) have been shown to be a promising cell candidate for cell-based therapy [6]. MSCs have attracted the interest of clinician-scientists not only because of the low immunogenicity and tissue regenerative properties that could enable their use in allogeneic settings [7], but also for their anti-tumorigenic, anti-fibrotic, anti-apoptotic, anti-inflammatory, pro-angiogenic, neuro-protective, anti-bacterial, and chemo-attractive effects [8–10]. Recently, it has become evident that the functional benefits exerted by MSCs upon transplantation are due to the release of paracrine factors and biologically relevant molecules to the neighboring diseased or injured tissue, referred to as the MSCs secretome [11–13]. The secretome of MSCs influences the microenvironment upon injury, promoting cytoprotection, angiogenesis, and tissue repair at the damaged area [14].

MSCs differentiated from induced pluripotent stem cells (iPSCs), called induced mesenchymal stem cell-like cells (iMSCs), represent an alternative to primary MSCs isolated from different tissues [1]. Due to the limited availability, in vitro proliferation capacity, and differentiation potential of primary MSCs, impeding their application in the clinical routine [15], iMSCs have been identified as a promising source of transplantable donor cells with similar capacities.

A variety of studies have demonstrated that MSCs and iMSCs have comparable properties either in their morphology, marker expression, differentiation potential, or immunomodulatory properties [16,17]. The advantages of the use of iMSCs are that they have been characterized as rejuvenated MSCs [18–20] and that they show no risk of tumor formation as they do not express oncogenic pluripotency-associated genes such as OCT4 [21]. In addition, iMSCs outperformed native MSCs in the treatment of multiple sclerosis in a rodent model [22]. Another animal study demonstrated that iMSCs could successfully improve in vivo bone regeneration by their direct differentiation into bone cells and by the recruitment of host cells in a radial defect model in mice [23].

In previous studies, our lab succeeded in the generation of iPSCs from jaw periosteal cells (JPCs) and the resulting differentiation of iPSCs to iMSCs [24,25]. We provided evidence for tri-lineage differentiation of generated iMSCs compared to that of the parental JPCs by histologic staining and gene expression patterns. Furthermore, we gave information about their morphology and telomere lengths, about their proliferative and mitochondrial activities, and about cellular senescence compared to that detected in parental JPCs.

Angiogenesis and bone formation are coupled processes during skeletal development and fracture healing [26]. New blood vessels bring oxygen and nutrients to the metabolically highly active regenerating callus, and serve as a route for inflammatory cells, cartilage, and bone precursor cells to reach the injury site [27]. It has been suggested that the MSC secretome induces angiogenesis [28].

To the best of our knowledge, the present study is the first study examining the influence of iMSCs secretomes on angiogenesis. The aim of our study was to compare the angiogenic potential of the secretome from differently pre-conditioned iMSCs in order to find a suitable cell source with pro-angiogenic capacities for the generation of impactful bone tissue engineering constructs.

2. Materials and Methods

2.1. Cell Culture

iMSCs derived from three donors were grown in hPL10-medium (DMEM/F12 (Gibco) + 10% human platelet lysate (hPL; PL BioScience GmbH, Aachen, Germany), 100 U/mL penicillin-streptomycin (Pen-Strep; Lonza, Basel, Switzerland), and 2.5 µg/mL amphotericin B (Biochrom, Berlin, Germany). Passages 5-6 iMSCs were used for this study [25].

2.2. iMSC Stimulation and Secretome Collection

First, 2.3×10^6 iMSCs were seeded into T-175 cell culture flasks (Greiner Bio-One GmbH, Frickenhausen, Germany). For pro-inflammatory conditions (Figure 1), iMSCs were stimulated with IFN-γ (200 ng/mL, Sigma-Aldrich, St. Louis, MO, USA) for 5 days under normoxia (21% O_2, 5% CO_2, IFN-N group) and hypoxia (5% O_2, 5% CO_2, IFN-H group).

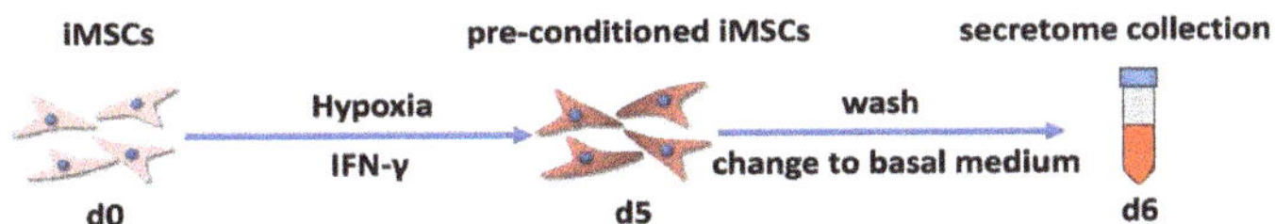

Figure 1. Timeline of IFN-γ stimulation and secretome collection: iMSCs were seeded into T-175 cell culture flasks (d0) and were stimulated with IFN-γ under hypoxic condition for 5 days (d5). Cells were washed and medium change to basal medium followed. After 24 h culture with basal medium, iMSCs secretomes were collected at day 6 (d6).

As control groups, iMSCs were cultured only in both normoxic and hypoxic conditions without IFN-γ stimulation. After 5 days, the medium was removed, the cells were washed three times with 10 mL PBS and 10 mL basal medium (DMEM/F12, 100 U/mL penicillin-streptomycin, and 2.5 μg/mL amphotericin B), then subsequently starved in 37 mL basal medium without hPL supplementation. After 24 h, 37 mL of iMSCs secretome was collected in 50 mL tubes (Greiner Bio-One GmbH, Frickenhausen, Germany). After centrifuging these tubes at 600 g for 7 min to discard the cell debris, 34 mL of supernatant was recollected in fresh 50 mL tubes. After rapid freezing in liquid nitrogen, the secretome was stored at −80 °C in a freezer until the process of secretome enrichment.

2.3. Flow Cytometric Analyses of HLA-I and HLA-II Expression of iMSCs

As the parental JPCs were shown to be very sensitive to IFN-γ stimulation, we determined the optimal duration of IFN-γ treatment in pre-experiments. JPCs as well as other MSCs respond to IFN-γ through the up-regulation of their HLA-II expression. We treated iMSCs with IFN-γ for 3, 5, and 7 days, and detected the highest HLA-II response at day 5. Therefore, IFN-γ stimulation for 5 days was chosen for all of the experiments performed for this study. Stimulated iMSCs and cells from the control groups were collected for flow cytometric analyses. Then, 2×10^5 cells per sample were incubated in 20 μL blocking buffer (PBS, 0.1% BSA, 1 mg/mL sodium azide (Sigma-Aldrich) and 10% Gamunex (human immune globulin solution, Talecris Biotherapeutics, Frankfurt, Germany) for 15 min on ice. Then, the cells were incubated on ice with a FACS buffer (DPBS, 0.1% BSA, and 0.1% sodium azide) as well as mouse APC-conjugated anti-HLA-I and anti-HLA-II antibodies (MACS Miltenyi Biotec, Bergisch Gladbach, Germany) for 20 min in the dark. For the isotype controls, APC-labeled IgG2a antibodies (Biolegend, San Diego, CA, USA) were used. For the analysis of the iMSC and iPSC surface marker expression, the APC- and PE-labeled antibodies listed in Table 1 were used. Representative histograms are shown in Supplementary Figure S1. After two washing steps with a FACS buffer, the cell pellets were resuspended in 200 μL FACS buffer and were analyzed by flow cytometry using the Guava easyCyte 6HT-2L (Merck Millipore, Billerica, MA, USA). For data evaluation, the guavaSoft 2.2.3 (InCyte 2.2.2, Luminex Corporation, Chicago, IL, USA) software was used.

Table 1. List of the antibodies used for the detection of MSC and iPSC marker expression on iMSCs; * Biolegend, ** Miltenyi. Rec-recombinant.

	PE-Labeled Antibodies	Volume [μL]	APC-Labeled Antibodies	Volume [μL]
1	Rec. Tra-1–0	1 **	CD90 (IgG1)	5 *
2	Rec. Tra-1–81	1 **	CD73 (IgG1)	5 *
3	Rec. SSEA4	5 **	CD105 (IgG1)	5 *
4	CD44 (IgG1)	5 *	CD326 (IgG2a *)	5 *
5	Rec.-PE (control)	5 *	IgG2a-APC (control)	5 *
6	IgG1-PE (control)	5 **	IgG1-APC (control)	5 *

2.4. Secretome Enrichment

According to the instruction of Vivaspin® 20 centrifugal concentrators (5 kDa cut off, Sartorius, Goettingen, Germany), iMSC secretome was concentrated to 100-fold. First, 14 mL of iMSC supernatant was pipetted into the Vivaspin tubes and centrifuged at 6000× *g* (6831 rpm). After ca. 2 h of centrifugation, the volume from the upper compartment of the tube was checked and further volume was filled for secretome enrichment until the total volume of the added supernatant reached 34 mL (14 mL + 10 mL + 10 mL). After a total centrifugation time of 9 h, less than 340 μL of the concentrated iMSC secretome were collected and filled to exactly 340 μL with basal medium (DMEM/F12, 100 U/mL penicillin-streptomycin, and 2.5 μg/mL amphotericin B). After the enrichment of the secretomes from three different donors, those obtained under the same condition were mixed completely in order to minimize the interindividual variations and to obtain more reliability of the data. The optimal iMSC secretome concentration was determined by dose response kinetics. Therefore, tube formation assays were performed in the presence of 1×, 5×, and 10× concentrated secretomes. As illustrated in Supplementary Figure S2, 10× secretomes showed the highest angiogenic effect. Based on this result from the pre-experiments, all other experiments were performed with mixed 10-fold concentrated iMSC secretomes.

2.5. Detection of IL-8 and VEGF-A Protein Concentrations in iMSC Secretomes by Enzyme-Linked Immunosorbent Assay (ELISA)

The secretomes from different conditions were collected and IL-8 and VEGF-A protein secretions were quantified using the human IL-8 ELISA kit II (Invitrogen, Thermo Fisher, Darmstadt, Germany) and the human VEGF ELISA kit (R&D Systems, Minneapolis, USA) according to the manufacturer's instructions. All measurements were performed in duplicate. ELISA plates were read immediately with a microplate ELISA reader (BioTek, Friedrichshall, Germany) at a wavelength of 450 nm. IL-8 and VEGF concentrations were quantified using a standard curve of known concentrations. The lowest detection limit was in the range of 9–15.63 pg/mL.

2.6. Cell Culture of Human Umbilical Vein Endothelial Cells

Human umbilical vein endothelial cells (HUVECs) were purchased from PromoCell (Heidelberg, Germany) and cultured in endothelial cell growth medium 2 (EGM-2 kit, PromoCell, Heidelberg, Germany) with 1% amphotericin B and penicillin/streptomycin (Biochrom, Berlin, Germany) at 37 °C and 5% CO_2. Cells of passages 5–7 were used for all experiments and medium change was performed three times per week.

2.7. Endothelial Tube Formation

Tube formation assays were performed with HUVECs using a method adapted from Wang and co-authors [29]. Briefly, 100 μL/well of GeltrexTM LDEV-free reduced growth factor basement membrane matrix (Invitrogen/Thermo Fisher Scientific, Waltham, MA, USA) was incubated in a 24 well plate for 45 min at 37 °C for matrix gel polymerization. Then, 5×10^4 HUVECs were seeded onto GeltrexTM matrices and cultured for 8 h with EGM-2 medium containing all the supplements (+GF, positive group) and EGM-2 medium containing 10× secretome of different conditions (IFN-H, IFN-N, Co-H, and Co-N). EGM-2 medium without growth factors served as the negative control (-GF). After 8 h of cultivation, 1 μM calcein-AM dye (Invitrogen/Thermo Fisher Scientific, Waltham, MA, USA) was added to the plate and incubated for 20 min. Fluorescence images were captured from at least three wells per culture condition at a 1.25× magnification using the Axio Observer Z1 fluorescence microscope (Zeiss, Oberkochen, Germany). Network branches, meshes, and nodes were counted from the collected images using the Image J software, in order to quantify the angiogenic network formation.

2.8. Detection of Wound Closure by Using the Endothelial Scratch Assay

An in vitro scratch assay was performed as described previously [30]. HUVECs were seeded in 24-well plates at a density of 1×10^5. After incubation for 24 h, each colonialized well was manually scratched with a 200 µL pipette tip, washed with PBS three times, and incubated at 37 °C under different conditions (positive, IFN-H, IFN-N, Co-H, Co-N, and negative groups). Three randomly selected views along the scraped line were photographed from each well immediately after manual scratching and 8 h later with a 5× objective in a brightfield microscope. The area of the wound gap was calculated using the ImageJ software:

$$\% \text{ wound closure} = \frac{A(0) - A(t)}{A(0)} \times 100\%$$

where A(t) is the wound area at 8 h and A(0) is its initial area. Cell migration was quantified and expressed as the average percentage of closure of the scratch area [31].

2.9. Spheroid Sprouting Assay in Collagen/Hydroxyapatite Scaffolds

A spheroid sprouting assay was adapted from the method previously described by Maracle and co-authors [32]. The principle of this assay is based on the sprout formation originating from aggregated and gel-embedded HUVECs. In brief, methocel solution was prepared by dissolving 6 g methylcellulose (Sigma-Aldrich, USA) in 500 mL EGM-2 medium (PromoCell, Heidelberg, Germany). The HUVECs were then harvested. A total of 8×10^3 HUVECs were added to each well of a 96-well polypropylene plate (Corning, Sigma-Aldrich, USA) in 200 µL EGM-2 medium containing 20% methocel. Spheroids formed overnight at 37 °C. Then, 30 spheroids were resuspended in 300 µL of the solution for the preparation of collagen/hydroxyapatite composites. After incubation at 37 °C for 30 min for polymerization of the collagen/hydroxyapatite composites, 500 µL of the different pre-conditioned iMSCs secretomes containing 10% FBS (IFN-H, IFN-N, CO-H, and CO-N groups) were added to the wells. Sprout formation by the HUVEC spheroids was detected after 48 h of incubation.

2.10. Fluorescence Staining

Different groups of collagen/hydroxyapatite scaffolds containing HUVEC spheroids were washed three times with PBS and were fixed with 4% paraformaldehyde for 1 h. After cell permeabilization with PBS + 1% Triton-X100 (Sigma), the cells were washed with PBS and stained with Alexa488-Phalloidin (10 µg/mL in bovine serum albumin, Sigma-Aldrich, USA) and Hoechst 33342 (1 µg/mL, Promocell, Heidelberg, Germany) at room temperature for 1 h. After three wash steps with PBS, images were taken using an Axio Observer Z1 fluorescence microscope (Zeiss, Oberkochen, Germany) at 10× magnifications. Spheroids were quantified using the ImageJ software in order to measure the length of the sprouts and to calculate the cumulative sprout length (CSL). Data from at least 10 spheroids per experimental group were calculated.

2.11. RNA Isolation and Quantitative Gene Expression Analyses in iMSCs and HUVECs

The total mRNA was isolated from HUVECs using the NucleoSpin RNA kit (Macherey-Nagel, Hoerd, France) according to the manufacturer's guidelines. After isolation, RNA was quantified using a Nanodrop micro-volume spectrophotometer (Invitrogen/Thermo Fisher Scientific, Waltham, MA, USA). Then, cDNA synthesis was performed using the LunaScript® RT SuperMix kit according to the instructions of the manufacturer (New England Biolabs, Ipswich, MA, USA). To quantify the mRNA expression levels, the Applied Biosystems® QuantStudio® 5 Real-Time PCR System (Thermo Fisher Scientific, Waltham, MA, USA) was used. For the PCR reactions, DEPC-treated water, Luna® Universal Probe qPCR Master Mix (New England Biolabs, Ipswich, MA, USA), and primer kits (FLT-1, KDR, HGF, MET, IL-8, HIF-1α, MMP-1, TIMP-1, TIMP-4, IGFBP-1, IGFBP-2, and VCAM-1) from Thermo Fisher Scientific (Waltham, MA, USA) were used for 40 amplification cycles of the target cDNA following the manufacturer's instructions. The target gene transcript levels were normalized to those of the housekeeping gene GAPDH. X-fold induction values were calculated by the quotient of the sample and the corresponding control. All cDNA samples were analyzed in triplicate.

2.12. Statistical Analysis

The data of all measurements are expressed as means ± standard error of means (SEM). All statistical analyses were carried out using the GraphPad Prism software (La Jolla, CA, USA). The two-tailed Student's t-test or one-way analysis of variance (ANOVA) for repeated measurements followed by Tukey's multiple comparisons tests were used. Values were considered significant with a p-value < 0.05.

3. Results

3.1. HLA-I and HLA-II Expression by iMSCs Cultured under Pro-Inflammatory (IFN-γ) and Hypoxic/Normoxic Condition

In a previous study, we provided evidence of tri-lineage differentiation, telomere lengths, and proliferative and mitochondrial capacities of generated iMSCs, compared to the functions of the parental JPCs [25]. The MSC and iPSC marker expressions of the iMSCs used in this work are given in the Supplementary Figure S1.

To characterize the response of iMSCs to IFN-γ stimulation, the expression of HLA-I and HLA-II surface markers was analyzed via flow cytometry. iMSCs derived from three donors expressed high levels of HLA-I, under both hypoxic and normoxic conditions (Figure 2A). After iMSC stimulation with IFN-γ for 5 days (IFN-γ), the expression of HLA-II (Hypoxia: IFN-γ: 62.52 ± 25.21%; Normoxia: IFN-γ: 65.37 ± 19.79%) was significantly upregulated compared to that of the untreated iMSCs (CO) (Normoxia: CO: 0.12 ± 0.01%; Hypoxia: CO: 0.27 ± 0.09%) under both hypoxic and normoxic conditions (Figure 2B).

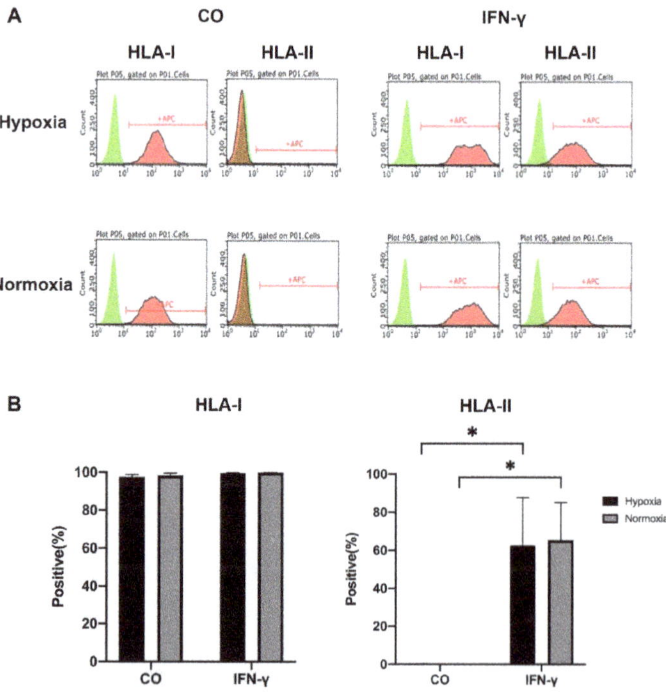

Figure 2. HLA-I and HLA-II expression of iMSCs of untreated (CO) and stimulated (with 200 ng/mL IFN-γ (IFN-γ groups) for 5 days under hypoxic and normoxic conditions. (**A**) Representative

histograms of the surface marker expression were detected by flow cytometry. Positive cells (red), unstained control (green). (**B**) Amount of HLA-I and -II positive cells under the indicated conditions. Differences in surface marker expression were compared using two-way ANOVA (n = 3 patients, * $p < 0.05$).

3.2. Gene Expression of the Pro-Angiogenic Genes IL-8 and VEGF-A by iMSCs Cultured under Pro-Inflammatory (IFN-γ) and Hypoxic/Normoxic Conditions

To investigate the angiogenic potential of the treated and untreated iMSCs, the angiogenesis-related genes in iMSCs were analyzed. It is well known that interleukin-(IL-)-8 and VEGF-A are critical regulators of angiogenesis [33,34]. In our present study, it was evident that the IL-8 gene expression in iMSCs from the IFN-H group (IFN-H: 102.77 ± 1.89) was significantly higher than in the other three groups (IFN-N group: 7.26 ± 0.24, CO-H group: 1.36 ± 0.13 and CO-N group: 1 ± 0.25), and the expression of the IL-8 gene in iMSCs from the IFN-N group was also significantly higher than the IL-8 levels from the CO-H and CO-N groups (Figure 3A). Concerning the VEGF-A gene expression, the levels in the IFN-H and IFN-N groups (IFN-H: 3.93 ± 0.12; IFN-N: 4.31 ± 0.11) were found to be significantly higher compared to those detected in the CO-H and CO-N groups (CO-H: 0.75 ± 0.25; CO-N: 1 ± 0.22). No significant differences were detected between the stimulated and unstimulated groups (Figure 3B). In summary, these results indicate that with IFN-γ pre-conditioned iMSCs possess a higher angiogenic potential compared to the control iMSCs.

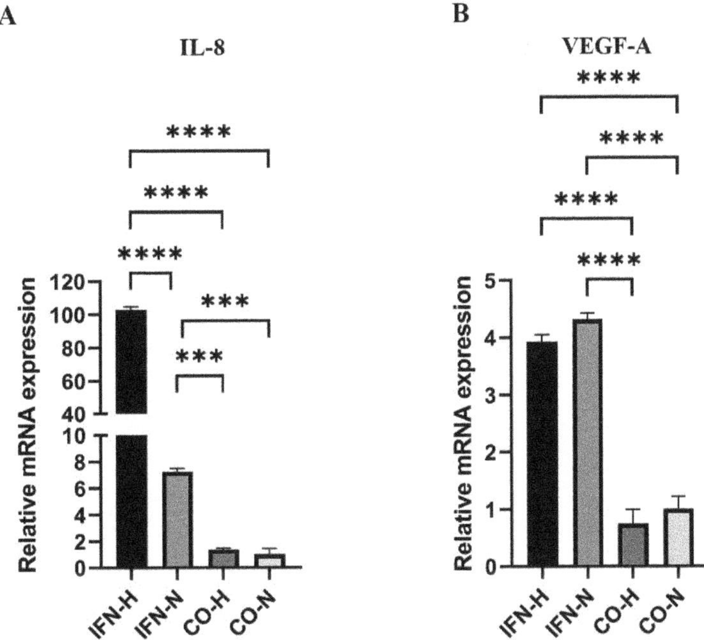

Figure 3. Expression of angiogenesis-related genes by iMSCs pre-conditioned with IFN-γ under hypoxic and normoxic conditions. Gene expression levels of IL-8 (**A**) and VEGF-A (**B**) in iMSCs were quantified using a Thermo Fisher Scientific PCR instrument, and ratios of the target mRNA copy numbers related to copy numbers of the housekeeping gene (GAPDH) were calculated. Gene expressions mean values ± SEM in iMSCs from the IFN-H, IFN-N, CO-H, and CO-N groups displayed as x-fold induction values relative to the CO-N (control normoxic) group. Data were collected from three independent experiments (*** $p < 0.001$; **** $p < 0.0001$).

3.3. Examination of IL-8 and VEGF-A Protein Concentrations in Pre-Conditioned iMSCs Secretomes

To quantify the IL-8 and VEGF protein release in secretomes from different pre-conditioned iMSCs groups, we used human IL-8 and human VEGF ELISA kits. As shown in Figure 4, VEGF secretion was significantly increased in both IFN-stimulated groups (IFN-H: 9.75 ± 0.22; IFN-N: 8.68 ± 0.50) compared to levels detected in the unstimulated groups (CO-H: 3.51 ± 0.14; CO-N: 2.75 ± 0.11). After quantification of IL-8 protein secretion, a significant upregulation was detected in the IFN-H group (IFN-H: 52.48 ± 0.30) compared to the normoxic and untreated groups (IFN-N: 0.002 ± 0.001; CO-H: 0.01 ± 0.003; CO-N: 0.07 ± 0.004).

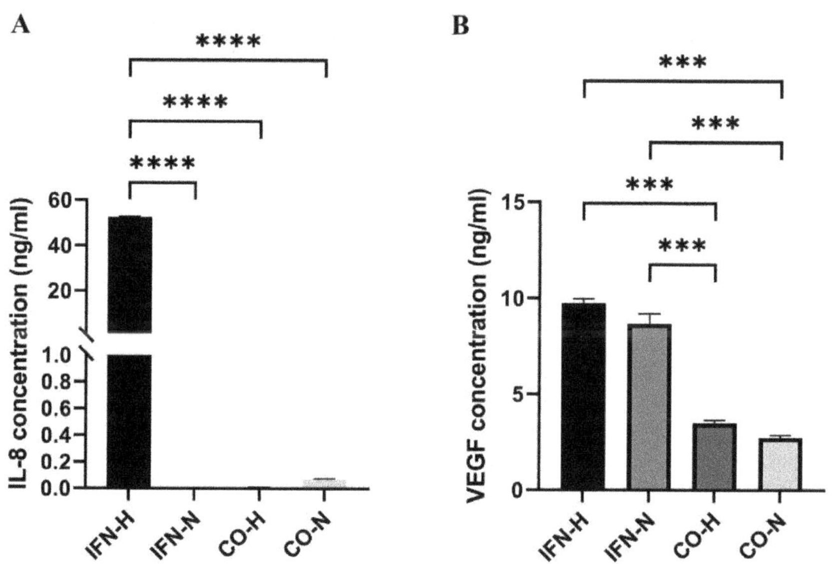

Figure 4. Quantification of IL-8 (**A**) and VEGF (**B**) protein concentration in secretomes from iMSCs pre-conditioned with IFN-γ under hypoxic and normoxic conditions. Values represent means ± SEM values from three independent experiments (*** $p < 0.001$; **** $p < 0.0001$).

3.4. Tube Formation Assays with HUVECs Cultured in the Presence of Secretomes from Pre-Conditioned iMSCs

In order to evaluate the angiogenic potential of the secretomes obtained from differently pre-conditioned iMSCs, endothelial tube formation assays were performed. After imaging, different amounts of tube-like structures were formed. For the quantification using the Image J software, we used three indicators to determine the angiogenic effects of the different iMSC secretomes: the number of nodes, number of meshes, and the number of branches. As illustrated in Figure 5, HUVECs from the positive group (+GF) formed significantly more nodes, meshes, and branches (Figure 5B) than the cells from all other groups, except the number of branches and nodes compared to HUVECs from the IFN-H group. HUVECs from the IFN-H group formed significantly higher numbers of nodes, branches, and meshes compared to the cells from the negative control group (−GF) and in the tendency compared to all other groups, without reaching statistical significance.

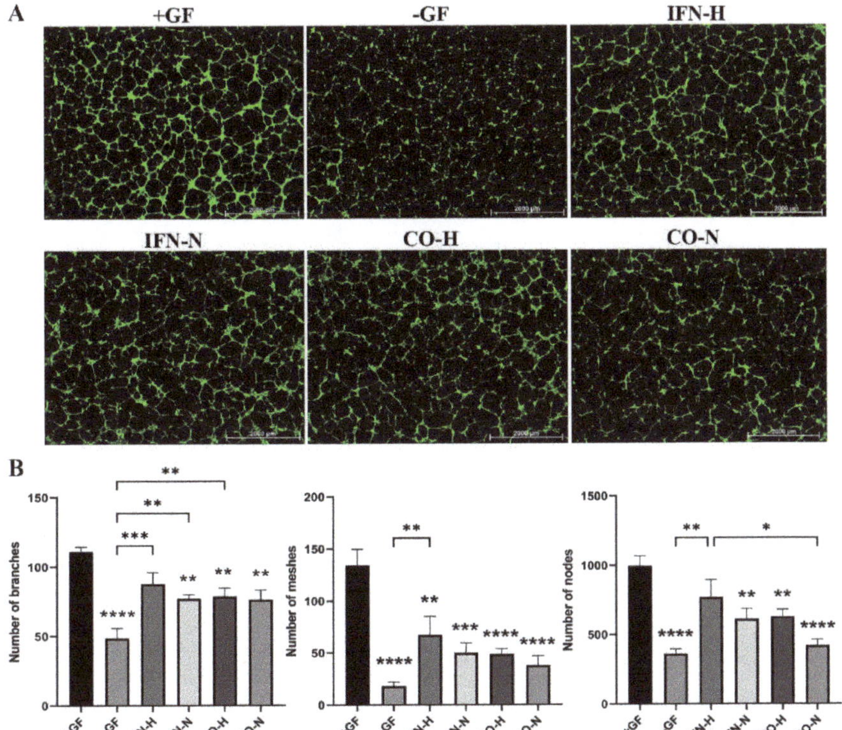

Figure 5. Tube formation of HUVECs incubated with secretomes obtained from differently pre-conditioned iMSCs. HUVECs seeded onto the Geltrex matrix in a medium containing secretomes from differently pre-conditioned iMSCs, growth factors (+GF, positive control group), or without growth factors (−GF, negative control group). (**A**) Representative images (1.25× magnification, scale bar = 2000 µm) of tube formation were taken using fluorescent microscopy (calcein-AM green staining), 8 h after cell seeding. (**B**) Number of branches, number of meshes, and number of nodes were quantified with the ImageJ software. Values represent means ±SEM from three independent experiments (*, **, ***, **** without bar indicates significant differences to the +GF group; * $p < 0.05$, ** $p < 0.01$, *** $p < 0.001$, **** $p < 0.0001$).

3.5. Wound Healing Assays with HUVECs Cultured in the Presence of Pre-Conditioned iMSCs Secretomes

The migration capability of the endothelial cells represents the first step in the angiogenesis process [35]. In order to evaluate the wound closure by the migration capacity of HUVECs cultured in the presence of pre-conditioned iMSCs secretomes, wound healing assays were performed. After imaging, the wound closure area in the different groups was calculated by ImageJ. As shown in Figure 6, the wound closure percentage of the positive cell group (+GF: 44.83 ± 1.77%) was significantly higher (Figure 6B) than in the other groups (−GF: 19.68 ± 3.97%; IFN-H: 38.82 ± 3.50%; IFN-N: 28.54 ± 1.02%; CO-H: 32.31 ± 3.86%; CO-N: 29.27 ± 3.55%), except when compared to cells from the IFN-H group. HUVECs from the IFN-H group showed the highest percentages of wound closure, which reached statistical significance compared to the negative control group, which was cultured in the absence of all growth factors (−GF).

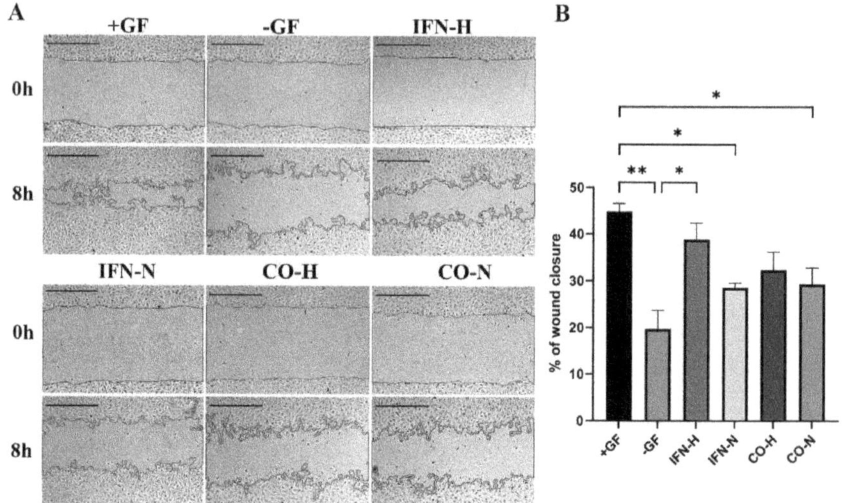

Figure 6. Wound healing assay of HUVECs incubated with differently pre-conditioned iMSCs secretomes. HUVECs cultured with differently pre-conditioned iMSCs secretomes, growth factors (+GF, positive group), or without growth factors (−GF, negative control group). (**A**) Representative images (5× magnification, scale bar = 500 μm) of the scratched area were taken using brightfield microscopy at 0 and 8 h after cell seeding. (**B**) Quantification of the wound closure by HUVECs cultured with pre-conditioned iMSCs secretomes. Values represent means ± SEM from three independent experiments (* $p < 0.05$, ** $p < 0.01$).

3.6. Gene Expression Analysis of Angiogenic Markers by HUVECs Cultured in the Presence of Pre-Conditioned iMSC Secretomes

Further investigation of the biofunctionality of the differently pre-conditioned iMSCs secretomes includes the analysis of their effects on the angiogenic gene expression in HUVECs (Figure 6). HUVECs cultured in the presence of iMSC secretome from the IFN-γ (for 5 days) stimulation under hypoxic condition group (IFN-H) showed an up-regulation of seven pro-angiogenic genes (FLT-1, KDR, MET, TIMP-1, HIF-1α, IL-8, and VCAM-1) and a down-regulation of two anti-angiogenic genes (TIMP-4 and IGFBP-1) when compared to all other groups. The MMP-1 gene expression in the IFN-H HUVEC group was higher than that detected in the IFN-N, CO-H, CO-N, and −GF groups and lower than that of the +GF group, however, there was no significant difference among these groups (Figure 7E). The gene expression of IGFBP-2 in the IFN-H, IFN-N, CO-H, and CO-N groups was higher than in the −GF group and lower than that of the +GF HUVEC group (Figure 7J). Concerning the expression of the HGF gene, higher mRNA levels were detected in the IFN-H group compared to those detected in the +GF and -GF groups, but they were lower compared to those of the IFN-N, CO-H, and CO-N groups. The statistical significances are illustrated in Figure 7. The expression pattern of the 12 analyzed genes gives an overview of the iMSC secretome with the highest angiogenic potential under hypoxic and pro-inflammatory (IFN-γ) pre-treatment.

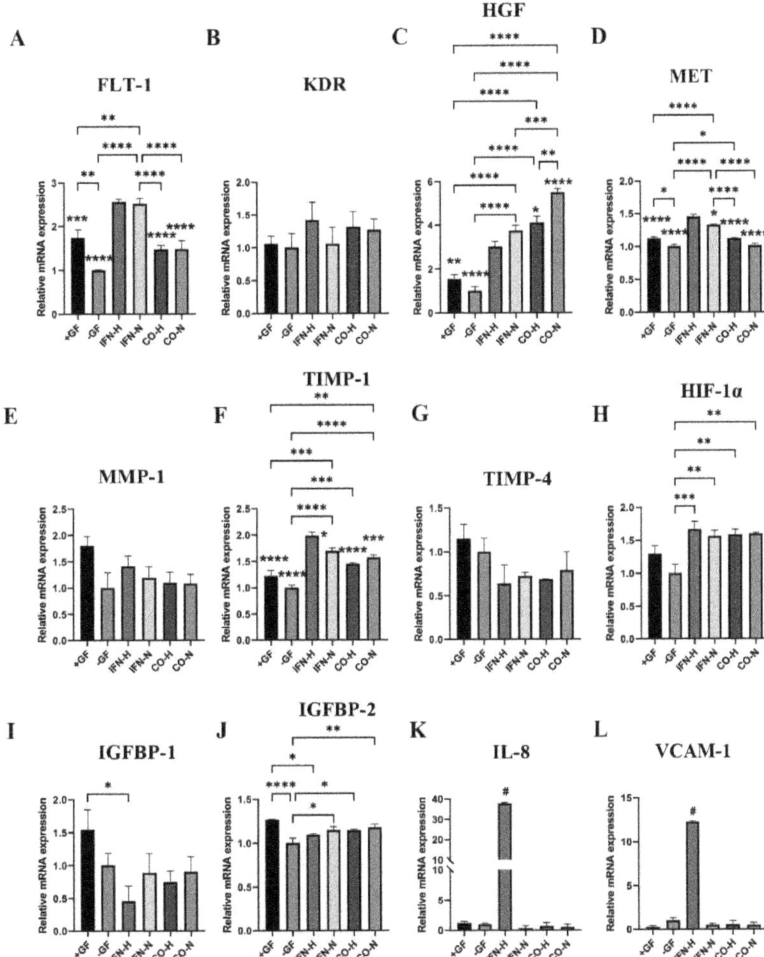

Figure 7. Expression of angiogenesis-related genes by HUVECs cultivated in the presence of pre-conditioned iMSCs secretomes. Gene expression levels of FLT-1 (**A**), KDR (**B**), HGF (**C**), MET (**D**), MMP-1 (**E**), TIMP-1 (**F**), TIMP-4 (**G**), HIF-1α (**H**), IGFBP-1 (**I**), IGFBP-2 (**J**), IL-8 (**K**), and VCAM-1 (**L**) in HUVECs cultivated in the presence of pre-conditioned iMSCs secretomes. Quantification of mRNA levels was performed using a Thermo Fisher Scientific PCR instrument, and ratios of the target mRNA copy numbers to copy numbers of the housekeeping gene (GAPDH) were calculated in all samples. Mean values ± SEM in HUVECs from different iMSCs secretome groups are displayed as x-fold induction values relative to the negative control group (−GF). Data were collected from three independent experiments (*, **, ***, **** without bar indicates significant differences compared to the IFN-H group, # indicates that the IFN-H group show significant differences to all other groups, * $p < 0.05$; ** $p < 0.01$; *** $p < 0.001$; **** $p < 0.0001$).

3.7. Sprouting Assays by HUVEC Spheroids Seeded on 3D Collagen/Hydroxyapatite Composites in the Presence of Pre-conditioned iMSC Secretomes

To evaluate the functionality of differently conditioned iMSCs secretomes also in the 3D-culture, sprouting assays with HUVEC spheroids were performed within collagen/hydroxyapatite (coll/HA) composite scaffolds. We described the generation of these scaffolds in a previous work [36]. HUVEC spheroids were incorporated into coll/HA scaffolds and incubated in the presence of differently pre-conditioned iMSCs secretomes (IFN-H, IFN-N, CO-H, and CO-N groups). Figure 8 gives an overview of the merged images; the respective cytoplasmic and nuclear staining is illustrated in the Supplementary Figure S3. Sprout lengths were analyzed by ImageJ software after 48 h of incubation (Figure 8A). By quantifying the cumulative sprout length (CSL), we detected a significantly higher CSL in the IFN-H scaffold group (IFN-H: 2263.25 ± 228.68), compared to the CSL detected for all other groups (+GF: 1347.29 ± 122.98; −GF: 342.35 ± 49.52; IFN-N: 1303.20 ± 122.75; CO-H: 584.57 ± 76.12; CO-N: 492.75 ± 57.85). The CSL calculated in the scaffolds from the IFN-N and the +GF groups were significantly higher than those detected in the scaffolds from the CO-H, CO-N, and −GF groups. (Figure 8B). The CSL calculated in scaffolds from the CO-H and CO-N groups was higher than that calculated for the −GF group; however, without statistical significance.

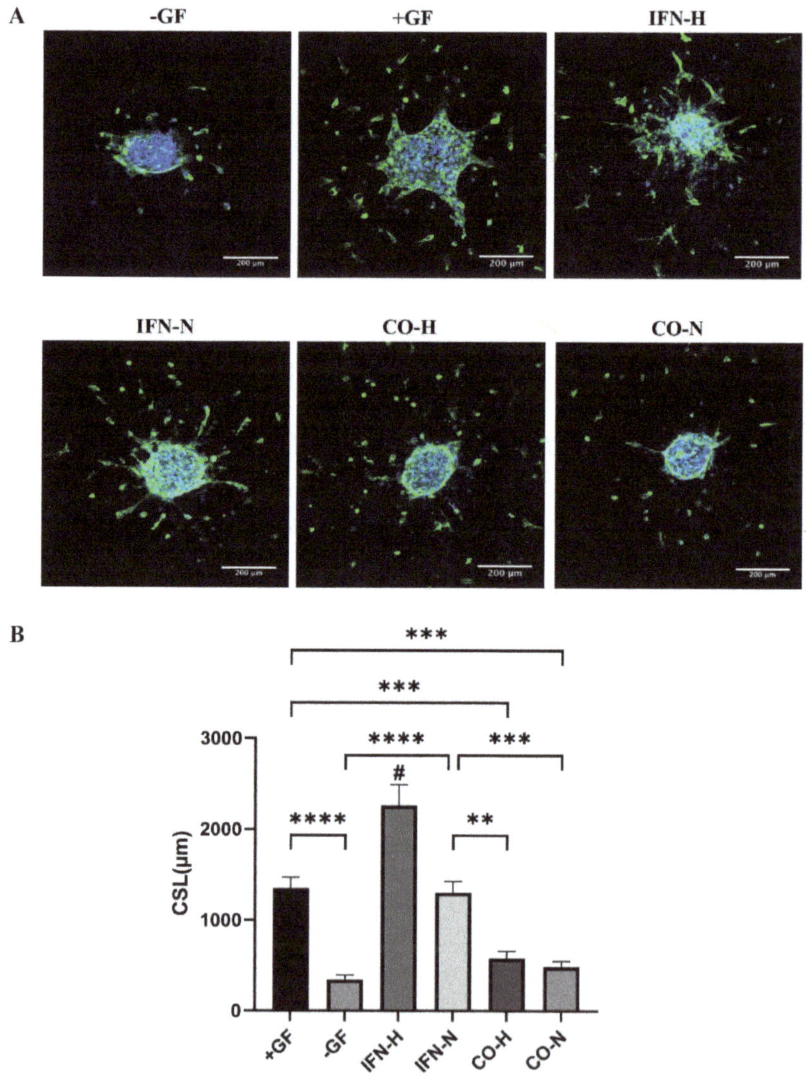

Figure 8. Sprout formation by 3D HUVEC spheroids cultured within coll/HA composites in the

presence of iMSC secretomes for 48 h. (**A**) Representative confocal images (Alexa488-Phalloidin staining) of HUVEC spheroids cultured within coll/HA composites in the presence of differently pre-conditioned iMSCs secretomes in a 10× magnification (scale bar = 200 µm). (**B**) Quantification of cumulative sprouts length (CSL) analyzed using the ImageJ software. Mean CSL was calculated for at least 10 randomly selected spheroids per experimental group. Values represent means ± SEM from three independent experiments (# indicates that the IFN-H HUVEC group showed significant differences to all other groups, ** $p < 0.01$; *** $p < 0.001$; **** $p < 0.0001$).

4. Discussion

MSCs have been widely explored for cell-based therapy in the field of tissue regeneration due to their remarkable immunosuppressive and immunomodulatory properties [37], and their ability to enhance angiogenesis and accelerate tissue healing [38]. The similarity of iMSCs to primary MSCs and their regenerative potential in vivo have already been demonstrated in initial studies [39,40], but further investigation is needed. MSC-derived secretomes contain many bioactive molecules, such as growth factors, cytokines, chemokines, free nucleic acids, lipids, and extracellular vesicles, which carry proteins and/or miRNAs to target cells [41–43]. Detailed transcriptome analysis of iMSCs showed a rejuvenation-associated gene signature, as well as more genes in common with fetal MSCs than with adult MSCs [20]. The same study demonstrated that protein composition of iMSC secretomes is similar to that of both fetal and adult MSCs. However, the secretome composition can be regulated by preconditioning strategies during the in vitro iMSC culture [44]. The influence of different conditions has been investigated so far, including hypoxic and inflammatory conditions, addition of pharmacological agents, and 3D culture conditions [45]. A recent study investigated the influence of interferon-γ and hypoxia on the proteome and metabolome of therapeutic adipose-derived mesenchymal stem cells [46]. Pro-inflammatory and hypoxic conditions coexist in settings of chronic diseases, acute injury, and adipositas. The authors demonstrated that dual priming (IFN-γ/hypoxia) of MSCs intensified their immunomodulatory capacity, promoted their own survival, prevented them from clearance, and led to an anti-fibrotic MSC phenotype. Hypoxic conditions switch MSCs to glycolysis, causing fast consumption of glucose and fast production of lactate, which has inhibitory effects on T-cells. For the manufacturing of clinical-grade MSCs, there is a need to develop standardized assays to prove their potency. Guan and co-authors [47], detected after IFN exposure of MSCs, elevated expression levels of IDO and PD-L1 (programmed death ligand 1), which correlated with their suppressive potential on third-party T cell proliferation. The authors concluded that flow cytometric measurement of intracellular IDO and cell surface protein PD-L1 represents a potential and rapid assay for the assessment of their immunosuppressive potential.

Far less is known about the phenotype and features of iMSCs, as well as about their therapeutic capacities. Our study shows for the first time the effects of a pro-inflammatory IFN-γ activation under a hypoxic condition (5% O_2) on the angiogenic potential of the obtained iMSCs secretome. The obtained angiogenic potency was compared to that induced by the normoxic (20% O_2) and non-inflammatory environment of in vitro cultured iMSCs.

Pro-angiogenic proteins secreted by MSCs are mediated by growth factors (such as VEGF) and chemokines (such as IL-8) [48]. Many studies report that MSC secretomes can exert different effects in the context of angiogenesis, and many of these differences largely depend on the preconditioning of MSC cultures [49]. By the results described in the present study, we completely agree with this finding. Furthermore, the tissue origin seems to also influence the angiogenic profile of the human MSC secretome [50]. The use of rejuvenated iMSCs could bypass tissue- and age-related heterogeneities, which are associated with primary MSCs. In our study, we demonstrate for the first time that pro-inflammatory and hypoxic conditions enhance the angiogenic potency of the released iMSC secretome. As IL-8 and VEGF are both potent regulators of angiogenesis [33,34], we analyzed the gene and protein expression of both factors and detected the highest levels in the IFN-γ and hypoxia pre-conditioned iMSC group.

In recent years, priming approaches have been investigated to empower MSC efficacy. Among them, preconditioning with IFN-γ under hypoxic conditions seems to enhance the immunosuppressive properties of MSCs [45]. Under IFN-γ priming, MSCs increase the expression of class II histocompatibility leukocyte antigen (HLA) molecules [51], and MCSs preconditioning with IFN-γ and TNF-α in combination promoted angiogenesis and accelerated tumor growth [52]. Under hypoxic conditions, MSCs have been shown to possess a higher angiogenic and regenerative potential than under normoxic conditions [53]. This result is exactly in line with our obtained findings. Exosomes released by hypoxia-treated adipose tissue derived MSCs have been shown to enhance angiogenesis through the protein kinase A (PKA) signaling pathway in HUVECs [54]. Low oxygen tension is thought to be an integral component of the endosteal niche microenvironment. As IL-8 secretion by human primary MSCs is clearly increased under hypoxic conditions and IL-8 in turn possesses strong pro-angiogenic and chemotactic abilities, MSCs can enhance their migratory capacity in an autocrine manner [55] and promote osteogenesis at the same time [56]. However, IFN-γ exposure alone led to the suppression of VEGF and IL-8 by adipose-derived stromal cells. Conditioned medium of IFN-γ primed ASCs was not able to activate in vivo vessel formation [57]. Dual primed MSCs downregulated thrombospondin 1 and 2 expression, both factors that inhibit angiogenesis [46]. We did not analyze the thrombospondin expression, but its suppression might be a reason for the pro-angiogenic effect of the iMSC secretome from the IFN-H group, in addition to the suppression of the anti-angiogenic TIMP-4 and IGFBP-1, as described below.

Currently, we developed collagen/hydroxyapatite composites with good angiogenic properties by incorporating VEGF-mimetic peptides [36]. However, a cell type with pro-angiogenic properties could further support construct endothelialization in situ. Due to the limited proliferation and differentiation capacity of the primary jaw periosteal cells we usually work with, we found with iMSCs an alternative cell source that we derived from reprogrammed JPCs. As this MSC-like cell type is new, we drew our attention to a detailed analysis of its phenotype and functions, and demonstrated that iMSC priming with IFN-γ under hypoxic conditions induced pro-angiogenic properties of their secretome.

Besides evaluating the effect of iMSCs secretome on HUVEC tube formation, we also investigated the angiogenic gene expression in HUVECs. The analyzed genes play an important role during angiogenesis. VEGFR-1 (FLT-1) and VEGFR-2 (KDR) are both receptors of VEGF, representing principal initiators of the angiogenesis process [58]. For angiogenesis amplification and vascular stabilization, other factors like IL-8, IGFBP-2, MMP-1, VCAM-1, HIF-1α, HGF, and MET come into play [59–63]. Angiogenesis inhibitory factors like TIMP-4, endostatin, angiostatin, or thrombospondin suppress vascular growth [64]. Although several studies demonstrated that the inhibition of TIMP1 promotes angiogenesis [65], there are also contradictory works reporting on enhanced tumor angiogenesis in the brain metastasis of lung carcinoma associated with TIMP-1 overexpression [66]. In our study, HUVECs cultured in the presence of IFN-γ and hypoxia pre-conditioned iMSCs secretome showed an up-regulation of pro-angiogenic genes (FLT-1, KDR, MET, TIMP-1, HIF-1α, IL-8, and VCAM-1) and a down-regulation of anti-angiogenic genes such as TIMP-4, IGFBP-1, and IGFBP-2, compared to all other groups. Relatively little is actually known about the function of TIMP-4 and its role in angiogenesis. However, TIMP-4 has been shown to inhibit platelet aggregation and to decrease the migration and invasive potential of cancer cell lines. Further studies could demonstrate the inhibitory effect of TIMP-4 on capillary tube formation only in high doses and on capillary endothelial cell migration [67]. We suggest that TIMP-4 inhibition in the HUVEC group cultivated with the pre-conditioned (IFN-γ and hypoxia) iMSCs secretome was based on its anti-angiogenic effect. IGFBPs usually inhibit the metabolic and proliferative actions of IGFs by binding them, prolonging their half-lives, and altering their interactions with cell surface receptors. Hypoxia induces IGFBP-1 hyperphosphorylation, leading to decreased IGF-I bioavailability in fetal growth [68]. However, the interactions between IGFs and IGFBPs are mutual, resulting in reciprocal regulation dependent on the cellular environment [66].

Additionally, the HGF gene expression has been shown to be inversely proportional to detected c-MET levels in the pre-conditioned iMSCs secretome HUVEC group, compared to the other groups. HGF unfolds its angiogenic effect via tyrosine phosphorylation of its specific receptor, c-Met, expressed on blood vessels including endothelial cells (ECs) and vascular smooth muscle cells (VSMCs) [69]. Based on maximal c-MET transcription levels detected in the IFN-H group, we suppose higher HGF-responsiveness in this group compared to the other IFN-treated or untreated groups. The pre-conditioned iMSCs secretome obtained upon IFN-γ and hypoxia treatment could also include TGF-beta [51], which is able to inhibit the expression of HGF [70]. We suppose this might be the reason for reduced HGF levels in the IFN-H group compared to the other three secretome groups.

Many studies have shown the relevance of VCAM-1 in angiogenesis [71], and the chemokine interleukin-8 exerts potent pro-angiogenic effects through binding to the CXCR2 receptor of intestinal microvascular endothelial cells and downstream signaling by phosphorylation of the extracellular signal-regulated protein kinase 1/2 (ERK 1/2) [72]. As IL-8 and VCAM-1 gene expressions have been shown to be strongly and significantly upregulated in treated HUVECs, we suggest that these are key responder factors induced by the strong pro-angiogenic effect starting from the IFN-γ/hypoxia pre-conditioned iMSC secretome. We did not analyze the gene expression pattern of 3D-cultured HUVECs (as shown in Figure 8), but sprout formation assays showed very clearly similar results to the performed tube formation assays with 2D-cultured HUVECs (as shown in Figure 5). These results showed that regardless of the cultivation approach in 2D on a standard growth-factor reduced matrigel preparation composed of laminin I, type IV collagen, entactin, and heparan sulfate proteoglycans (information from the company), or in 3D within the type I collagen/hydroxyapatite constructs, HUVECs were able to be activated by the IFN-γ and hypoxia pre-conditioned iMSCs secretome.

Taken together, in this study we provide insight into the excellent pro-angiogenic functional capacity starting from IFN-γ and hypoxia pre-conditioned iMSCs secretome.

5. Conclusions

In this study, we present a cell-free approach for the development of collagen/hydroxyapatite scaffolds exhibiting angiogenic properties. Secondly, we demonstrate that dual priming with hypoxia and IFN-γ significantly improved the pro-angiogenic properties of iMSCs. Based on this result, we conclude that iMSCs priming before clinical application can activate neovascularization and improve the therapeutic efficacy of these stem cells.

Supplementary Materials: The following supporting information can be downloaded at: https://www.mdpi.com/article/10.3390/cells11060988/s1, Figure S1: Representative histograms of MSC- and iPSC-surface marker expression on iMSCs of passage 3. Figure S2: Tube formation of HUVECs incubated with different concentrations of iMSCs secretome; Figure S3: Sprout formation by 3D HUVEC spheroids cultured within coll/HA composites for 48 h.

Author Contributions: Conceptualization, S.W., F.U. and D.A.; methodology, S.W. and W.C.; validation, S.W. and F.U.; investigation, S.W.; data curation, S.W.; writing—original draft preparation, S.W.; writing—review and editing, S.W., F.U., A.J.S., S.R. and D.A.; visualization, S.W.; supervision, F.U. and D.A.; project administration, S.R.; funding acquisition, S.W. and D.A. All authors have read and agreed to the published version of the manuscript.

Funding: S.W. was financed by the China Scholarship Council (CSC) (grant 201908080045).

Institutional Review Board Statement: Not applicable.

Informed Consent Statement: Not applicable.

Data Availability Statement: Obtained data for this study are available from the corresponding author upon reasonable request.

Acknowledgments: We thank Hui Liu for the technical support. Additionally, we acknowledge support by Deutsche Forschungsgemeinschaft and the Open Access Publishing Fund of University of Tuebingen.

Conflicts of Interest: The authors declare no conflict of interest.

References

1. Jungbluth, P.; Spitzhorn, L.-S.; Grassmann, J.; Tanner, S.; Latz, D.; Rahman, M.S.; Bohndorf, M.; Wruck, W.; Sager, M.; Grotheer, V.; et al. Human IPSC-Derived IMSCs Improve Bone Regeneration in Mini-Pigs. *Bone Res.* **2019**, *7*, 32. [CrossRef]
2. Sakkas, A.; Wilde, F.; Heufelder, M.; Winter, K.; Schramm, A. Autogenous Bone Grafts in Oral Implantology-Is It Still a "Gold Standard"? A Consecutive Review of 279 Patients with 456 Clinical Procedures. *Int. J. Implant Dent.* **2017**, *3*, 23. [CrossRef] [PubMed]
3. Rai, R.; Raval, R.; Khandeparker, R.V.S.; Chidrawar, S.K.; Khan, A.A.; Ganpat, M.S. Tissue Engineering: Step Ahead in Maxillofacial Reconstruction. *J. Int. Oral Health* **2015**, *7*, 138–142. [PubMed]
4. Costello, B.J.; Shah, G.; Kumta, P.; Sfeir, C.S. Regenerative Medicine for Craniomaxillofacial Surgery. *Oral Maxillofac. Surg. Clin. N. Am.* **2010**, *22*, 33–42. [CrossRef] [PubMed]
5. Amini, A.R.; Laurencin, C.T.; Nukavarapu, S.P. Bone Tissue Engineering: Recent Advances and Challenges. *Crit. Rev. Biomed. Eng.* **2012**, *40*, 363–408. [CrossRef] [PubMed]
6. Praveen, K.L.; Kandoi, S.; Misra, R.; Vijayalakshmi, S.; Rajagopal, K.; Verma, R.S. The Mesenchymal Stem Cell Secretome: A New Paradigm towards Cell-Free Therapeutic Mode in Regenerative Medicine. *Cytokine Growth Factor Rev.* **2019**, *46*, 1–9.
7. Stagg, J.; Galipeau, J. Mechanisms of Immune Modulation by Mesenchymal Stromal Cells and Clinical Translation. *Curr. Mol. Med.* **2013**, *13*, 856–867. [CrossRef] [PubMed]
8. Bartosh, T.J.; Ullah, M.; Zeitouni, S.; Beaver, J.; Prockop, D.J. Cancer Cells Enter Dormancy after Cannibalizing Mesenchymal Stem/Stromal Cells (MSCs). *Proc. Natl. Acad. Sci. USA* **2016**, *113*, E6447–E6456. [CrossRef]
9. Galipeau, J.; Sensébé, L. Mesenchymal Stromal Cells: Clinical Challenges and Therapeutic Opportunities. *Cell Stem Cell* **2018**, *22*, 824–833. [CrossRef]
10. Timaner, M.; Tsai, K.K.; Shaked, Y. The Multifaceted Role of Mesenchymal Stem Cells in Cancer. *Semin. Cancer Biol.* **2020**, *60*, 225–237. [CrossRef]
11. Ahangar, P.; Mills, S.J.; Cowin, A.J. Mesenchymal Stem Cell Secretome as an Emerging Cell-Free Alternative for Improving Wound Repair. *Int. J. Mol. Sci.* **2020**, *21*, 7038. [CrossRef] [PubMed]
12. Baraniak, P.R.; McDevitt, T.C. Stem Cell Paracrine Actions and Tissue Regeneration. *Regen. Med.* **2010**, *5*, 121–143. [CrossRef] [PubMed]
13. Murphy, M.B.; Moncivais, K.; Caplan, A.I. Mesenchymal Stem Cells: Environmentally Responsive Therapeutics for Regenerative Medicine. *Exp. Mol. Med.* **2013**, *45*, e54. [CrossRef] [PubMed]
14. Maumus, M.; Jorgensen, C.; Noël, D. Mesenchymal Stem Cells in Regenerative Medicine Applied to Rheumatic Diseases: Role of Secretome and Exosomes. *Biochimie* **2013**, *95*, 2229–2234. [CrossRef] [PubMed]
15. Wagner, W.; Ho, A.D. Mesenchymal Stem Cell Preparations–Comparing Apples and Oranges. *Stem Cell Rev.* **2007**, *3*, 239–248. [CrossRef] [PubMed]
16. Jung, Y.; Bauer, G.; Nolta, J.A. Concise Review: Induced Pluripotent Stem Cell-Derived Mesenchymal Stem Cells: Progress toward Safe Clinical Products. *Stem Cells* **2012**, *30*, 42–47. [CrossRef]
17. Ng, J.; Hynes, K.; White, G.; Sivanathan, K.N.; Vandyke, K.; Bartold, P.M.; Gronthos, S. Immunomodulatory Properties of Induced Pluripotent Stem Cell-Derived Mesenchymal Cells. *J. Cell. Biochem.* **2016**, *117*, 2844–2853. [CrossRef]
18. Frobel, J.; Hemeda, H.; Lenz, M.; Abagnale, G.; Joussen, S.; Denecke, B.; Šarić, T.; Zenke, M.; Wagner, W. Epigenetic Rejuvenation of Mesenchymal Stromal Cells Derived from Induced Pluripotent Stem Cells. *Stem Cell Rep.* **2014**, *3*, 414–422. [CrossRef]
19. Lapasset, L.; Milhavet, O.; Prieur, A.; Besnard, E.; Babled, A.; Aït-Hamou, N.; Leschik, J.; Pellestor, F.; Ramirez, J.-M.; De Vos, J.; et al. Rejuvenating Senescent and Centenarian Human Cells by Reprogramming through the Pluripotent State. *Genes Dev.* **2011**, *25*, 2248–2253. [CrossRef]
20. Spitzhorn, L.-S.; Megges, M.; Wruck, W.; Rahman, M.S.; Otte, J.; Degistirici, Ö.; Meisel, R.; Sorg, R.V.; Oreffo, R.O.C.; Adjaye, J. Human IPSC-Derived MSCs (IMSCs) from Aged Individuals Acquire a Rejuvenation Signature. *Stem Cell Res. Ther.* **2019**, *10*, 100. [CrossRef]
21. Spitzhorn, L.-S.; Kordes, C.; Megges, M.; Sawitza, I.; Götze, S.; Reichert, D.; Schulze-Matz, P.; Graffmann, N.; Bohndorf, M.; Wruck, W.; et al. Transplanted Human Pluripotent Stem Cell-Derived Mesenchymal Stem Cells Support Liver Regeneration in Gunn Rats. *Stem Cells Dev.* **2018**, *27*, 1702–1714. [CrossRef]
22. Wang, X.; Kimbrel, E.A.; Ijichi, K.; Paul, D.; Lazorchak, A.S.; Chu, J.; Kouris, N.A.; Yavanian, G.J.; Lu, S.-J.; Pachter, J.S.; et al. Human ESC-Derived MSCs Outperform Bone Marrow MSCs in the Treatment of an EAE Model of Multiple Sclerosis. *Stem Cell Rep.* **2014**, *3*, 115–130. [CrossRef]
23. Sheyn, D.; Ben-David, S.; Shapiro, G.; De Mel, S.; Bez, M.; Ornelas, L.; Sahabian, A.; Sareen, D.; Da, X.; Pelled, G.; et al. Human Induced Pluripotent Stem Cells Differentiate Into Functional Mesenchymal Stem Cells and Repair Bone Defects. *Stem Cells Transl. Med.* **2016**, *5*, 1447–1460. [CrossRef]
24. Umrath, F.; Steinle, H.; Weber, M.; Wendel, H.-P.; Reinert, S.; Alexander, D.; Avci-Adali, M. Generation of IPSCs from Jaw Periosteal Cells Using Self-Replicating RNA. *Int. J. Mol. Sci.* **2019**, *20*, 1648. [CrossRef]

25. Umrath, F.; Weber, M.; Reinert, S.; Wendel, H.-P.; Avci-Adali, M.; Alexander, D. IPSC-Derived MSCs Versus Originating Jaw Periosteal Cells: Comparison of Resulting Phenotype and Stem Cell Potential. *Int. J. Mol. Sci.* **2020**, *21*, 587. [CrossRef]
26. Portal-Núñez, S.; Lozano, D.; Esbrit, P. Role of Angiogenesis on Bone Formation. *Histol. Histopathol.* **2012**, *27*, 559–566.
27. Hankenson, K.D.; Dishowitz, M.; Gray, C.; Schenker, M. Angiogenesis in Bone Regeneration. *Injury* **2011**, *42*, 556–561. [CrossRef]
28. Estrada, R.; Li, N.; Sarojini, H.; An, J.; Lee, M.-J.; Wang, E. Secretome from Mesenchymal Stem Cells Induces Angiogenesis via Cyr61. *J. Cell. Physiol.* **2009**, *219*, 563–571. [CrossRef]
29. Wang, W.; Zhang, X.; Chao, N.-N.; Qin, T.-W.; Ding, W.; Zhang, Y.; Sang, J.-W.; Luo, J.-C. Preparation and Characterization of Pro-Angiogenic Gel Derived from Small Intestinal Submucosa. *Acta Biomater.* **2016**, *29*, 135–148. [CrossRef]
30. Jonkman, J.E.N.; Cathcart, J.A.; Xu, F.; Bartolini, M.E.; Amon, J.E.; Stevens, K.M.; Colarusso, P. An Introduction to the Wound Healing Assay Using Live-Cell Microscopy. *Cell Adh. Migr.* **2014**, *8*, 440–451. [CrossRef]
31. Rameshbabu, A.P.; Datta, S.; Bankoti, K.; Subramani, E.; Chaudhury, K.; Lalzawmliana, V.; Nandi, S.K.; Dhara, S. Polycaprolactone Nanofibers Functionalized with Placental Derived Extracellular Matrix for Stimulating Wound Healing Activity. *J. Mater. Chem. B* **2018**, *6*, 6767–6780. [CrossRef]
32. Maracle, C.X.; Kucharzewska, P.; Helder, B.; van der Horst, C.; Correa de Sampaio, P.; Noort, A.-R.; van Zoest, K.; Griffioen, A.W.; Olsson, H.; Tas, S.W. Targeting Non-Canonical Nuclear Factor-KB Signalling Attenuates Neovascularization in a Novel 3D Model of Rheumatoid Arthritis Synovial Angiogenesis. *Rheumatology* **2017**, *56*, 294–302. [CrossRef]
33. SHI, J.; WEI, P.-K. Interleukin-8: A Potent Promoter of Angiogenesis in Gastric Cancer. *Oncol. Lett.* **2016**, *11*, 1043–1050. [CrossRef]
34. Carmeliet, P. VEGF as a Key Mediator of Angiogenesis in Cancer. *Oncology* **2005**, *69*, 4–10. [CrossRef]
35. Clemente, N.; Argenziano, M.; Gigliotti, C.L.; Ferrara, B.; Boggio, E.; Chiocchetti, A.; Caldera, F.; Trotta, F.; Benetti, E.; Annaratone, L.; et al. Paclitaxel-Loaded Nanosponges Inhibit Growth and Angiogenesis in Melanoma Cell Models. *Front. Pharmacol.* **2019**, *10*, 776. [CrossRef]
36. Wang, S.; Umrath, F.; Cen, W.; Reinert, S.; Alexander, D. Angiogenic Potential of VEGF Mimetic Peptides for the Biofunctionalization of Collagen/Hydroxyapatite Composites. *Biomolecules* **2021**, *11*, 1538. [CrossRef]
37. Parekkadan, B.; Milwid, J.M. Mesenchymal Stem Cells as Therapeutics. *Annu. Rev. Biomed. Eng.* **2010**, *12*, 87–117. [CrossRef]
38. Caplan, A.I.; Correa, D. The MSC: An Injury Drugstore. *Cell Stem Cell* **2011**, *9*, 11–15. [CrossRef]
39. Chen, Y.S.; Pelekanos, R.A.; Ellis, R.L.; Horne, R.; Wolvetang, E.J.; Fisk, N.M. Small Molecule Mesengenic Induction of Human Induced Pluripotent Stem Cells to Generate Mesenchymal Stem/Stromal Cells. *Stem Cells Transl. Med.* **2012**, *1*, 83–95. [CrossRef]
40. Kimbrel, E.A.; Kouris, N.A.; Yavanian, G.J.; Chu, J.; Qin, Y.; Chan, A.; Singh, R.P.; McCurdy, D.; Gordon, L.; Levinson, R.D.; et al. Mesenchymal Stem Cell Population Derived from Human Pluripotent Stem Cells Displays Potent Immunomodulatory and Therapeutic Properties. *Stem Cells Dev.* **2014**, *23*, 1611–1624. [CrossRef]
41. Munn, D.H.; Mellor, A.L. Indoleamine 2,3 Dioxygenase and Metabolic Control of Immune Responses. *Trends Immunol.* **2013**, *34*, 137–143. [CrossRef]
42. Wangler, S.; Kamali, A.; Wapp, C.; Wuertz-Kozak, K.; Häckel, S.; Fortes, C.; Benneker, L.M.; Haglund, L.; Richards, R.G.; Alini, M.; et al. Uncovering the Secretome of Mesenchymal Stromal Cells Exposed to Healthy, Traumatic, and Degenerative Intervertebral Discs: A Proteomic Analysis. *Stem Cell Res. Ther.* **2021**, *12*, 11. [CrossRef]
43. Beer, L.; Mildner, M.; Ankersmit, H.J. Cell Secretome Based Drug Substances in Regenerative Medicine: When Regulatory Affairs Meet Basic Science. *Ann. Transl. Med.* **2017**, *5*, 170. [CrossRef]
44. Madrigal, M.; Rao, K.S.; Riordan, N.H. A Review of Therapeutic Effects of Mesenchymal Stem Cell Secretions and Induction of Secretory Modification by Different Culture Methods. *J. Transl. Med.* **2014**, *12*, 260. [CrossRef]
45. de Cassia Noronha, N.; Mizukami, A.; Caliári-Oliveira, C.; Cominal, J.G.; Rocha, J.L.M.; Covas, D.T.; Swiech, K.; Malmegrim, K.C.R. Priming Approaches to Improve the Efficacy of Mesenchymal Stromal Cell-Based Therapies. *Stem Cell Res. Ther.* **2019**, *10*, 131. [CrossRef]
46. Wobma, H.M.; Tamargo, M.A.; Goeta, S.; Brown, L.M.; Duran-Struuck, R.; Vunjak-Novakovic, G. The Influence of Hypoxia and IFN-γ on the Proteome and Metabolome of Therapeutic Mesenchymal Stem Cells. *Biomaterials* **2018**, *167*, 226–234. [CrossRef]
47. Guan, Q.; Li, Y.; Shpiruk, T.; Bhagwat, S.; Wall, D.A. Inducible Indoleamine 2,3-Dioxygenase 1 and Programmed Death Ligand 1 Expression as the Potency Marker for Mesenchymal Stromal Cells. *Cytotherapy* **2018**, *20*, 639–649. [CrossRef]
48. Watt, S.M.; Gullo, F.; van der Garde, M.; Markeson, D.; Camicia, R.; Khoo, C.P.; Zwaginga, J.J. The Angiogenic Properties of Mesenchymal Stem/Stromal Cells and Their Therapeutic Potential. *Br. Med. Bull.* **2013**, *108*, 25–53. [CrossRef]
49. Kwon, H.M.; Hur, S.-M.; Park, K.-Y.; Kim, C.-K.; Kim, Y.-M.; Kim, H.-S.; Shin, H.-C.; Won, M.-H.; Ha, K.-S.; Kwon, Y.-G.; et al. Multiple Paracrine Factors Secreted by Mesenchymal Stem Cells Contribute to Angiogenesis. *Vascul. Pharmacol.* **2014**, *63*, 19–28. [CrossRef]
50. Kehl, D.; Generali, M.; Mallone, A.; Heller, M.; Uldry, A.-C.; Cheng, P.; Gantenbein, B.; Hoerstrup, S.P.; Weber, B. Proteomic Analysis of Human Mesenchymal Stromal Cell Secretomes: A Systematic Comparison of the Angiogenic Potential. *NPJ Regen. Med.* **2019**, *4*, 8. [CrossRef]
51. de Witte, S.F.H.; Franquesa, M.; Baan, C.C.; Hoogduijn, M.J. Toward Development of IMesenchymal Stem Cells for Immunomodulatory Therapy. *Front. Immunol.* **2016**, *6*, 648. [CrossRef]
52. Liu, Y.; Han, Z.; Zhang, S.; Jing, Y.; Bu, X.; Wang, C.; Sun, K.; Jiang, G.; Zhao, X.; Li, R.; et al. Effects of Inflammatory Factors on Mesenchymal Stem Cells and Their Role in the Promotion of Tumor Angiogenesis in Colon Cancer. *J. Biol. Chem.* **2011**, *286*, 25007–25015. [CrossRef] [PubMed]

53. Leroux, L.; Descamps, B.; Tojais, N.F.; Séguy, B.; Oses, P.; Moreau, C.; Daret, D.; Ivanovic, Z.; Boiron, J.-M.; Lamazière, J.-M.D.; et al. Hypoxia Preconditioned Mesenchymal Stem Cells Improve Vascular and Skeletal Muscle Fiber Regeneration After Ischemia Through a Wnt4-Dependent Pathway. *Mol. Ther.* **2010**, *18*, 1545–1552. [CrossRef]
54. Xue, C.; Shen, Y.; Li, X.; Li, B.; Zhao, S.; Gu, J.; Chen, Y.; Ma, B.; Wei, J.; Han, Q.; et al. Exosomes Derived from Hypoxia-Treated Human Adipose Mesenchymal Stem Cells Enhance Angiogenesis Through the PKA Signaling Pathway. *Stem Cells Dev.* **2018**, *27*, 456–465. [CrossRef]
55. Wobus, M.; Mueller, K.; Ehninger, G.; Bornhaeuser, M. Hypoxia Increases IL-8 Secretion of Mesenchymal Stroma Cells Affecting Migratory Capacity in An Autocrine Manner. *Blood* **2008**, *112*, 4752. [CrossRef]
56. Yang, A.; Lu, Y.; Xing, J.; Li, Z.; Yin, X.; Dou, C.; Dong, S.; Luo, F.; Xie, Z.; Hou, T.; et al. IL-8 Enhances Therapeutic Effects of BMSCs on Bone Regeneration via CXCR2-Mediated PI3k/Akt Signaling Pathway. *Cell Physiol. Biochem.* **2018**, *48*, 361–370. [CrossRef]
57. Andreeva, E.R.; Udartseva, O.O.; Zhidkova, O.V.; Buravkov, S.V.; Ezdakova, M.I.; Buravkova, L.B. IFN-Gamma Priming of Adipose-Derived Stromal Cells at "Physiological" Hypoxia. *J. Cell. Physiol.* **2018**, *233*, 1535–1547. [CrossRef]
58. Gerber, H.P.; Condorelli, F.; Park, J.; Ferrara, N. Differential Transcriptional Regulation of the Two Vascular Endothelial Growth Factor Receptor Genes. Flt-1, but Not Flk-1/KDR, Is up-Regulated by Hypoxia. *J. Biol. Chem.* **1997**, *272*, 23659–23667. [CrossRef]
59. Ema, M.; Taya, S.; Yokotani, N.; Sogawa, K.; Matsuda, Y.; Fujii-Kuriyama, Y. A Novel BHLH-PAS Factor with Close Sequence Similarity to Hypoxia-Inducible Factor 1α Regulates the VEGF Expression and Is Potentially Involved in Lung and Vascular Development. *Proc. Natl. Acad. Sci. USA* **1997**, *94*, 4273–4278. [CrossRef]
60. Ferrara, N. Vascular Endothelial Growth Factor: Basic Science and Clinical Progress. *Endocr. Rev.* **2004**, *25*, 581–611. [CrossRef]
61. Cheng, F.; Guo, D. MET in Glioma: Signaling Pathways and Targeted Therapies. *J. Exp. Clin. Cancer Res.* **2019**, *38*, 270. [CrossRef]
62. Azar, W.J.; Azar, S.H.X.; Higgins, S.; Hu, J.-F.; Hoffman, A.R.; Newgreen, D.F.; Werther, G.A.; Russo, V.C. IGFBP-2 Enhances VEGF Gene Promoter Activity and Consequent Promotion of Angiogenesis by Neuroblastoma Cells. *Endocrinology* **2011**, *152*, 3332–3342. [CrossRef]
63. Ding, Y.-B.; Chen, G.-Y.; Xia, J.-G.; Zang, X.-W.; Yang, H.-Y.; Yang, L. Association of VCAM-1 Overexpression with Oncogenesis, Tumor Angiogenesis and Metastasis of Gastric Carcinoma. *World J. Gastroenterol.* **2003**, *9*, 1409–1414. [CrossRef]
64. Fernández, C.A.; Louis, G.; Moses, M.A. Characterization of the Anti-Angiogenic Effects of TIMP-4. *Cancer Res.* **2004**, *64*, 530.
65. Reed, M.J.; Koike, T.; Sadoun, E.; Sage, E.H.; Puolakkainen, P. Inhibition of TIMP1 Enhances Angiogenesis in Vivo and Cell Migration in Vitro. *Microvasc. Res.* **2003**, *65*, 9–17. [CrossRef]
66. Rojiani, M.V.; Ghoshal-Gupta, S.; Kutiyanawalla, A.; Mathur, S.; Rojiani, A.M. TIMP-1 Overexpression in Lung Carcinoma Enhances Tumor Kinetics and Angiogenesis in Brain Metastasis. *J. Neuropathol. Exp. Neurol.* **2015**, *74*, 293–304. [CrossRef]
67. Fernández, C.A.; Moses, M.A. Modulation of Angiogenesis by Tissue Inhibitor of Metalloproteinase-4. *Biochem. Biophys. Res. Commun.* **2006**, *345*, 523–529. [CrossRef]
68. Damerill, I.; Biggar, K.K.; Abu Shehab, M.; Li, S.S.-C.; Jansson, T.; Gupta, M.B. Hypoxia Increases IGFBP-1 Phosphorylation Mediated by MTOR Inhibition. *Mol. Endocrinol.* **2016**, *30*, 201–216. [CrossRef]
69. Kaga, T.; Kawano, H.; Sakaguchi, M.; Nakazawa, T.; Taniyama, Y.; Morishita, R. Hepatocyte Growth Factor Stimulated Angiogenesis without Inflammation: Differential Actions between Hepatocyte Growth Factor, Vascular Endothelial Growth Factor and Basic Fibroblast Growth Factor. *Vascul. Pharmacol.* **2012**, *57*, 3–9. [CrossRef]
70. Correll, K.A.; Edeen, K.E.; Redente, E.F.; Zemans, R.L.; Edelman, B.L.; Danhorn, T.; Curran-Everett, D.; Mikels-Vigdal, A.; Mason, R.J. TGF Beta Inhibits HGF, FGF7, and FGF10 Expression in Normal and IPF Lung Fibroblasts. *Physiol. Rep.* **2018**, *6*, e13794. [CrossRef]
71. Kong, D.-H.; Kim, Y.K.; Kim, M.R.; Jang, J.H.; Lee, S. Emerging Roles of Vascular Cell Adhesion Molecule-1 (VCAM-1) in Immunological Disorders and Cancer. *Int. J. Mol. Sci.* **2018**, *19*, 1057. [CrossRef]
72. Heidemann, J.; Ogawa, H.; Dwinell, M.B.; Rafiee, P.; Maaser, C.; Gockel, H.R.; Otterson, M.F.; Ota, D.M.; Lugering, N.; Domschke, W.; et al. Angiogenic Effects of Interleukin 8 (CXCL8) in Human Intestinal Microvascular Endothelial Cells Are Mediated by CXCR2. *J. Biol. Chem.* **2003**, *278*, 8508–8515. [CrossRef]

 cells

Article

Assessment of the Neuroprotective and Stemness Properties of Human Wharton's Jelly-Derived Mesenchymal Stem Cells under Variable (5% vs. 21%) Aerobic Conditions

Ewelina Tomecka [1,†], Wioletta Lech [2,†], Marzena Zychowicz [2], Anna Sarnowska [2], Magdalena Murzyn [1], Tomasz Oldak [1], Krystyna Domanska-Janik [2], Leonora Buzanska [2,*] and Natalia Rozwadowska [3]

1 Polish Stem Cell Bank, FamiCord Group, 00-867 Warsaw, Poland; ewelina.tomecka@pbkm.pl (E.T.); magdalena.murzyn@pbkm.pl (M.M.); tomasz.oldak@pbkm.pl (T.O.)
2 Department of Stem Cell Bioengineering, Mossakowski Medical Research Institute, Polish Academy of Sciences, 02-106 Warsaw, Poland; wlech@imdik.pan.pl (W.L.); mzychowicz@imdik.pan.pl (M.Z.); asarnowska@imdik.pan.pl (A.S.); kd-j@imdik.pan.pl (K.D.-J.)
3 Institute of Human Genetics, Polish Academy of Sciences, 60-479 Poznan, Poland; nrozwad@man.poznan.pl
* Correspondence: buzanska@imdik.pan.pl; Tel.: +48-22-668-52-50
† These authors contributed equally to this paper as the first authors.

Citation: Tomecka, E.; Lech, W.; Zychowicz, M.; Sarnowska, A.; Murzyn, M.; Oldak, T.; Domanska-Janik, K.; Buzanska, L.; Rozwadowska, N. Assessment of the Neuroprotective and Stemness Properties of Human Wharton's Jelly-Derived Mesenchymal Stem Cells under Variable (5% vs. 21%) Aerobic Conditions. *Cells* **2021**, *10*, 717. https://doi.org/10.3390/cells10040717

Academic Editor: Joni H. Ylostalo

Received: 28 February 2021
Accepted: 21 March 2021
Published: 24 March 2021

Abstract: To optimise the culture conditions for human Wharton's jelly-derived mesenchymal stem cells (hWJ-MSCs) intended for clinical use, we investigated ten different properties of these cells cultured under 21% (atmospheric) and 5% (physiological normoxia) oxygen concentrations. The obtained results indicate that 5% O_2 has beneficial effects on the proliferation rate, clonogenicity, and slowdown of senescence of hWJ-MSCs; however, the oxygen level did not have an influence on the cell morphology, immunophenotype, or neuroprotective effect of the hWJ-MSCs. Nonetheless, the potential to differentiate into adipocytes, osteocytes, and chondrocytes was comparable under both oxygen conditions. However, spontaneous differentiation of hWJ-MSCs into neuronal lineages was observed and enhanced under atmospheric oxygen conditions. The cells relied more on mitochondrial respiration than glycolysis, regardless of the oxygen conditions. Based on these results, we can conclude that hWJ-MSCs could be effectively cultured and prepared under both oxygen conditions for cell-based therapy. However, the 5% oxygen level seemed to create a more balanced and appropriate environment for hWJ-MSCs.

Keywords: Wharton's jelly mesenchymal stem cells; umbilical cord; oxygen conditions; secretory profile; neuroprotection

1. Introduction

Mesenchymal stromal/stem cells (MSCs) are promising tools in regenerative therapy and other clinical applications. According to the position statement of the International Society for Cellular Therapy [1], MSCs are characterised by (1) adherence to plastic under standard culture conditions; (2) expression of the surface markers CD73, CD90, and CD105 in the absence of CD11b or CD14, CD19, CD34, CD45, CD79a, and HLA-DR; and (3) the ability to differentiate into adipocytes, osteocytes, and chondrocytes in vitro. Other key features of these cells are self-renewal; multipotency; high proliferative potential; and immunomodulatory, paracrine, and anti-inflammatory properties [2,3]. MSCs can be isolated from various types of adult tissue, namely, bone marrow, adipogenic tissue, and dental pulp and foetal tissue, namely, umbilical cord, umbilical cord blood, amniotic fluid, placenta, and Wharton's jelly (WJ) [2,4–6]. The most common sources of mesenchymal stem cells used for clinical application are bone marrow and adipogenic tissue, which have some limitations [7–11]. For example, collection of these tissues is a painful and invasive procedure. Moreover, the proliferation ability and differentiation potential of human bone marrow-derived MSCs (hBM-MSCs) decrease with age [12,13].

A relevant source of MSCs is the umbilical cord matrix, or Wharton's jelly, which is the embryonic mucous connective tissue between the amniotic epithelium and umbilical vessels. hWJ-MSCs also show similar characteristics to adult MSCs, as determined by the International Society for Cellular Therapy [1]. They exhibit a higher proliferation rate and differentiation capabilities and lower immunogenicity than adult tissue-derived MSCs [14–18]. hWJ-MSCs also have unique immunomodulatory effects [19]. Moreover, the isolation of hWJ-MSCs is easy and noninvasive [20]. In contrast to bone marrow, the umbilical cord is considered medical waste and is obtained/utilised with no harm to the patient. The expression level of neural/neuronal markers (Nestin, NF-200, GFAP) is higher in hWJ-MSCs than in hBM-MSCs [18], and mesenchymal stem cells derived from Wharton's jelly exhibit neuroprotective properties, which were defined after indirect coculture of hWJ-MSCs with injured neuronal cells or tissue [21,22]. Due to these features of hWJ-MSCs, there is a growing interest in using these cells in cell-based therapies. Therefore, it is important to thoroughly investigate the properties of hWJ-MSCs.

Additionally, conventional in vitro cell cultures are carried out under atmospheric oxygen conditions (21% O_2), which do not correspond to the in vivo situation. In living organisms, the average physiological oxygen concentration is much lower and varies from 2–9% O_2 ("physiological normoxia") [23–25] depending on the vascularization of tissue and its metabolic activity [26]. The oxygen level can be a significant environmental factor that may affect mesenchymal stem cell properties. Moreover, it has been hypothesised that 21% O_2 culture conditions could lead to a reduction in MSCs' therapeutic potential [27]. Therefore, it is important to determine the influence of different oxygen concentrations on the biological activity of MSCs to optimise their culture conditions before clinical application.

The role of different oxygen conditions in mesenchymal stem cell biology has been studied by other researchers, who investigated the influence of low oxygen tension (1–5%) on mesenchymal stem cells derived from bone marrow [27–41], adipose tissue [28,35,36,42–49], dental pulp [50–52], cord blood [53,54], the umbilical cord [45,55,56], and Wharton's jelly [3,34,40,57,58]. However, studies on the effect of low oxygen levels on human WJ-MSCs were not comprehensive and have focused on a few cell properties. These works also reported contradictive results; thus, the impact of oxygen conditions on hWJ-MSCs has not been clarified.

The aim of the present work was to investigate and compare human Wharton's jelly-derived MSCs cultured under atmospheric (21% O_2) and low (5% O_2) oxygen conditions. We analysed the different properties of hWJ-MSCs, such as the morphology, phenotype, proliferation ability, clonogenicity, senescence, mesodermal differentiation potential, secretory profile, metabolic activity, neural differentiation potential, and neuroprotective effects. The results of this work highlight the need to determine optimal oxygen culture conditions for the expansion of hWJ-MSCs intended for clinical application.

We would like to emphasise that the potential therapeutic application of MSCs requires careful verification of their properties prior to their transplantation. Standard tests based only on the expression of surface markers (CD73, CD90, CD105) and the capacity for mesodermal differentiation are insufficient. Depending on their subsequent use for treating specific neural disorders, the characteristics of mesenchymal stem cells should also include neural differentiation potential, neuroprotective properties, or the ability to secrete specific neurotrophins.

2. Materials and Methods

2.1. Isolation and Culture of Human WJ-MSCs

Umbilical cord (UC) fragments were collected from four human donors after caesarean section or natural delivery with the consent of mothers and approval by the Bioethics Committee. To minimise the risk of contamination, the umbilical cords were immediately transferred to the laboratory in a sterile transportation container filled with transport liquid consisting of 0.9% sodium chloride solution (Fresenius Kabi, Sevres, France) supplemented with antibiotic/antimycotic solution (Gibco, New York, NY, USA). After washing with fresh sterile transport liquid, the umbilical cords were cut into approximately 3 cm pieces, and the blood vessels were mechanically removed. Vessel-free Wharton's jelly was sliced into 1–2 mm^3 scraps, which were placed into 525 cm^2 (triple layers, Falcon, Oxnard, CA, USA) culture flasks. The culture flasks were previously covered with MSC Attachment Solution Xeno Free (Biological Industries, Beit Ilaemek, Israel). For this purpose, culture surface of one culture flask was covered with 90 mL of attachment solution, prepared in a ratio of 1:100 in CTS DPBS (Gibco, New York, NY, USA), and incubated at 37 °C under a humidified atmosphere with 5% CO_2 and 21% O_2 for 30 min. After this time the culture flasks were rinsed with 90 mL of CTS DPBS (Gibco, New York, NY, USA). The explants were cultured with growth medium consisting of MSC NutriStem® XF basal medium (Biological Industries, Beit Ilaemek, Israel) supplemented with MSC NutriStem® XF Supplement Mix (Biological Industries, Beit Ilaemek, Israel) and 1% (*v*/*v*) antibiotic/antimycotic solution (Gibco, New York, NY, USA) at 37 °C under a humidified atmosphere with 5% CO_2 and 21% O_2. After up to 4 weeks (during this time culture medium was not exchange), the cultures were analysed for the presence of adherent, fibroblast-like cells. If nonadherent cells were present in the cultures, they were washed out. Subsequently, adherent cells isolated from Wharton's jelly were detached with TrypLE Express Enzyme (Gibco, Denmark) and cryopreserved (passage 0) in 5% human serum albumin (CSL Behring, Margburg, Germany) containing 10% DMSO (WAK-Chemie Medical, Steinbach, Germany).

For all experiments, cells derived from four human donors (WJ-MSCs (1), WJ-MSCs (2), WJ-MSCs (3), and WJ-MSCs (4)) were used. First, the cells were thawed, centrifuged (112× *g*, 3 min) and seeded into 75 cm^2 culture flasks at an initial density of $2 \times 10^3/\text{cm}^2$. The culture medium consisted of Dulbecco's modified Eagle's medium (DMEM, Macopharma, Mouvaux, France), 10% (*v*/*v*) human platelet lysate (Macopharma, Mouvaux, France), 2 U/mL heparin (Sigma-Aldrich, St. Louis, MO, USA), 1 mg/mL glucose (Sigma-Aldrich, St. Louis, MO, USA), and 1% (*v*/*v*) antibiotic/antimycotic solution (Gibco, New York, NY, USA). Before collection for subsequent passages, the cells were cultured to 70% confluence at 37 °C under atmospheric (21% O_2 and 5% CO_2) and "physiological normoxia" (5% O_2 and 5% CO_2) conditions. Experiments were performed with the use of cells at passages 1 to 5.

2.2. Flow Cytometry Analysis

An evaluation of the expression profile of surface markers was carried out using flow cytometry analysis. The cell phenotype was determined using a Human MSC Analysis Kit (Becton Dickinson, BD, Franklin Lakes, NJ, USA). For this purpose, the cells at passages 1–2 were detached with Accutase cell detachment solution (BD Bioscience, Franklin Lakes, NJ, USA) and resuspended in cold stain buffer (BD Pharmingen, Franklin Lakes, NJ, USA) at a minimal density of 1×10^6 cells/mL. The WJ-derived cells were incubated with fluorochrome-conjugated (FITC, PerCP-Cy5.5, APC, PE) antibodies against CD105, CD90, and CD73 (mesenchymal cell surface markers) and CD11b, CD19, CD34, CD45, and HLA-DR (haematopoietic markers) for 30 min at room temperature in the dark. To exclude nonspecific binding, corresponding isotype antibodies (IgG1/IgG2) were used as a control. Cell analysis was performed using a FACSCalibur II flow cytometer and FACSDiva software (Becton Dickinson, Franklin Lakes, USA). The results are presented as the percentage of positive cells for suitable markers in relation to the isotype control.

2.3. Cell Proliferation Assays

The proliferation rate of the hWJ-MSCs was evaluated by a WST-1 assay (Roche, Mannheim, Germany) and expression analysis of the proliferation marker Ki67 (Abcam, Cambridge, MA, USA).

2.3.1. WST-1 Assay

Quantitative analyses of cell proliferation were performed using a commercially available WST-1 assay. The principle of this test is based on the reduction of colourless tetrazolium salt to colourful formazan by cellular dehydrogenases. The amount of colourful product was then colourimetrically measured and directly correlated to the viable cell number. For this purpose, the cells at passages 3–5 were seeded into 96-well plates at a density of 2×10^3 cells/cm^2. For both analysed 7-day cultures (under 21% O_2 and 5% O_2), WST-1 reagent was added every day to the appropriate wells of 96-well plates at a volume ratio of 1:10 and incubated at 37 °C for 2 h. The absorbance was measured using a multiwell plate reader (FLUOstar Omega, BMG LABTECH, Ortenberg, Germany) at a wavelength of 420 nm. Based on the prepared standard curves, the obtained absorbance values were converted to the number of viable, metabolically active cells.

2.3.2. Analysis of Ki67 Marker Expression

The proliferation ability of hWJ-MSCs cultured under 21% O_2 and 5% O_2 was evaluated by analysing the expression of the proliferation marker Ki67. The Ki67 protein is a nuclear antigen, and its expression is closely related to cell cycle activity. The expression of the Ki67 marker was determined by immunofluorescence staining analysis. For experiments, hWJ-MSCs at early (p1–p2) and late passages (p4–p5) were used. The cells were seeded into plates previously covered with poly-L-lysine (Sigma-Aldrich, St. Louis, MO, USA) coverslips placed in 24-well plates at a density of 2.5×10^3 cells/cm^2. When the cells reached 70% confluence, they were fixed with 4% paraformaldehyde solution (Sigma Aldrich, St. Louis, MO, USA) for 20 min at room temperature. To permeabilise the cell membrane, the cells were incubated with 0.1% Triton X-100 (Sigma Aldrich, St. Louis, MO, USA) for 15 min at room temperature. The nonspecific reaction was blocked with 10% goat serum (Sigma Aldrich, St. Louis, MO, USA) for 1 h at room temperature. The cells prepared in this way were incubated overnight at 4 °C with the primary rabbit polyclonal antibody anti-Ki67 (1:500, Abcam, Cambridge, MA, USA). Subsequently, the cells were incubated with Alexa Fluor 488 secondary antibody (1:1000, Thermo Fisher Scientific, Waltham, MA, USA) for 1 h at room temperature in the dark. Additionally, cell nuclei were stained with Hoechst dye (Sigma-Aldrich, St. Louis, MO, USA). The expression of the proliferation marker Ki67 was evaluated by microscopic observations. The labelled cells were counted from at least 6 independent images (~800 cells per image). The proliferation ability of hWJ-MSCs was expressed as the percentage of Ki67-positive cells to all cell nuclei.

2.4. Colony Forming Unit (CFU) Assay

For the colony forming unit assay, hWJ-MSCs at passage 2 were seeded into 6-well plates at a density of 10 cells/well. The cells were cultured for 14 days in complete medium under 21% O_2 and 5% O_2. To visualise the cell colonies, the cells were washed with PBS (Gibco, Bleiswijk, The Netherlands), fixed with 4% paraformaldehyde solution (Sigma Aldrich, St. Louis, MO, USA), and stained with 0.5% toluidine blue solution (Sigma-Aldrich, St. Louis, MO, USA). Colony forming potency was defined as the percentage of the colony number to the number of seeded cells.

2.5. Senescence-Associated β-Galactosidase Assay

A β-galactosidase assay was performed to evaluate cellular senescence. The activity of the β-galactosidase enzyme (cellular senescence biomarker) was detected with the use of a Senescence Cells Histochemical Staining Kit (Sigma-Aldrich, St. Louis, MO, USA). For the experiment, hWJ-MSCs at passage 5 were used. The cells were seeded into 6-well plates at a density of 2.5×10^3 cells/cm^2 and cultured until they reached 70% confluence. Subsequently, hWJ-MSCs were washed with PBS, fixed with fixation buffer, and then stained with staining solution according to the manufacturers' protocol. The cells prepared in this way were incubated overnight at 37 °C, and then analysed by microscopic observations. The results of this study are presented as the percentage of blue-stained, senescent cells to the general number of analysed cells.

2.6. Mesodermal Differentiation Ability of hWJ-MSCs

The potential of human WJ-MSCs for differentiation into adipogenic, osteogenic, and chondrogenic lineages under 21% O_2 and 5% O_2 was evaluated. For the experiments, the cells at passage 3 were used. In the case of adipogenic and osteogenic differentiation, the cells were seeded into 24-well plates at an initial density of 2×10^3 cells/cm^2 in growth medium. After reaching 50–70% confluence, the growth medium was replaced by appropriate differentiation medium. Differentiation into cartilage tissue cells was performed by the hanging drop method (8×10^4 cells/drop). After 1 h of incubation at 37 °C, the obtained cell aggregates were transferred into 24-well plates with appropriate differentiation medium.

Adipogenesis was induced by culturing hWJ-MSCs for 14 days in commercial adipogenic differentiation medium (Gibco, New York, NY, USA). Differentiation was confirmed by the staining method with Oil Red O (Sigma-Aldrich, St. Louis, MO, USA). Briefly, the cells were fixed with 4% paraformaldehyde solution (Sigma-Aldrich, St. Louis, MO, USA) for 30 min, washed 2 times with distilled water, and then with 60% isopropanol for 5 min (Merck, Darmstad, Germany). Then, the cells were stained for 5 min with Oil Red O (Sigma-Aldrich, St. Louis, MO, USA) to detect the presence of lipid droplets.

For the evaluation of osteogenesis, hWJ-MSCs were cultured for 21 days in commercial osteogenic differentiation medium (Gibco, New York, NY, USA). To verify osteogenic differentiation, the cells were fixed with 4% paraformaldehyde solution (Sigma-Aldrich, St. Louis, MO, USA) for 30 min, washed 2 times with distilled water, and stained with 2% Alizarin Red S (Sigma-Aldrich, St. Louis, MO, USA) for 3 min. Staining with Alizarin Red S allowed observation of the formation of calcium deposits.

Chondrogenic differentiation was performed using a commercial chondrogenic differentiation medium (Gibco, New York, NY, USA) for 14-day cultures. For verification of the differentiation of hWJ-MSCs into chondrocytes, the cells were fixed with 4% paraformaldehyde solution (Sigma Aldrich, St. Louis, MO, USA) for 30 min, washed 2 times with distilled water, and stained with 1% Alcian Blue (Sigma-Aldrich, St. Louis, MO, USA) for 30 min to observe protoglycans.

2.7. Analysis of hWJ-MSCs Secretome

Evaluation of the cytokine profile of Wharton's jelly-derived MSCs cultured under different oxygen conditions (21% O_2 and 5% O_2) was carried out using ELISA and Luminex Multiplex assays (R&D Systems, Minneapolis, MN, USA). The secretion of the following factors was analysed: BDNF, INF-γ, TNF-α, IL-4, IL-6, IL-8, IL-10, IL-13, IL-17, VEGF, HGF, TGF-β, IGF-1, HLA-G, IDO, and PGE2. The following kits were used: LXSAHM-10 (for BDNF, CXCL8/IL-8, HGF, IFN-γ, IL-10, IL-17A, IL-4, IL-6, TNF-α, VEGF-A), DY6030-05 Human Indoleamine 2,3-dioxygenase/IDO DuoSet ELISA, DG100 Human IGF-I Quantikine ELISA Kit, DB100B Human TGF-beta 1 Quantikine ELISA Kit, KGE004B Prostaglandin E2 Parameter Assay Kit, D1300B Human IL-13 Quantikine ELISA Kit, and NBP2-62174 Human HLA G ELISA Kit. Culture supernatants at passages 1 and 5 were used for the experiment. When the cultures reached 70% confluence, the growth medium was replaced with serum-free medium to eliminate factors present in the platelet lysate. After the cells were incubated for 24 h with the new medium, the culture supernatants were harvested, centrifuged and thickened using filtering-thickening columns (Vivaspin 20, 5000 MWCO, Sartorius, Merck, Gloucestershire, United Kingdom). Thickened supernatants were cryopreserved at −80 °C until analysis. Quantitative analysis of TGF-β, IGF-1, IDO, and PGE2 was performed by ELISA (R&D Systems, Minneapolis, MN, USA), while other cytokines were analysed using a Luminex multiplex assay; both assays were performed according to the manufacturers' protocol. Based on the absorbance measurements of protein quantity standards and tested samples (Bradford method), the obtained cytokine levels were calculated in relation to 1 μg of total protein. The results were expressed as pg/1 μg of total protein.

2.8. Determination of the Oxygen Consumption Rate (OCR) and Extracellular Acidification Rate (ECAR)

The metabolic potential of the hWJ-MSCs was evaluated using an Agilent Seahorse XF Energy Cell Phenotype Test Kit (Agilent Technologies, Santa Clara, CA, USA). This method allows one to simultaneously measure the two major energy-producing pathways of the cell, namely, mitochondrial respiration (determination of oxygen consumption rate (OCR)) and glycolysis (determination of extracellular acidification rate (ECAR)), under basal and induced stressed conditions. For this experiment, the cells at passage 2 were seeded into 8-well plates at a density of 5×10^3 cells/well and incubated overnight under 21% O_2 and 5% O_2. Subsequently, the growth medium was replaced with assay medium consisting of Agilent Seahorse XF basal medium, 1 mM pyruvate, 2 mM glutamine, and 10 mM glucose and incubated with the cells for 1 h at 37 °C without CO_2. Then, the OCR and ECAR were determined without the addition (baseline) and with the addition of electron transfer chain inhibitors (stressed). To generate stress conditions, the cells were treated with oligomycin (inhibitor of ATP synthase) and FCCP (mitochondrial uncoupling agent). The cells prepared in this way were analysed using an Agilent Seahorse XFe/XF Analyzer (Agilent Technologies, Santa Clara, CA, USA). Metabolic potential was expressed as the percentage of stressed OCR to baseline OCR and stressed ECAR to baseline ECAR.

2.9. Neural Differentiation Ability of hWJ-MSCs

Evaluation of the spontaneous differentiation potential of hWJ-MSCs towards neural progenitors (NG2, A2B5), glial (GFAP), and neuronal (β-tubIII, DCX, NF-200) cells was performed by immunofluorescence staining. For the experiment, the cells at passage 2 were seeded on poly-L-lysine-coated 24-well plates at a density of 2.5×10^3 cells/cm^2 and cultured under different oxygen conditions (21% O_2 and 5% O_2). After 48 h, the cells were fixed with 4% paraformaldehyde solution (Sigma Aldrich, St. Louis, MO, USA) for 20 min at room temperature and incubated with 0.1% Triton X-100 (Sigma Aldrich, St. Louis, MO, USA) for 15 min at room temperature to permeabilise cell membranes. The nonspecific reaction was blocked with 10% goat serum (Sigma-Aldrich, St. Louis, MO, USA) for 1 h at room temperature. The cells prepared in this way were incubated overnight at 4 °C with the following primary antibodies: polyclonal rabbit anti-NG2 chondroitin sulfate proteoglycan (1:300, Millipore, Burlington, MA, USA), mouse monoclonal anti-β-tubIII, IgG2b (1:1000, Sigma-Aldrich, St. Louis, MO, USA), polyclonal rabbit anti glial fibrillary acidic protein (GFAP, 1:500, Dako, Glostrup, Denmark), mouse monoclonal anti neurofilament 200, IgG1 (NF-200, 1:400, Sigma-Aldrich, St. Louis, MO, USA), mouse monoclonal anti A2B5, IgM (1:500, Millipore, Burlington, MA, USA), and polyclonal rabbit anti-doublecortin (DCX, 1:500, Cell Signaling Technology, Danvers, MA, USA). Subsequently, the antibodies were washed out and the cells were incubated with appropriate secondary antibodies conjugated with Alexa Fluor 488 or Alexa Fluor 546 fluorochromes (1:1000, Thermo Fisher Scientific, Waltham, MA, USA). Additionally, cell nuclei were stained with Hoechst dye (Sigma-Aldrich, St. Louis, MO, USA). Then, the labelled cells were analysed by microscopic observations.

2.10. Neuroprotective Properties of hWJ-MSCs

To analyse the neuroprotective abilities of hWJ-MSCs, coculture of human WJ-MSCs with organotypic rat hippocampal slices (obtained from 7-day-old rat pups) subjected to the transient oxygen-glucose deprivation (OGD) procedure was performed. Rat hippocampi were cut into 400 μm slices using a tissue chopper (McIlwan, Stoelting, Wood Dale, IL, USA) and placed onto the semipermeable membranes (Millicell CM, Millipore, Burlington, MA, USA) located in 6-well plates. A total of 900 μL of culture medium was added in each well. The composition of the medium was as follows: DMEM (Gibco, New York, NY, USA), horse serum (25%; Sigma-Aldrich, St. Louis, MO, USA), HEPES-Buffered Hanks Balanced

Salt Solution (HHBSS, 25%; Gibco, New York, NY, USA), 1 M HEPES (Gibco, New York, NY, USA), 5 mg/mL glucose (Sigma-Aldrich, St. Louis, MO, USA), 1% amphotericin B, and 0.4% penicillin–streptomycin (Gibco, New York, NY, USA). Organotypic hippocampal slices were cultured at 35 °C. After 5 days of culture, when the concentration of the serum was lowered until it achieved a serum-free medium consisting of DMEM (Gibco, New York, NY, USA), HBSS (Gibco, New York, NY, USA), 1 M HEPES (Gibco, New York, NY, USA), glucose (Sigma-Aldrich, St. Louis, MO, USA), and antibiotics (Gibco, New York, NY, USA), the OGD procedure was performed. OGD causes early neuronal death in the CA1 region of the hippocampus. To mimic an ischemic injury, hippocampal slices were kept for 40 min at 35 °C in oxygen-free conditions that were prepared by placing the slices in an anaerobic chamber and saturating with 95% N_2 and 5% CO_2. The neuroprotective effect induced by hWJ-MSCs was determined as the ratio of neuronal death in the CA1 region of the hippocampus after OGD to neuronal death in the CA1 region of OGD-treated hippocampal slices after indirect coculture with hWJ-MSCs. For this purpose, human mesenchymal stem cells at passages 2–3 cultured under both oxygen conditions were seeded into 6-well plates and cocultured with OGD-treated hippocampal slices. After 24 h of coculture, the cells from the hippocampus were stained with 1.4 µg/mL propidium iodide (Sigma-Aldrich, St. Louis, MO, USA) to determine the number of dead cells in the CA1 region. Preparations were evaluated by microscopic observations.

2.11. *Microscopic Observations*

Analysis of the cell morphology, cellular senescence, and mesodermal differentiation potential of the hWJ-MSCs was carried out using a light microscope (Axio Vert. A1, Zeiss, Oberkochen, Germany). The expression of the Ki67 marker was analysed using an inverted fluorescence microscope (AxioVert 200, Zeiss, Oberkochen, Germany). All microscopes were coupled with a CCD camera. The neural differentiation ability and neuroprotective properties of the hWJ-MSCs were evaluated in the Laboratory of Advanced Microscopy Techniques, Mossakowski Medical Research Centre, Polish Academy of Sciences using a confocal microscope (LSM510, Zeiss, Oberkochen, Germany). For data acquisition and analysis, ZEN 2.3 (Zeiss) image analysis software was used.

2.12. *Statistical Analysis*

Each experiment was repeated independently 3–12 times. All obtained data are expressed as the mean ± standard deviation (SD). Statistical analyses were performed using GraphPad Prism 8 software (GraphPad Software, San Diego, CA, USA) and assessed using a one-way analysis of variance (ANOVA). Here, p-values of less than 0.05, 0.001, 0.0001, and 0.00001 were considered statistically significant.

3. Results

3.1. *Morphology and Immunophenotype of hWJ-MSCs*

Human mesenchymal stem cells derived from Wharton's jelly cultured under 21% O_2 and 5% O_2 were characterised by the typical morphology of MSCs (Figure 1A). Differences in morphology were not observed between the culture conditions or between passages and donors. The cells formed an adherent heterogeneous cell population in terms of the size and cytoplasm to nucleus ratio. They also exhibited fibroblast-like shapes during culture. The hWJ-MSCs cultured under both analysed oxygen conditions were positive for mesenchymal markers, such as CD105, CD90, and CD73, and negative for haematopoietic markers, such as CD11b, CD19, CD34, CD45, and HLA-DR (Figure 1B). Significant immunophenotypic differences were not observed between 21% O_2 and 5% O_2.

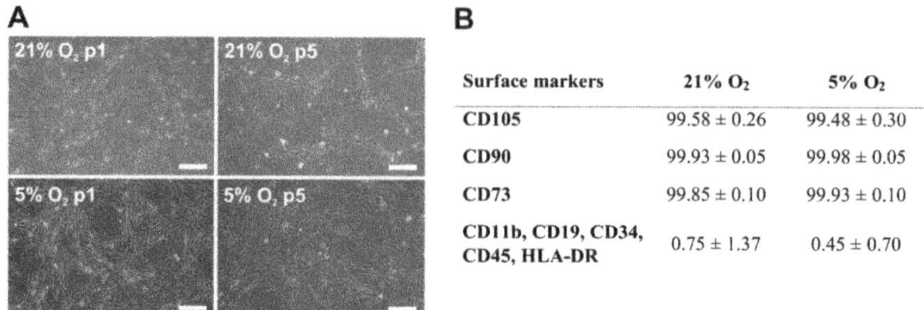

Surface markers	21% O₂	5% O₂
CD105	99.58 ± 0.26	99.48 ± 0.30
CD90	99.93 ± 0.05	99.98 ± 0.05
CD73	99.85 ± 0.10	99.93 ± 0.10
CD11b, CD19, CD34, CD45, HLA-DR	0.75 ± 1.37	0.45 ± 0.70

Figure 1. Morphology of the human Wharton's jelly-derived mesenchymal stem cells (hWJ-MSCs) cultured under 21% O_2 and 5% O_2 (passages 1 and 5). (**A**) Phase contrast images for hWJ-MSC (3) donors are presented as a representative (scale bars: 100 μm). (**B**) Cell immunophenotype results are expressed as the mean ± SD of 4 investigated donors.

3.2. Cell Proliferation

The proliferation of hWJ-MSCs was evaluated by two different methods: WST-1 assay and expression analysis of the proliferation marker Ki67. The results of the WST-1 assay (Figure 2A) showed that the number of cells cultured under different oxygen conditions increased each day of the culture for each investigated passage. Moreover, the cell proliferation rate was higher under 5% O_2 than under 21% O_2 for the cells at passages 4 and 5. At passage 3, a slightly higher cell number was observed under 5% O_2 at days 5–7 of the cultures. However, the differences between 21% O_2 and 5% O_2 were not significant. For each following passage, the differences between the number of cells cultured under 21% and 5% oxygen levels were higher. In the case of passages 4 and 5, the cell proliferation rate was significantly higher under 5% O_2 than under 21% O_2 at days 3–7 of culture. At passage 5, the differences between the number of cells cultured under both oxygen conditions were the highest. Expression analyses of the Ki67 proliferation marker (Figure 2B) indicated that cells at passages 4 and 5 showed higher proliferation ability under 5% O_2 than under 21% O_2. At these passages, a significantly higher percentage of Ki67-positive cells under 5% O_2 was observed. At passages 1–2, the cells showed a similar ability to proliferate under both investigated culture conditions.

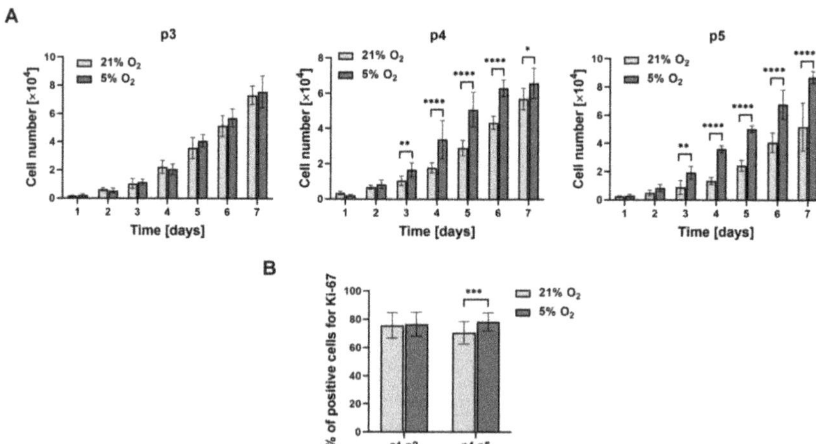

Figure 2. Results of proliferation of the hWJ-MSCs cultured under 21% O_2 and 5% O_2: (**A**) WST-1 assay (passages 3–5) and (**B**) expression of the Ki67 proliferation marker (passages 1–2 and 4–5). All results are expressed as the mean ± SD of 4 investigated donors at 6 replications ($n = 24$). The asterisks denote significant differences (* $p < 0.05$, ** $p < 0.001$, *** $p < 0.0001$, **** $p < 0.00001$) between 21% and 5% oxygen concentrations.

3.3. Clonogenicity and Cellular Senescence of hWJ-MSCs

Analysis of the colony forming potency (Figure 3A) indicated that cells cultured under 5% O_2 exhibited higher clonogenic potential than cells cultured under 21% O_2. For three of the investigated donors (WJ-MSCs (1), WJ-MSCs (3), and WJ-MSCs (4)), the differences between the two oxygen conditions were significant. In the case of one donor (WJ-MSCs (2)), the results were not significantly different. To evaluate cellular senescence, a β-galactosidase assay was performed. The results of this test are presented in Figure 3B and indicated that the hWJ-MSCs showed slight signs of cellular senescence. Moreover, the cultures under 5% O_2 exhibited a significantly lower percentage of senescent cells than the cultures under atmospheric oxygen conditions. The same dependency was observed for all investigated cell donors.

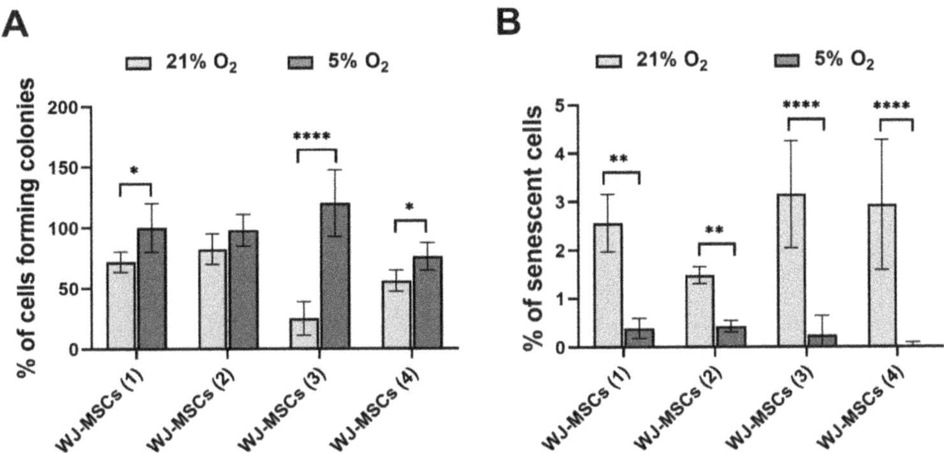

Figure 3. Results of the (**A**) colony forming unit (CFU passage 2) and (**B**) β-galactosidase assays (passage 5) for hWJ-MSCs cultured under 21% O_2 and 5% O_2. All results are expressed as the mean ± SD for the individual donors ($n = 6$ for the CFU assay and $n = 12$ for the β-galactosidase assay). The asterisks denote significant differences (* $p < 0.05$, ** $p < 0.001$, **** $p < 0.00001$) between the 21% and 5% oxygen concentrations.

3.4. Adipogenic, Osteogenic, and Chondrogenic Differentiation Potential of hWJ-MSCs

To evaluate the multilineage differentiation potential of hWJ-MSCs cultured under different oxygen conditions, adipogenesis, osteogenesis, and chondrogenesis were induced. The results (Figure 4) showed that the human mesenchymal stem cells derived from Wharton's jelly exhibited the ability to differentiate into adipocytes, osteocytes, and chondrocytes under both 21% and 5% oxygen concentrations, and differences were not detected between the two culture conditions. Staining with Oil Red O indicated the presence of intracellular lipid droplets, which confirmed adipogenic differentiation. Alizarin Red S staining indicated the formation of calcium deposits, which are characteristic of osteocytes. Alcian Blue staining indicated the occurrence of proteoglycans, which confirmed the ability of the hWJ-MSCs to differentiate into chondrocytes. The same results were observed for all investigated donors.

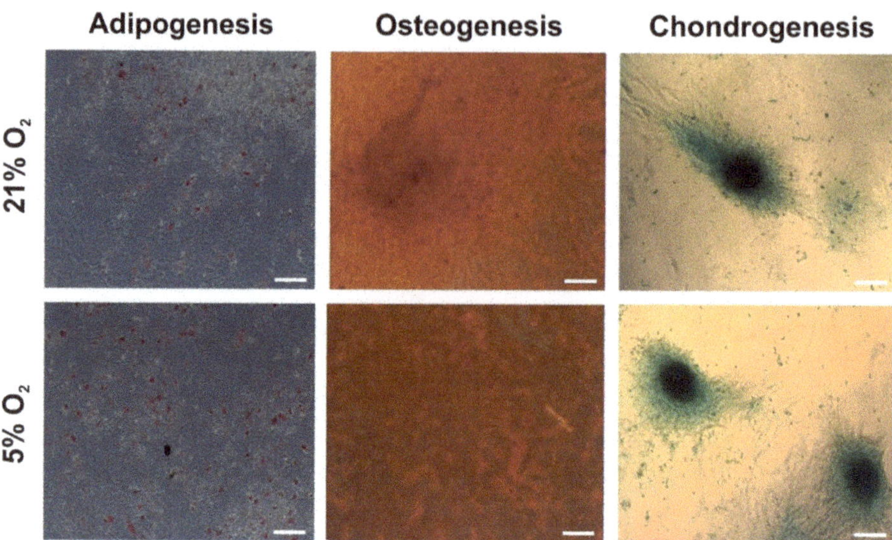

Figure 4. Multilineage differentiation potential of hWJ-MSCs cultured under 21% O_2 and 5% O_2 (passage 3). Phase contrast images of adipogenesis, osteogenesis, and chondrogenesis for WJ-MSC (1) donors are presented as representative (scale bars: 100 μm).

3.5. Secretory Profile of hWJ-MSCs

The influence of different oxygen conditions on cytokine secretion by hWJ-MSCs was also analysed. The analysis of the secretory profile, which is presented in Figure 5, indicated that 5% O_2 promoted an increase in IFN-γ, IL-4, IL-6, HGF, TGF-β, and PGE-2 secretion for all or most investigated donors. The augmented secretion of these cytokines was observed for the cells at passage 1 as well as at passage 5. However, in the case of BDNF, a decrease in secretion under 5% oxygen was detected. This relationship was similar for most analysed donors and both investigated passages. For other analysed cytokines, depending on the cell donor, both an increase and a decrease in their secretion under 5% O_2 were observed. However, due to large differences between donors, the dependence of the secretory profile of the hWJ-MSCs on the oxygen conditions of the cell culture could not be clarified. The results of the hWJ-MSC secretome also showed that cells isolated from all investigated donors secreted predominantly cytokines, such as HGF, IDO, and IL-6. In contrast, TNF-α, IL-10, IL-17, and VEGF secretion by hWJ-MSCs was at the lowest level.

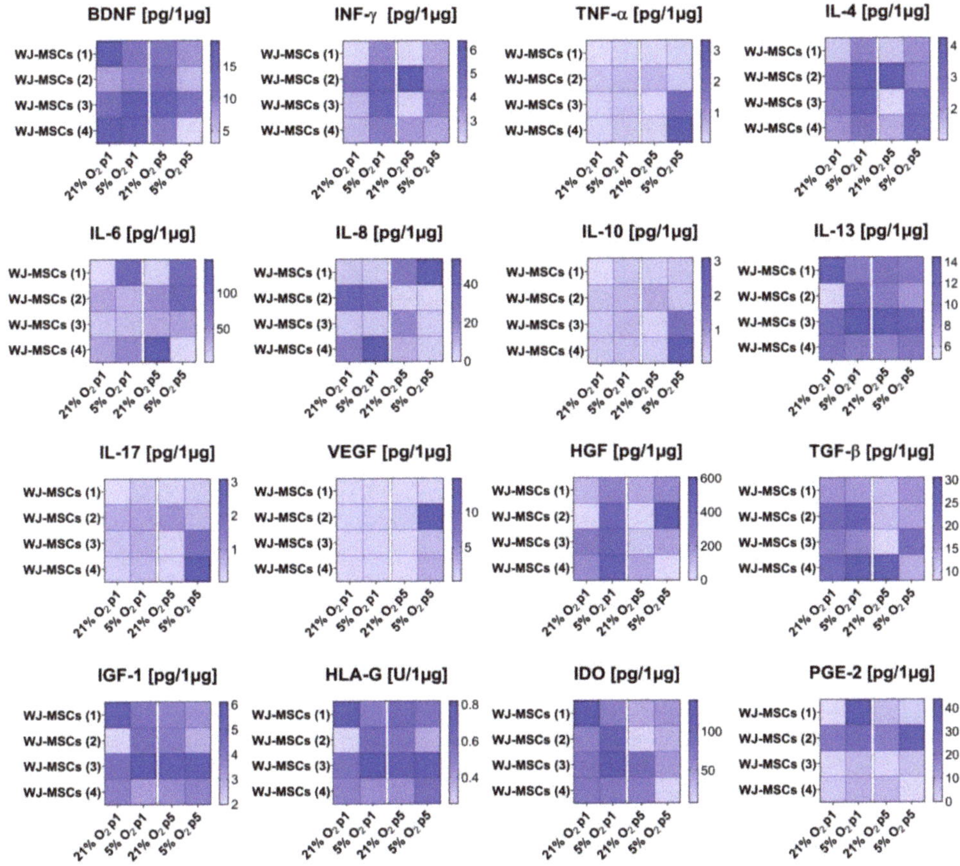

Figure 5. Secretory profile of hWJ-MSCs cultured under 21% O_2 and 5% O_2 (passages 1 and 5). The results are expressed as the mean for the individual donors (n = 4). The colour scale illustrates the relative expression of the analysed cytokines.

3.6. Metabolic Potential of hWJ-MSCs

The analysis of the metabolic potential of the hWJ-MSCs cultured under 21% O_2 and 5% O_2 was based on the oxygen consumption rate (OCR) and extracellular acidification rate (ECAR). The results (Figure 6) indicated a significantly higher oxygen consumption rate than extracellular acidification rate under both oxygen conditions for most of the investigated donors (except for WJ-MSCs (2) and WJ-MSCs (4) donors under 5% O_2). A higher OCR than ECAR suggests that the cells relied more on mitochondrial respiration than on glycolysis. Moreover, mitochondrial respiration was slightly higher under 21% O_2 than under 5% O_2 for most of the investigated donors (in the case of WJ-MSCs donor (1), the difference was significant). Opposite results were detected for glycolysis, which was slightly enhanced under 5% O_2 (for WJ-MSC donor (4), the difference was significant).

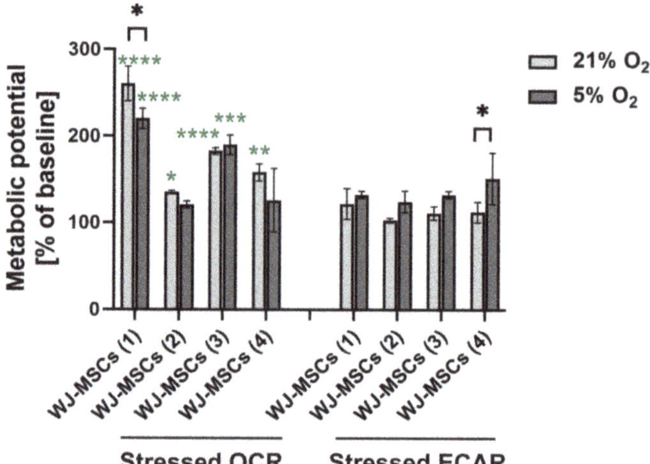

Figure 6. Metabolic potential of the hWJ-MSCs cultured under 21% O_2 and 5% O_2 (passage 2). The results are expressed as the mean ± SD for the individual donors (n = 4). The asterisks denote significant differences (* $p < 0.05$, ** $p < 0.001$, *** $p < 0.0001$, **** $p < 0.00001$) between 21% and 5% oxygen concentrations (black asterisks) as well as between stressed oxygen consumption rate (OCR) and extracellular acidification rate (ECAR) (green asterisks).

3.7. Neural Differentiation Potential of hWJ-MSCs

The potential of human WJ-MSCs for differentiation into neural progenitors and glial and neuronal cells was evaluated. The results (Figure 7) showed that human mesenchymal stem cells derived from Wharton's jelly expressed neural markers under both 21% and 5% oxygen conditions. The cells expressed markers of neural progenitors (NG2, A2B5), glia (GFAP), and neurons (β-tubIII, DCX, NF-200), which proves the ability of hWJ-MSCs to spontaneously differentiate into neurons. An immunocytochemical analysis performed on hWJ-MSCs isolated from four donors indicated a similar tendency to promote neuronal lineage differentiation (enhanced expression of NF-200 and DCX) of the hWJ-MSCs under atmospheric oxygen conditions. The expression of early neural progenitor markers (NG2 and A2B5) and glial marker (GFAP) was comparable under both oxygen conditions.

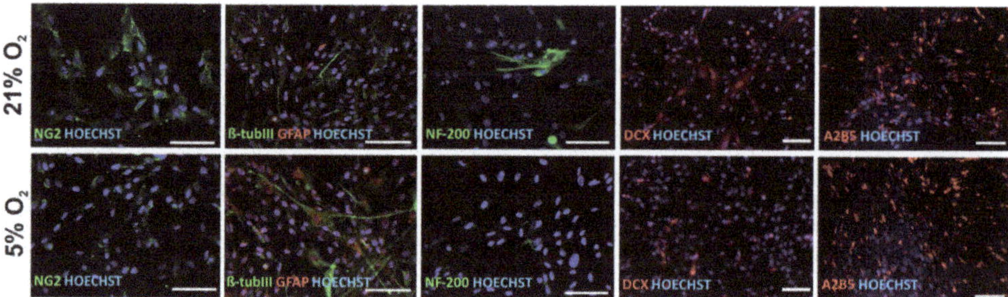

Figure 7. Neural differentiation potential of hWJ-MSCs cultured under 21% O_2 and 5% O_2 (passage 2). Fluorescent images showing the expression of markers of neural progenitors (NG2, A2B5), glial (GFAP), and neuronal (β-tubIII, DCX, NF-200) cells in Wharton's jelly-derived human mesenchymal stem cells. Representative images of the cells isolated from WJ-MSC (2) donor are presented (scale bars: 100 μm).

3.8. Neuroprotective Effect of hWJ-MSCs

To evaluate the neuroprotective properties of hWJ-MSCs, they were co-cultured with organotypic rat hippocampal slices subjected to the transient oxygen-glucose deprivation (OGD) procedure. The analysis of the results, presented in Figure 8, indicated that the hWJ-MSCs exhibited high protective potential in relation to OGD-treated nerve tissue. The percentage of dead cells in the CA1 region was significantly lower in hippocampal slices treated with OGD and cocultured with hWJ-MSCs than in hippocampal slices treated with OGD procedure and cultured without WJ-derived mesenchymal stem cells. No significant differences between the culture oxygen conditions were detected.

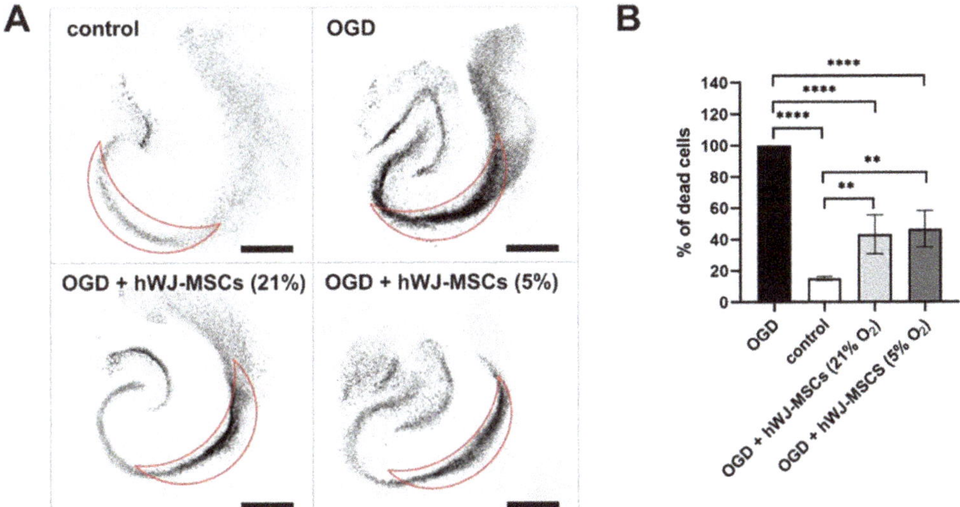

Figure 8. Neuroprotective effect of hWJ-MSCs cultured under 21% and 5% O_2. (**A**) Confocal images of hippocampal slices for 4 investigated groups: control (without OGD), after OGD, after OGD and coculture with hWJ-MSCs from 21% O_2, after OGD and coculture with hWJ-MSCs from 5% O_2 (scale bars: 500 μm). The results for cells isolated from WJ-MSC (3) donor are presented as representative. (**B**) Quantitative evaluation of neuronal death in the CA1 region. The results are expressed as the mean ± SD of 4 investigated donors at 4 replications ($n = 16$). The asterisks denote significant differences (** $p < 0.001$, **** $p < 0.00001$) between the investigated groups.

4. Discussion

Mesenchymal stromal/stem cells (MSCs) have regenerative properties and thus are increasingly used in cell-based therapies. However, the most commonly used sources of MSCs, such as bone marrow and adipogenic tissue, have some limitations [7,8]. MSCs isolated from Wharton's jelly, which is considered medical waste and obtained with no harm to the patient, are becoming a popular source of cells for future applications in regenerative medicine. Therefore, it is important to thoroughly investigate the properties of these cells. Additionally, it is believed that oxygen conditions play an important role in stem cell physiology. Standard in vitro cell cultures under 21% O_2 do not refer to a physiological stem cell niche. In vivo oxygen levels are much lower and range from 2% to 9%, thus representing "physiological normoxia" [23–25]. In our study, we investigated and evaluated human Wharton's jelly-derived MSCs cultured in vitro under different oxygen concentrations (21% O_2 and 5% O_2) and performed analysis of ten different hallmarks: morphology, phenotype, proliferation ability, clonogenicity, senescence, mesodermal differentiation potential, secretory profile, metabolic activity, neural differentiation potential, and neuroprotective effect. Previous studies on human WJ-MSCs' activity under atmospheric and low oxygen levels [3,34,40,57–66] considered only a few cell properties. Moreover, the obtained results are often contradictory.

Similar to previous works, the results of our experiments showed that oxygen levels do not influence the morphology [31,48,53,67] or phenotype of MSCs [27,34,35,42,46,48–54, 57,60,61,65], and these features did not differ between cells cultured under 21% O_2 and 5% O_2. The hWJ-MSCs exhibited similar morphology and expressed surface antigens at similar levels under both investigated oxygen levels. The cells were positive for mesenchymal markers (CD105, CD90, CD73) and negative for haematopoietic markers (CD11b, CD19, CD34, CD45, HLA-DR). However, changes in cell morphology [34,50,65] and surface marker expression [62,67] in MSCs cultured under low oxygen concentrations have been reported. Nekanti et al. [34] indicated that hWJ-MSCs at early and late passages showed higher amounts of large and flattened cells under low oxygen relative to atmospheric oxygen. They explained that enlarging the cell surface probably promotes an increase in the oxygen diffusion rate to the cell. Different results were presented by Drela et al. [64], where hWJ-MSCs cultured under 5% O_2 were smaller in size, had a round shape, and showed a tendency to form colonies compared to cells cultured under 21% O_2. Similar morphological changes were indicated by Ahmed et al. [50]. Kwon et al. [67] observed in MSCs under 21% O_2 enhanced expression of mesenchymal markers and reduced expression of haematopoietic markers under low oxygen. On the other hand, Majumdar et al. [62] detected reduced expression of mesenchymal markers under 21% O_2.

Many studies proved the higher proliferation rate under low oxygen relative to atmospheric oxygen for different types of mesenchymal stem cells derived from: bone marrow (BM-MSCs) [31,36,38,62], adipose tissue (AD-MSCs) [36,44,45,48,49,68], cord blood (CB-MSCs) [28,45,53–55], amniotic fluid (AF-MSCs) [28], and Wharton's jelly (WJ-MSCs) [3,34,40,58,60,65]. Our studies confirmed the previously reported results based on WST-1 assays and expression analyses of the Ki67 marker. They showed that the cell proliferation rate was significantly higher under 5% O_2 than 21% O_2 for the cells at passages 4 and 5. For the cells at passage 3, the differences between 5% O_2 and 21% O_2 were not significant. The analysis of the expression of the Ki67 proliferation marker indicated similar results. At passages 4 and 5, a significantly higher percentage of Ki67-positive cells (higher proliferation ability) was observed under 5% O_2. In the case of early passages (1 and 2), the cells exhibited similar proliferative potential under both investigated culture conditions. In contrast, several works reported a lower proliferation rate of MSCs under low oxygen levels [35,42,44,46], while Roemeling-van Rhijn et al. reported that low oxygen does not influence the proliferation of AD-MSCs [47]. Such differences could result from the different concentrations of low oxygen and the duration of these conditions [23].

We also investigated the clonogenicity and cellular senescence of hWJ-MSCs. Analysis of the colony forming unit (CFU) assay results indicated that cells cultured under "physiological normoxia" exhibited higher clonogenic potential than cells cultured under atmospheric conditions. Such dependency was revealed for hWJ-MSCs [40,61] and for other mesenchymal stem cells [27,30,33,53,54,69]. However, Lee at el. [38] observed reduced colony forming efficiency of hBM-MSCs under low oxygen concentration.

Ageing cells show β-galactosidase enzyme activity, which is regarded to be a biomarker of cellular senescence [70,71]. Our results indicated that senescence (β-galactosidase activity) was inhibited under "physiological normoxia". We observed a significantly lower percentage of senescent cells under 5% O_2 in comparison to cultures under 21% O_2, which suggests that "physiological normoxia" prevents senescence of hWJ-MSCs. Other studies on MSCs [53,67,72–76] confirmed our observations.

The multilineage differentiation potential of hWJ-MSCs is one of the reasons underlying the use of these cells in regenerative medicine [2]. In previous works, several groups demonstrated that low oxygen concentrations stimulated adipogenesis [27,40,43,45,47,63,77], osteogenesis [27,31,41,42,45,61,67,78], and chondrogenesis [30,39–41,46,49,61,69,79] of different types of MSCs. In most of these studies, the influence of 5% O_2 on the differentiation potential of MSCs was investigated. Some of these works evaluated the differentiation ability of hWJ-MSCs under lower (e.g., 3%) oxygen conditions [40,61]. However, few studies have presented opposite results in which low oxygen levels (1–3%) suppress the differentiation of MSCs into adipogenic [37,49,79], osteogenic [37,40,46,49,61], and chondrogenic cells [37,80]. Furthermore, in other works [29,34,39,57], no significant differences in efficient mesodermal differentiation of MSCs between low (1–5%) and atmospheric oxygen levels were found. Similar to these results, our work also showed that human mesenchymal stem cells derived from Wharton's jelly have comparable potential to differentiate into adipogenic, osteogenic, and chondrogenic lineages. No impact of oxygen level on the differentiation potential of the hWJ-MSCs was detected. Such heterogeneous results obtained in the abovementioned studies could be caused by the application of various levels of low oxygen (1–5%) tension and different exposure times to these conditions (72 h–30 days) [23]. It is worth noting the inconsistency across the publications in the terminology used to indicate oxygen level conditions. While the term "hypoxia" should be applied to describe oxygen concentrations lower than 1%, it also frequently refers to oxygen conditions, which are termed "physiological normoxia" in our study [23].

The immunomodulatory properties of MSCs are one of the main factors for using these cells in therapies [23]. MSCs secrete a variety of cytokines and growth factors that can influence tissue homeostasis and repair. According to several researchers [31,33,62], low oxygen increases the secretion level of several of these factors. Moreover, Teixeira et al. [59] indicated that low oxygen conditions led to an increased secretion profile of hWJ-MSCs compared to atmospheric conditions. They identified 104 factors for oxygen levels under 21%, 166 proteins for oxygen levels under 5%, and 81 proteins that were common for both oxygen conditions. In our work, we investigated the secretion of 16 factors. We demonstrated that 5% O_2 promotes the growth of IFN-γ, IL-4, IL-6, HGF, TGF-β, and PGE-2 secretion. In the case of BDNF, a decrease in secretion under 5% oxygen was observed, which is similar to the work of Majumdar et al. [62]. In the work of Lech et al. [81], the mRNA expression of BDNF in hWJ-MSCs cultured in 3D hydrogel scaffolds was also decreased under 5% O_2 in comparison to atmospheric conditions. However, due to the large differences between investigated donors, the dependence of the secretion of other factors on oxygen conditions could not be clarified. Detailed studies should be performed to investigate the influence of individual factors on hWJ-MSC properties.

Stem cells in their physiological environment (low oxygen conditions) rely more on glycolysis than on mitochondrial oxidative phosphorylation, although this relationship changes in favour of oxidative phosphorylation during their differentiation [23,82–84]. However, oxygen conditions significantly influence the process of respiration. When hWJ-MSCs were exposed to atmospheric oxygen conditions, glycolysis decreased in favour of mitochondrial respiration. Our analysis of the metabolic activity of the hWJ-MSCs confirmed that the cells cultured under 21% O_2 relied more on mitochondrial respiration than on glycolysis. Moreover, mitochondrial respiration was slightly higher under 21% O_2 than under 5% O_2 for most of the investigated donors. Opposite results were detected for glycolysis, which was slightly enhanced under low oxygen concentrations. Lavrentieva et al. [55] and Dos Santos et al. [38] also observed higher consumption of glucose by MSCs under low oxygen than atmospheric conditions.

The spontaneous differentiation of hWJ-MSCs towards neural lineages observed in the current study confirmed our previous findings [18,22,64,66,81]. Drela et al. [18] described enhanced neuronal and glial differentiation of hWJ-MSCs as compared to hBM-MSCs cultured under 21% O_2. In the current study early neural (NG2, A2B5), neuronal (β-Tubulin III, NF-200, DCX) and glial (GFAP) markers were expressed with no significant differences between oxygen conditions. However, an enhanced tendency for neuronal differentiation under 21% O_2 was observed. This was in line with observed tendency of hWJ-MSCs neuronal lineage commitment in 21% O_2 and 3D hydrogel scaffolds, but only at the protein level. The transcriptomic data previously obtained by our group [81] proved that lowered oxygen together with 3D hydrogel scaffolds together enhance the expression of neuronal markers. These results clearly show the importance of setting up appropriate in vitro biomimetic conditions (3D and oxygen tension) to obtain conclusive results regarding hWJ-MSC differentiation potential.

This unique feature of spontaneous differentiation of mesenchymal stem cells derived from afterbirth tissue towards neural cells favours the hypothesis about the regenerative potential of hWJ-MSCs; however, the ability of MSCs to differentiate into fully mature neurons and to functionally integrate with injured nerve tissue is still not proved.

The therapeutic effect of hWJ-MSCs is related mostly to their ability to secrete trophic and protective factors [22,85]. In this work, we confirmed this effect via the indirect (noncontact) coculture of these cells with damaged nervous tissue. Despite the lack of cell–cell contact, hWJ-MSCs significantly reduced apoptosis of neurons in the hippocampal CA1 region via secretion of neuroprotective factors in both tested oxygen conditions (21% O_2 and 5% O_2). A similar neuroprotective effect independent of the oxygen concentration was also observed by Lech et al. [81] for hWJ-MSCs cultured in 3D hydrogel scaffolds. Puig-Pijuan et al. also proved potential neuroprotective and antioxidants effects of WJ-MSCs on hippocampal cultures [86].

5. Conclusions

We characterised human mesenchymal stem cells derived from Wharton's jelly (hWJ-MSCs) and evaluated the influence of different oxygen conditions on their properties. The results indicated that regardless of the oxygen level, hWJ-MSCs maintained the typical morphology and phenotype of mesenchymal stem cells according to the position statement of the International Society for Cellular Therapy [1]. Moreover, they have a higher proliferation rate and clonogenic potential under 5% O_2 than under atmospheric conditions. The hWJ-MSCs displayed signs of cellular senescence during in vitro cultures; however, this process was reduced under "physiological normoxia". The investigated cells have a similar ability to differentiate towards adipogenic, osteogenic, and chondrogenic lineages, and they present similar neuroprotective effects under both oxygen culture conditions. The ability of hWJ-MSCs to spontaneously differentiate into neuronal lineages was observed under 21% and 5% O_2. Furthermore, cells rely more on mitochondrial respiration than on glycolysis under both oxygen conditions. However, the significance of these findings should be confirmed in preclinical trials conducted under defined conditions.

Based on our results, we can conclude that hWJ-MSCs are suitable as a cell source for application in regenerative medicine and display neuroprotective effect regardless of oxygen culture conditions. However, "physiological normoxia" favourably influences some regenerative properties of hWJ-MSCs and may enhance their therapeutic potential. Therefore, preconditioning cells under 5% O_2 before transplantation might be beneficial. However, preclinical and clinical validation of our findings is required.

Author Contributions: E.T. and W.L. are equivalent first authors. Conceptualization: K.D.-J. and L.B.; methodology: W.L., M.Z., A.S.; formal analysis: M.Z., E.T., N.R.; investigation: W.L, M.Z. and A.S.; resources: M.M., T.O. and N.R.; writing—original draft preparation, E.T., W.L., M.Z.; writing—review and editing, L.B., N.R. and T.O.; visualization, E.T., M.Z.; supervision: K.D.-J., L.B. and N.R.; project administration: T.O.; funding acquisition: T.O. and L.B. All authors have read and agreed to the published version of the manuscript.

Funding: This work was financially supported by National Center for Research and Development within a frame of project "Development of the advanced therapy product based on the mesenchymal cells from Wharton's jelly in the treatment of amyotrophic lateral sclerosis" No. POIR.01.02.00-00-00024/17-00 (T.O). Statutory funds to MMRI PAS (L.B.); MM is supported by Ministry of Science and Higher Education Project No. DWD/3/27/2019.

Institutional Review Board Statement: The study was conducted according to the guidelines of the Declaration of Helsinki and approved by the Ministry of Health and National Center for Tissues and Cells Banking (SZTR.4061.9.2019.MN).

Informed Consent Statement: Informed consent was obtained from all subjects involved in the study.

Data Availability Statement: The data presented in this study are available in the article.

Conflicts of Interest: T.O. is a board member and an employee of Polish Stem Cell Bank, FamiCord; ET and MM are an employees of Polish Stem Cell Bank, FamiCord Group; NR has been involved as a consultant and expert in Polish Stem Cell Bank, FamiCord Group.

References

1. Dominici, M.; Blanc, K.L.; Mueller, I.; Slaper-Cortenbach, I.; Marini, F.; Krause, D.; Deans, R.; Keating, A.; Prockop, D.; Horwitz, E. Minimal criteria for defining multipotent mesenchymal stromal cells. The International Society for Cellular Therapy position statement. *Cytotherapy* **2006**, *8*, 315–317. [CrossRef] [PubMed]
2. Widowati, W.; Rihibiha, D.D.; Khiong, K.; Widodo, M.A.; Sumitro, S.B.; Bachtiar, I. Hypoxia in Mesenchymal Stem Cell. In *Hypoxia and Human Diseases*; InTech: London, UK, 2017; pp. 91–115.
3. Shuvalova, N.S.; Kordium, V.A. Comparison of proliferative activity of Wharton jelly mesenchymal stem cells in cultures under various gas conditions. *Biopolym. Cell* **2015**, *31*, 233–239. [CrossRef]
4. Abumaree, M.H.; Abomaray, F.M.; Alshehri, N.A.; Almutairi, A.; AlAskar, A.S.; Kalionis, B.; Al Jumah, M.A. Phenotypic and Functional Characterization of Mesenchymal Stem/Multipotent Stromal Cells from Decidua Parietalis of Human Term Placenta. *Reprod. Sci.* **2016**, *23*, 1193–1207. [CrossRef]
5. Berebichez-Fridman, R.; Montero-Olvera, P.R. Sources and Clinical Applications of Mesenchymal Stem Cells: State-of-the-art review. *Sultan Qaboos Univ. Med. J.* **2018**, *18*, e264–e277. [CrossRef]
6. Mushahary, D.; Spittler, A.; Kasper, C.; Weber, V.; Charwat, V. Isolation, cultivation, and characterization of human mesenchymal stem cells. *Cytometry* **2018**, *93*, 19–31. [CrossRef] [PubMed]
7. Wang, Y.; Yuan, M.; Guo, Q.Y.; Lu, S.B.; Peng, J. Mesenchymal stem cells for treating articular cartilage defects and osteoarthritis. *Cell Transplant.* **2015**, *24*, 1661–1678. [CrossRef] [PubMed]
8. Heidari, B.; Shirazi, A.; Akhondi, M.M.; Hassanpour, H.; Behzadi, B.; Naderi, M.M.; Sarvari, A.; Borjian, S. Comparison of proliferative and multilineage differentiation potential of sheep mesenchymal stem cells derived from bone marrow, liver, and adipose tissue. *Avicenna J. Med. Biotechnol.* **2013**, *5*, 104–117. [PubMed]
9. Squillaro, T.; Peluso, G.; Galderisi, U. Clinical Trials with Mesenchymal Stem Cells: An Update. *Cell Transplant.* **2016**, *25*, 829–848. [CrossRef] [PubMed]
10. Brown, C.; McKee, C.; Bakshi, S.; Walker, K.; Hakman, E.; Halassy, S.; Svinarich, D.; Dodds, R.; Govind, C.K.; Chaudhry, G.R. Mesenchymal stem cells: Cell therapy and regeneration potential. *J. Tissue Eng. Regen. Med.* **2019**, *13*, 1738–1755. [CrossRef]
11. Xu, L.; Liu, Y.; Sun, Y.; Wang, B.; Xiong, Y.; Lin, W.; Wei, Q.; Wang, H.; He, W.; Wang, B.; et al. Tissue source determines the differentiation potentials of mesenchymal stem cells: A comparative study of human mesenchymal stem cells from bone marrow and adipose tissue. *Stem Cell Res. Ther.* **2017**, *8*, 275. [CrossRef]
12. Fossett, E.; Khan, W.S.; Pastides, P.; Adesida, A.B. The Effects of Ageing on Proliferation Potential, Differentiation Potential and Cell Surface Characterisation of Human Mesenchymal Stem Cells. *Curr. Stem Cell Res. Ther.* **2012**, *7*, 282–286. [CrossRef]
13. Zaim, M.; Karaman, S.; Cetin, G.; Isik, S. Donor age and long-term culture affect differentiation and proliferation of human bone marrow mesenchymal stem cells. *Ann. Hematol.* **2012**, *91*, 1175–1186. [CrossRef] [PubMed]
14. Troyer, D.L.; Weiss, M.L. Concise Review: Wharton's Jelly-Derived Cells Are a Primitive Stromal Cell Population. *Stem Cells* **2008**, *26*, 591–599. [CrossRef]
15. Can, A.; Karahuseyinoglu, S. Concise Review: Human Umbilical Cord Stroma with Regard to the Source of Fetus-Derived Stem Cells. *Stem Cells* **2007**, *25*, 2886–2895. [CrossRef] [PubMed]
16. Hass, R.; Kasper, C.; Böhm, S.; Jacobs, R. Different populations and sources of human mesenchymal stem cells (MSC): A comparison of adult and neonatal tissue-derived MSC. *Cell Commun. Signal.* **2011**, *9*, 1–14. [CrossRef]
17. Eve, D.J.; Sanberg, P.R.; Buzanska, L.; Sarnowska, A.; Domanska-Janik, K. Human Somatic Stem Cell Neural Differentiation Potential. In *Human Neural Stem Cells From Generation to Differentiation and Application*, 1st ed.; Buzanska, L., Ed.; Springer: Berlin/Heidelberg, Germany, 2018; Volume 66, pp. 21–87.
18. Drela, K.; Lech, W.; Figiel-Dabrowska, A.; Zychowicz, M.; Mikula, M.; Sarnowska, A.; Domanska-Janik, K. Enhanced neuro-therapeutic potential of Wharton's Jelly-derived mesenchymal stem cells in comparison with bone marrow mesenchymal stem cells culture. *Cytotherapy* **2016**, *18*, 497–509. [CrossRef] [PubMed]
19. Paladino, F.V.; de Moraes Rodrigues, J.; da Silva, A.; Goldberg, A.C. The Immunomodulatory Potential of Wharton's Jelly Mesenchymal Stem/Stromal Cells. *Stem Cells Int.* **2019**, *2019*, 3548917.
20. Forraz, N.; Mcguckin, C.P. The umbilical cord: A rich and ethical stem cell source to advance regenerative medicine. *Cell Prolif.* **2011**, *44*, 60–69. [CrossRef] [PubMed]
21. Li, F.; Zhang, K.; Liu, H.; Yang, T.; Xiao, D.J.; Wang, Y.S. The neuroprotective effect of mesenchymal stem cells is mediated through inhibition of apoptosis in hypoxic ischemic injury. *World J. Pediatrics* **2020**, *16*, 193–200. [CrossRef] [PubMed]
22. Dabrowska, S.; Sypecka, J.; Jablonska, A.; Strojek, L.; Wielgos, M.; Domanska-Janik, K.; Sarnowska, A. Neuroprotective Potential and Paracrine Activity of Stromal Vs. Culture Expanded hMSC Derived from Wharton Jelly under Co-Cultured with Hippocampal Organotypic Slices. *Mol. Neurobiol.* **2018**, *55*, 6021–6036. [CrossRef]
23. Mas-Bargues, C.; Sanz-Ros, J.; Román-Domínguez, A.; Inglés, M.; Gimeno-Mallench, L.; El Alami, M.; Viña-Almunia, J.; Gambini, J.; Viña, J.; Borrás, C. Relevance of oxygen concentration in stem cell culture for regenerative medicine. *Int. J. Mol. Sci.* **2019**, *20*, 1195. [CrossRef]

24. Kolf, C.M.; Cho, E.; Tuan, R.S. Mesenchymal stromal cells. Biology of adult mesenchymal stem cells: Regulation of niche, self-renewal and differentiation. *Arthritis Res. Ther.* **2007**, *9*, 1–10. [CrossRef]
25. Caplan, A.I. Adult mesenchymal stem cells for tissue engineering versus regenerative medicine. *J. Cell. Physiol.* **2007**, *213*, 341–347. [CrossRef]
26. Ward, J.P. Oxygen sensors in context. *Biochim. Biophys. Acta Bioenerg.* **2008**, *1777*, 1–14. [CrossRef]
27. Basciano, L.; Nemos, C.; Foliguet, B.; de Isla, N.; de Carvalho, M.; Tran, N.; Dalloul, A. Long term culture of mesenchymal stem cells in hypoxia promotes a genetic program maintaining their undifferentiated and multipotent status. *BMC Cell Biol.* **2011**, *12*, 1–12. [CrossRef]
28. Dionigi, B.; Ahmed, A.; Pennington, E.C.; Zurakowski, D.; Fauza, D.O. A comparative analysis of human mesenchymal stem cell response to hypoxia in vitro: Implications to translational strategies. *J. Pediatric Surg.* **2014**, *49*, 915–918. [CrossRef] [PubMed]
29. Fotia, C.; Massa, A.; Boriani, F.; Baldini, N.; Granchi, D. Hypoxia enhances proliferation and stemness of human adipose-derived mesenchymal stem cells. *Cytotechnology* **2015**, *67*, 1073–1084. [CrossRef]
30. Adesida, A.B.; Mulet-Sierra, A.; Jomha, N.M. Hypoxia mediated isolation and expansion enhances the chondrogenic capacity of bone marrow mesenchymal stromal cells. *Stem Cell Res. Ther.* **2012**, *3*, 1–12. [CrossRef] [PubMed]
31. Hung, S.P.; Ho, J.H.; Shih, Y.R.V.; Lo, T.; Lee, O.K. Hypoxia promotes proliferation and osteogenic differentiation potentials of human mesenchymal stem cells. *J. Orthop. Res.* **2012**, *30*, 260–266. [CrossRef] [PubMed]
32. Bornes, T.D.; Jomha, N.M.; Mulet-Sierra, A.; Adesida, A.B. Hypoxic culture of bone marrow-derived mesenchymal stromal stem cells differentially enhances in vitro chondrogenesis within cell-seeded collagen and hyaluronic acid porous scaffolds. *Stem Cell Res. Ther.* **2015**, *6*, 1–17. [CrossRef] [PubMed]
33. Antebi, B.; Rodriguez, L.A.; Walker, K.P.; Asher, A.M.; Kamucheka, R.M.; Alvarado, L.; Mohammadipoor, A.; Cancio, L.C. Short-term physiological hypoxia potentiates the therapeutic function of mesenchymal stem cells. *Stem Cell Res. Ther.* **2018**, *9*, 1–15. [CrossRef] [PubMed]
34. Nekanti, U.; Dastidar, S.; Venugopal, P.; Totey, S.; Ta, M. Increased proliferation and analysis of differential gene expression in human Wharton's jelly-derived mesenchymal stromal cells under hypoxia. *Int. J. Biol. Sci.* **2010**, *6*, 499–512. [CrossRef] [PubMed]
35. Ranera, B.; Remacha, A.R.; Álvarez-Arguedas, S.; Romero, A.; Vázquez, F.J.; Zaragoza, P.; Martín-Burriel, I.; Rodellar, C. Effect of hypoxia on equine mesenchymal stem cells derived from bone marrow and adipose tissue. *BMC Vet. Res.* **2012**, *8*, 2–13. [CrossRef] [PubMed]
36. Burian, E.; Probst, F.; Palla, B.; Riedel, C.; Saller, M.M.; Cornelsen, M.M.; König, F.; Schieker, M.; Otto, S. Effect of hypoxia on the proliferation of porcine bone marrow-derived mesenchymal stem cells and adipose-derived mesenchymal stem cells in 2- and 3-dimensional culture. *J. Cranio-Maxillofac. Surg.* **2017**, *45*, 414–419. [CrossRef]
37. Cicione, C.; Muiños-López, E.; Hermida-Gómez, T.; Fuentes-Boquete, I.; Díaz-Prado, S.; Blanco, F.J. Effects of severe hypoxia on bone marrow mesenchymal stem cells differentiation potential. *Stem Cells Int.* **2013**, *2013*, 1–11. [CrossRef]
38. Dos Santos, F.; Andrade, P.Z.; Boura, J.S.; Abecasis, M.M.; Da Silva, C.L.; Cabral, J.M.S. Ex vivo expansion of human mesenchymal stem cells: A more effective cell proliferation kinetics and metabolism under hypoxia. *J. Cell. Physiol.* **2010**, *223*, 27–35. [CrossRef] [PubMed]
39. Ranera, B.; Remacha, A.R.; Álvarez-Arguedas, S.; Romero, A.; Vázquez, F.J.; Zaragoza, P.; Martín-Burriel, I.; Rodellar, C. Expansion under hypoxic conditions enhances the chondrogenic potential of equine bone marrow-derived mesenchymal stem cells. *Vet. J.* **2013**, *195*, 248–251. [CrossRef] [PubMed]
40. Reppel, L.; Margossian, T.; Yaghi, L.; Moreau, P.; Mercier, N.; Leger, L.; Hupont, S.; Stoltz, J.-F.; Bensoussan, D.; Huselstein, C. Hypoxic Culture Conditions for Mesenchymal Stromal/Stem Cells from Wharton's Jelly: A Critical Parameter to Consider in a Therapeutic Context. *Curr. Stem Cell Res. Ther.* **2014**, *9*, 306–318. [CrossRef] [PubMed]
41. Lee, J.S.; Park, J.C.; Tae-Wan, K.; Kim, T.W.; Jung, B.J.; Lee, Y.; Shim, E.K.; Park, S.; Choi, E.Y.; Cho, K.S.; et al. Human bone marrow stem cells cultured under hypoxic conditions present altered characteristics and enhanced in vivo tissue regeneration. *Bone* **2015**, *78*, 34–45. [CrossRef] [PubMed]
42. Valorani, M.G.; Montelatici, E.; Germani, A.; Biddle, A.; Alessandro, D.D.; Strollo, R.; Patrizi, M.P.; Lazzari, L.; Nye, E.; Otto, W.R.; et al. Pre-culturing human adipose tissue mesenchymal stem cells under hypoxia increases their adipogenic and osteogenic differentiation potentials. *Cell Prolif.* **2012**, *45*, 225–238. [CrossRef] [PubMed]
43. Zubillaga, V.; Alonso-varona, A.; Fernandes, S.C.M.; Salaberria, A.M.; Palomares, T. Adipose-derived mesenchymal stem cell chondrospheroids cultured in hypoxia and a 3D porous chitosan/chitin nanocrystal scaffold as a platform for cartilage tissue engineering. *Int. J. Mol. Sci.* **2020**, *21*, 1004. [CrossRef] [PubMed]
44. Efimenko, A.; Starostina, E.; Kalinina, N.; Stolzing, A. Angiogenic properties of aged adipose derived mesenchymal stem cells after hypoxic conditioning. *J. Transl. Med.* **2011**, *9*, 1–13. [CrossRef]
45. Pan, Q.; Wang, D.; Chen, D.; Sun, Y.; Feng, X.; Shi, X.; Xu, Y.; Luo, X.; Yu, J.; Li, Y.; et al. Characterizing the effects of hypoxia on the metabolic profiles of mesenchymal stromal cells derived from three tissue sources using chemical isotope labeling liquid chromatography-mass spectrometry. *Cell Tissue Res.* **2019**, *380*, 79–91. [CrossRef]
46. Shell, K.; Raabe, O.; Freitag, C.; Ohrndorf, A.; Christb, H.-J.; Wenisch, S.; Arnhold, S. Comparison of Equine Adipose Tissue-Derived Stem Cell Behavior and Differentiation Potential Under the Influence of 3% and 21% Oxygen Tension. *J. Equine Vet. Sci.* **2013**, *33*, 74–82. [CrossRef]

47. Roemeling-van Rhijn, M.; Mensah, F.K.F.; Korevaar, S.S.; Leijs, M.J.; van Osch, G.J.V.M.; Ijzermans, J.M.N.; Betjes, M.G.H.; Baan, C.C.; Weimar, W.; Hoogduijn, M.J. Effects of hypoxia on the immunomodulatory properties of adipose tissue-derived mesenchymal stem cells. *Front. Immunol.* **2013**, *4*, 1–8. [CrossRef]
48. Choi, J.R.; Pingguan-Murphy, B.; Abas, W.A.B.W.; Azmi, M.A.N.; Omar, S.Z.; Chua, K.H.; Safwani, W.K.Z.W. Hypoxia Promotes Growth and Viability of Human Adipose-Derived Stem Cells with Increased Growth Factors Secretion. *J. Asian Sci. Res.* **2014**, *4*, 328–338.
49. Choi, J.R.; Pingguan-Murphy, B.; Abas, W.A.B.W.; Azmi, M.A.N.; Omar, S.Z.; Chua, K.H.; Safwani, W.K.Z.W. Impact of low oxygen tension on stemness, proliferation and differentiation potential of human adipose-derived stem cells. *Biochem. Biophys. Res. Commun.* **2014**, *448*, 218–224. [CrossRef] [PubMed]
50. Ahmed, N.E.M.B.; Murakami, M.; Kaneko, S.; Nakashima, M. The effects of hypoxia on the stemness properties of human dental pulp stem cells (DPSCs). *Sci. Rep.* **2016**, *6*, 35476. [CrossRef] [PubMed]
51. Iida, K.; Takeda-Kawaguchi, T.; Tezuka, Y.; Kunisada, T.; Shibata, T.; Tezuka, K. Hypoxia enhances colony formation and proliferation but inhibits differentiation of human dental pulp cells. *Arch. Oral Biol.* **2010**, *55*, 648–654. [CrossRef] [PubMed]
52. Aranha, A.M.F.; Zhang, Z.; Neiva, K.G.; Costa, C.A.S.; Hebling, J.; Nör, J.E. Hypoxia Enhances the Angiogenic Potential of Human Dental Pulp Cells. *J. Endod.* **2010**, *36*, 1633–1637. [CrossRef]
53. Peng, L.; Shu, X.; Lang, C.; Yu, X. Effects of hypoxia on proliferation of human cord blood-derived mesenchymal stem cells. *Cytotechnology* **2016**, *68*, 1615–1622. [CrossRef]
54. Kheirandish, M.; Gavgani, S.P.; Samiee, S. The effect of hypoxia preconditioning on the neural and stemness genes expression profiling in human umbilical cord blood mesenchymal stem cells. *Transfus. Apher. Sci.* **2017**, *56*, 392–399. [CrossRef]
55. Lavrentieva, A.; Majore, I.; Kasper, C.; Hass, R. Effects of hypoxic culture conditions on umbilical cord-derived human mesenchymal stem cells. *Cell Commun. Signal.* **2010**, *8*, 1–9. [CrossRef]
56. Naaldijk, Y.; Johnson, A.A.; Ishak, S.; Meisel, H.J.; Hohaus, C.; Stolzing, A. Migrational changes of mesenchymal stem cells in response to cytokines, growth factors, hypoxia, and aging. *Exp. Cell Res.* **2015**, *338*, 97–104. [CrossRef] [PubMed]
57. Bao, Y.; Huang, S.; Zhao, Z. Comparison of Different Culture Conditions for Mesenchymal Stem Cells from Human Umbilical Cord Wharton's Jelly for Stem Cell Therapy. *Turkish J. Hematol.* **2020**, *37*, 67–69. [CrossRef] [PubMed]
58. Balgi-Agarwal, S.; Winter, C.; Corral, A.; Mustafa, S.B.; Hornsby, P.; Moreira, A. Comparison of Preterm and Term Wharton's Jelly-Derived Mesenchymal Stem Cell Properties in Different Oxygen Tensions. *Cells Tissues Organs* **2018**, *205*, 137–150. [CrossRef]
59. Teixeira, F.G.; Panchalingam, K.M.; Anjo, S.I.; Manadas, B.; Pereira, R.; Sousa, N.; Salgado, A.J.; Behie, L.A. Do hypoxia/normoxia culturing conditions change the neuroregulatory profile of Wharton Jelly mesenchymal stem cell secretome? *Stem Cell Res. Ther.* **2015**, *6*, 1–14. [CrossRef] [PubMed]
60. Widowati, W.; Wijay, L.; Bachtiar, I.; Gunanegara, R.F.; Sugeng, S.U.; Irawan, Y.A.; Sumitrod, S.B.; Widodoe, M.A. Effect of oxygen tension on proliferation and characteristics of Wharton's jelly-derived mesenchymal stem cells. *Biomark. Genom. Med.* **2014**, *6*, 43–48. [CrossRef]
61. Obradovic, H.; Krstic, J.; Trivanovic, D.; Mojsilovic, S.; Okic, I.; Kukolj, T.; Ilic, V.; Jaukovic, A.; Terzic, M.; Bugarski, D. Improving stemness and functional features of mesenchymal stem cells from Wharton's jelly of a human umbilical cord by mimicking the native, low oxygen stem cell niche. *Placenta* **2019**, *82*, 25–34. [CrossRef] [PubMed]
62. Majumdar, D.; Bhonde, R.; Datta, I. Influence of ischemic microenvironment on human Wharton's Jelly mesenchymal stromal cells. *Placenta* **2013**, *34*, 642–649. [CrossRef] [PubMed]
63. Jiang, C.; Sun, J.; Dai, Y.; Cao, P.; Zhang, L.; Peng, S.; Zhou, Y.; Li, G.; Tang, J.; Xiang, J. HIF-1A and C/EBPs transcriptionally regulate adipogenic differentiation of bone marrow-derived MSCs in hypoxia. *Stem Cell Res. Ther.* **2015**, *6*, 1–14. [CrossRef] [PubMed]
64. Drela, K.; Sarnowska, A.; Siedlecka, P.; Szablowska-Gadomska, I.; Wielgos, M.; Jurga, M.; Lukomska, B.; Domanska-Janik, K. Low oxygen atmosphere facilitates proliferation and maintains undifferentiated state of umbilical cord mesenchymal stem cells in an hypoxia inducible factor-dependent manner. *Cytotherapy* **2014**, *16*, 881–892. [CrossRef]
65. Musial-Wysocka, A.; Kot, M.; Sukowski, M.; Badyra, B.; Majka, M. Molecular and functional verification of wharton's jelly mesenchymal stem cells (WJ-MSCs) pluripotency. *Int. J. Mol. Sci.* **2019**, *20*, 1807. [CrossRef]
66. Lech, W.; Figiel-Dabrowska, A.; Sarnowska, A.; Drela, K.; Obtulowicz, P.; Noszczyk, B.H.; Buzanska, L.; Domanska-Janik, K. Phenotypic, Functional, and Safety Control at Preimplantation Phase of MSC-Based Therapy. *Stem. Cell Int.* **2016**, *2016*, 2514917. [CrossRef]
67. Kwon, S.Y.; Chun, S.Y.; Ha, Y.-S.; Kim, D.H.; Kim, J.; Song, P.H.; Kim, H.T.; Yoo, E.S.; Kim, B.S.; Kwon, T.G. Hypoxia Enhances Cell Properties of Human Mesenchymal Stem Cells. *Tissue Eng. Regen. Med.* **2017**, *14*, 595–604. [CrossRef]
68. Kim, J.H.; Song, S.Y.; Park, S.G.; Song, S.U.; Xia, Y.; Sung, J.H. Primary involvement of NADPH oxidase 4 inhypoxia-induced generation of reactive oxygen species in adipose-derived stem cells. *Stem. Cells Dev.* **2012**, *21*, 2212–2221. [CrossRef]
69. Bae, H.C.; Park, H.J.; Wang, S.Y.; Yang, H.R.; Lee, M.C.; Han, H.S. Hypoxic condition enhances chondrogenesis in synovium-derived mesenchymal stem cells. *Biomater. Res.* **2018**, *26*, 22–28. [CrossRef]
70. Hall, B.M.; Balan, V.; Gleiberman, A.S.; Strom, E.; Krasnov, P.; Virtuoso, L.P.; Rydkina, E.; Vujcic, S.; Balan, K.; Gitlin, I.; et al. Aging of mice is associated with p16(Ink4a)- and β-galactosidase-positive macrophage accumulation that can be induced in young mice by senescent cells. *Aging* **2016**, *8*, 1294–1315. [CrossRef] [PubMed]

71. Hall, B.M.; Balan, V.; Gleiberman, A.S.; Strom, E.; Krasnov, P.; Virtuoso, L.P.; Rydkina, E.; Vujcic, S.; Balan, K.; Gitlin, I.; et al. p16(Ink4a) and senescence-associated β-galactosidase can be induced in macrophages as part of a reversible response to physiological stimuli. *Aging* **2017**, *9*, 1867. [CrossRef] [PubMed]
72. Kim, D.S.; Ko, Y.J.; Lee, M.W.; Park, H.J.; Park, Y.J.; Kim, D.I.; Sung, K.W.; Koo, H.H.; Yoo, K.H. Effect of lowoxygen tension on the biological characteristics of human bone marrow mesenchymal stem cells. *Cell Stress Chaperones* **2016**, *21*, 1089–1099. [CrossRef] [PubMed]
73. Tsai, C.C.; Chen, Y.J.; Yew, T.L.; Chen, L.L.; Wang, J.Y.; Chiu, C.H.; Hung, S.C. Hypoxia inhibits senescence and maintains mesenchymal stem cell properties through down-regulation of E2A-p21 by HIF-TWIS. *Blood* **2011**, *117*, 459–469. [CrossRef] [PubMed]
74. Vono, R.; Garcia, E.J.; Spinetti, G.; Madeddu, P. Oxidative Stress in Mesenchymal Stem Cell Senescence: Regulation by Coding and Noncoding RNAs. *Antioxid. Redox Signal.* **2018**, *29*, 864–879. [CrossRef]
75. Chen, C.; Tang, Q.; Zhang, Y.; Yu, M.; Jing, W.; Tian, W. Physioxia: A more effective approach for culturing human adipose-derived stem cells for cell transplantation. *Stem Cell Res. Ther.* **2018**, *9*, 148. [CrossRef]
76. Ratushnyy, A.; Lobanova, M.; Buravkova, L.B. Expansion of adipose tissue-derived stromal cells at"physiologic" hypoxia attenuates replicative senescence. *Cell Biochem. Funct.* **2017**, *35*, 232. [CrossRef]
77. Safwani, W.K.Z.W.; Choi, J.R.; Yong, K.W.; Ting, I.; Adenan, N.A.M.; Pingguan-Murphy, B. Hypoxia enhances the viability, growth and chondrogenic potential of cryopreserved human adipose-derived stem cells. *Cryobiology* **2017**, *75*, 91–99. [CrossRef] [PubMed]
78. Wagegg, M.; Gaber, T.; Lohanatha, F.L.; Hahne, M.; Strehl, C.; Fangradt, M.; Tran, C.L.; Schonbeck, K.; Hoff, P.; Ode, A.; et al. Hypoxia promotes osteogenesis but suppresses adipogenesis of human mesenchymal stromal cells in a hypoxia-inducible factor-1 dependent manner. *PLoS ONE* **2012**, *7*, e46483. [CrossRef]
79. Markway, B.D.; Tan, G.K.; Brooke, G.; Hudson, J.E.; Cooper-White, J.J.; Doran, M.R. Enhanced chondrogenic differentiation of human bone marrow-derived mesenchymal stem cells in low oxygen environment micropellet cultures. *Cell Transplant.* **2010**, *19*, 29–42. [CrossRef]
80. Kim, J.H.; Yoon, S.M.; Song, S.U.; Park, S.G.; Kim, W.S.; Park, I.G.; Lee, J.; Sung, J.H. Hypoxia suppresses spontaneous mineralization and osteogenic differentiation of mesenchymal stem cells via IGFBP3 Up-regulation. *Int. J. Mol. Sci.* **2016**, *17*, E1389. [CrossRef]
81. Lech, W.; Sarnowska, A.; Kuczynska, Z.; Dabrowski, F.; Figiel-Dabrowska, A.; Domanska-Janik, K.; Buzanska, L.; Zychowicz, M. Biomimetic microenvironmental preconditioning enhance neuroprotective properties of human mesenchymal stem cells derived from Wharton's Jelly (WJ-MSCs). *Sci. Rep.* **2020**, *10*, 16946. [CrossRef] [PubMed]
82. Ito, K.; Suda, T. Metabolic requirements for the maintenance of self-renewing stem cells. *Nat. Rev. Mol. Cell Biol.* **2014**, *15*, 243–256. [CrossRef] [PubMed]
83. Mohyeldin, A.; Garzon-Muvdi, T.; Quinones-Hinojosa, A. Oxygen in stem cell biology: A critical component of the stem cell niche. *Cell Stem Cell* **2010**, *7*, 150–161. [CrossRef] [PubMed]
84. Shyh-Chang, N.; Daley, G.Q.; Cantley, L.C. Stem cell metabolism in tissue development and aging. *Development* **2013**, *140*, 2535–2547. [CrossRef] [PubMed]
85. Marino, L.; Castaldi, M.A.; Rosamilio, R.; Ragni, E.; Vitolo, R.; Fulgione, C.; Castaldi, S.G.; Serio, B.; Bianco, R.; Guida, M.; et al. Mesenchymal Stem Cells from the Wharton's Jelly of the Human Umbilical Cord: Biological Properties and Therapeutic Potential. *Int. J. Stem Cells* **2019**, *12*, 218–226. [CrossRef]
86. Puig-Pijuan, T.; de Godoy, M. A, Pinheiro Carvalho, L.R.; Bodart-Santos, V.; Soares Lindoso, R.; Moreno Pimentel-Coelho, P.; Mendez-Otero, R. Human Wharton's jelly mesenchymal stem cells protect neural cells from oxidative stress through paracrine mechanisms. *Future Sci. OA* **2020**, *6*, FSO627. [CrossRef] [PubMed]

Article

Effect of Long-Term 3D Spheroid Culture on WJ-MSC

Agnieszka Kaminska [1], Aleksandra Wedzinska [1], Marta Kot [2] and Anna Sarnowska [1,2,*]

[1] Mossakowski Medical Research Centre, Translational Platform for Regenerative Medicine, Polish Academy of Science, 02-106 Warsaw, Poland; akaminska@imdik.pan.pl (A.K.); awedzinska@imdik.pan.pl (A.W.)
[2] Mossakowski Medical Research Centre, Department of Stem Cell Bioengineering, Polish Academy of Sciences, 02-106 Warsaw, Poland; mkot@imdik.pan.pl
* Correspondence: asarnowska@imdik.pan.pl; Tel.: +48-22-6086598

Abstract: The aim of our work was to develop a protocol enabling a derivation of mesenchymal stem/stromal cell (MSC) subpopulation with increased expression of pluripotent and neural genes. For this purpose we used a 3D spheroid culture system optimal for neural stem cells propagation. Although 2D culture conditions are typical and characteristic for MSC, under special treatment these cells can be cultured for a short time in 3D conditions. We examined the effects of prolonged 3D spheroid culture on MSC in hope to select cells with primitive features. Wharton Jelly derived MSC (WJ-MSC) were cultured in 3D neurosphere induction medium for about 20 days in vitro. Then, cells were transported to 2D conditions and confront to the initial population and population constantly cultured in 2D. 3D spheroids culture of WJ-MSC resulted in increased senescence, decreased stemness and proliferation. However long-termed 3D spheroid culture allowed for selection of cells exhibiting increased expression of early neural and SSEA4 markers what might indicate the survival of cell subpopulation with unique features.

Keywords: mesenchymal stem cells; mesenchymal stromal cells; 3D culture; neurospheres; spheroids; pluripotency; neural; quiescence

Citation: Kaminska, A.; Wedzinska, A.; Kot, M.; Sarnowska, A. Effect of Long-Term 3D Spheroid Culture on WJ-MSC. *Cells* **2021**, *10*, 719. https://doi.org/10.3390/cells10040719

Academic Editor: Joni H. Ylostalo

Received: 2 March 2021
Accepted: 22 March 2021
Published: 24 March 2021

1. Introduction

Mesenchymal stromal/stem cells (MSC) were discovered by Friedenstein in 1966 [1] and since that time most of the researchers have used 2D culture condition to expand this population. The adherence is listed as one of the criteria to revise cells as MSC [2]. Monolayer culture system allows MSC to attach to the surface just like in natural environment and to expand in two dimensions. In spite of its widespread, this method has multiple limitations, and it is discussed how close is to the natural cell environment of MSC [3,4].

Currently, cell cultures are cultivated more and more often in 3D conditions as an alternative for 2D conditions. Spheroid culture, one from the multiple solutions, provides cell-to-cell contacts and intercellular signaling what resembles the environment of tissue [5]. Moreover, such a method of culture is also supposed to imitate the natural cell niche with the stem cells preserved in it.

Spheroid is the floating aggregate of cells with visible changes across its structure. The core consists of proliferating cells, whereas cells from external layer might differentiate and migrate. In order to mimic the natural conditions, MSC has also started to be cultured as 3D aggregates. MSC spheres are described to be formed by using different protocols including low attachment surface [6,7], hanging drop culture [8,9], scaffolds [10], and even the bioreactors [11]. Few research groups tested also neurosphere assay, proposed for neural stem cells (NSC) [12,13] to achieve MSC spheroids. Culture media for neurospheres contain epithelial growth factors (EGF) and basal fibroblast growth factor (bFGF) but no serum.

Most of the experiments conducted with MSC-neurospheres were focused on acquisition of neural phenotype. it was suggested that 3D culture condition could improve neural differentiation of MSC. In spite of improvement of neural differentiation under 3D conditions, evidences of receiving fully functional neuronal cells from MSC populations are limited. Still there is a demand for efficient protocol, which could be used in clinic. Except analysis of neural phenotype, other aspects such as proliferation, senescence and stemness were not so broadly taken into consideration during research. Moreover, majority of MSC spheroids experiments were a short-term—3D cultured did not last up to seven days in vitro of culture (div) [4] and results were obtained usually during first three div. There were little evidences whether observed effects were transient or constant whether how cells would react for prolonged 3D conditions. Even less is known about the influence of 3D condition on stem cell niche: effect on surface markers expression, commitment of specific type of MSC or interaction between cells. That would explain why some cells survive in 3D conditions and how we could select with better properties.

In the present study, Wharton Jelly derived (WJ)-MSC were cultured in two different culture conditions as the standard, monolayer 2D culture or as 3D culture. Both cultures were conducted parallel for about 20 div—the time required to achieve three passages during standard 2D MSC culture. Cells derived from spheroid culture were compared to those cultured as monolayer regarding such properties as cell senescence, rate of proliferation, capability to form the colonies, and pluripotent and neural gene expression.

2. Materials and Methods

2.1. WJ-MSC Isolation and Primary Culture

Human umbilical cords were acquired from full-term deliveries with the written consent of mother according to the Ethics Committee of Warsaw Medical University guideline (KB/213/2016). Cords (15–20 cm) transported in phosphate buffer saline (PBS) solution (PBS; Sigma-Aldrich, Saint Louis, MO, USA) with mix of Penicillin-Streptomycin-Amphotericin B (1:100, Gibco, Thermo Fisher Scientific, Waltham, MA, USA) were cut with lancet to 2–3 mm in thickness slices. The cylindrical fragments of Wharton Jelly (WJ) of 2–3 mm diameter were obtained from the slices of umbilical cord using the diameter biopsy punch (Miltex, GmbH, Viernheim, Germany). Explants were transferred to six well cultured plates and culture in the standard cell culture medium for WJ-MSC: DMEM (Gibco), 10% human platelet cell lysate (Macopharma, Tourcoing, France), mix of penicillin, streptomycin amphotericin B (1:100; Gibco, Thermo Fisher Scientific), 2 µg/mL heparin (Sigma-Aldrich). Conditions for cell culture were following adherent surface, 37 °C temperature, 95% of humidity, 5% concentration of CO_2, and 5% concentration of O_2. The culture medium was replaced every 2 days for 14 div. When the cells migrated out of the explant and the culture reached semiconfluence, the cells were detached with Accutase Cell Detachment Solution (Beckton Dickinson, Franklin Lakes, NJ, USA) and counted.

WJ-MSC were cultured in conditions described above until the end of 3 passage (initial population of WJ-MSC). After the 3rd passage, cells were collected and divided into two group—part of them was cultured as a spheroids (3D cultured WJ-MSC) and the rest were continually cultured in previous cell culture conditions until the 7th passage (2D cultured WJ-MSC).

2.2. Spheroid Culture

WJ-MSC from the 3rd passage were collected and seeded on anti-adhesive 6 well plates (Nunclon Sphera, Thermo Fischer Scientific, Waltham, MA, USA) at the high density of $30 \times 10^3/\text{cm}^2$ in 5% O_2. Cells were cultured in medium DMEM/F12 (Gibco) containing mix of penicillin, streptomycin, amphotericin B (1:100) (Gibco), L-Glutamine 200 mM (1:100) (Gibco), N2 supplement (1:100; Biotechne, Minneapolis, MN, USA) and EGF (20 ng/mL; PeproTech, London, UK). In 3 div, Medium was replaced, and new medium was started to use from this point of culture—Neurobasal Medium (Gibco, Thermo Fischer Scientific) with mix of penicillin, streptomycin, amphotericin B (1:100; Gibco, Thermo Fischer Scientific), L-Glutamine 200 mM (1:100; Gibco, Thermo Fischer

Scientific), B27 supplement (1:50; Gibco, Thermo Fischer Scientific), EGF (20 ng/mL; PeproTech) and bFGF (20 ng/mL; PeproTech).

WJ-MSC spheres were cultured in parallel to standard 2D culture—until the standard culture acquire the 7th passage (about 20 div of culture). After that time spheroids were dissociated to obtain single cells with Accutase Cell Detachment Solution (Becton Dickinson) and seeded again to the 2D culture in standard medium. Reseeded cells were used to measure colony forming unit frequency (CFU-F), population doubling time (PDT), senescence processes, gene expression level, and to perform immunocytochemistry staining.

2.3. Flow Cytometry Analysis

Cells were detached with Accutase Cell Detachment Solution (Beckton Dickinson) and washed in PBS. Required cell number (1×10^6) was resuspend in cold Stain Buffer (Beckton Dickinson) and used for further flow cytometry analysis. Cell markers were analyzed with Human MSC Analysis Kit (Beckton Dickinson) containing antibodies conjugated with fluorochrome against following antigens: CD73, CD90, CD105 (positive markers), CD11b, CD19, CD34, CD45, and PE (negative markers) (Table 1). Cells were incubated in diluted antibodies in the dark for 30 min. After incubation, cells were washed twice with Stain Buffer (Beckton Dickinson) and resuspend in Stain Buffer. Resuspended cells were analyzed using FACS Canto II (Beckton Dickinson) with FACSDiva Software (Beckton Dickinson) and FlowJo 10 (Beckton Dickinson).

Table 1. List of antibodies used for flow cytometry—Human mesenchymal stem/stromal cell (MSC) Analysis Kit (Beckton Dickinson) (cat. nr 562245).

	Antigen	Fluorochrome
Positive cocktail	CD73 CD90 CD105	APC FITC PerCP-Cy5.5
Negative cocktail	CD11b CD19 CD34 CD49 HLA-DR	PE

In 3rd and 10th div of 3D culture, spheroids were dissociated and resuspended in Stain Buffer. Before running the sample, cells were filtered through 30 µm filter (Miltenyi Biotec, Bergisch Gladbach, Germany) to avoid dublets. Cells were analyzed to compare the change in size with flow cytometry FACS Canto II (Beckton Dickinson) with FACSDiva Software (Beckton Dickinson) and FlowJo 10 (Beckton Dickinson).

2.4. Live-Dead Staining

Aggregates were stained with mix of ethidium homodimer-1 (8 µM, EthD-1) and Calcein AM (Cal-AM) (0.1 µM, Invitrogen, Thermo Fischer Scientific, Waltham, MA, USA) to confirm the viability of cells in spheroids. Spheroids or single cells derived from spheroids were incubated with staining mixture for 45 min in room temperature in a darkness. Stained cells were observed in fluorescence microscope Axio Vert.A1 (Carl Zeiss, Oberkochen, Germany).

2.5. CFU Assay

WJ-MSC from initial population, 2D culture and 3D culture were seeded on 6-well plate in the amount of 100 cells per well. Cells were cultured for 10 div in standard conditions. Then, cells were washed with PBS, fixed with 4% PFA for 15 min and again washed with PBS. Fixed cells were stained with 0.5% toluidine blue for 20 min and washed with distilled water after staining. The number of colonies containing 50 cells or more were counted, and CFU-F was calculated as a percentage of seeded cells.

2.6. Senescence Assay

The cells senescence was analyzed with Senescence Cells Histochemical Staining Kit (Sigma-Aldrich) in initial population, 2D culture, and 3D culture of WJ-MSC. Cells from 2D cultures grew until the confluence reached 50–60% while the cells from dissociated spheroids were cultured 48 h in standard conditions and then the assay was performed.

Cells were washed with 1×PBS and then fixed with Fixation Buffer (provided with kit) for 6–7 min in room temperature. Then, cells were washed 3 times with 1×PBS. After washing, cells were incubated overnight in 37 °C with Staining Solution (prepared according to the protocol). Next day, total cell number was count as well as blue-stained cell number. The percentage of β-galactosidase positive cells was calculated.

2.7. Proliferation Analysis

The cell proliferation was analyzed in the initial population, 2D culture, and 3D culture. WJ-MSC were seeded at a density 2000 cells/cm^2 and cultured in standard conditions until the 80% confluence was reached. Then cells were collected, counted, and re-seeded again at initial density. PDT and cumulative population doublings (cPD) were calculated for the next 4 passages, based on total cell number at each passage. The PDT value was calculated with the following formula:

$$\mathrm{PDT} = ((\mathrm{t} - \mathrm{t_0}) \times \log 2)/(\log \mathrm{N} - \log \mathrm{N_0})$$

where N is the number of cells obtained at the end of the passage, N_0 is the initial number of seeded cells, and $t - t_0$ is the duration of passage (counted in days).

2.8. Cryostat Sectioning

Spheroids were collected, fixed in 4% PFA for 15 min and washed twice with PBS. Then, PBS was replaced with 7.5% sucrose solution and incubated overnight. Next day, solution was replaced with 15% sucrose solution and 30% sucrose solution. Then, spheroids were embedded in medium for frozen tissue specimen (OCT Sakura Tissue-Tek, Sakura Finetek Europe, Alphen aan den Rijn, The Netherlands) and moved to −80 °C. Spheroids were cut with cryostat for the 20–30 µm thickness sections. Sections were collected on APTEX (3-Aminopropyl) triethoxysilane) coated glass microscope slides, stored in −20 °C and used for immunocytochemical staining.

2.9. Immunocytochemistry

Immunocytochemistry was performed to detect pluripotency and early neural markers in initial population, 2D culture and 3D culture of WJ-MSC. WJ-MSC were washed with PBS and fixed in 4% PFA for 15 min. Samples were permeabilized with 0.2% Triton X-100 (Sigma-Aldrich) for 15 min and then washed with PBS. After incubation with 10% Goat Serum (Sigma-Aldrich) for 1 h primary antibodies were applied for 24 h in 4 °C (Table 2). Next day, cells were washed with PBS and then incubated with the secondary antibodies conjugated with fluorochrome for 1 h (Supplementary Table S1, Supplementary Figure S1). Cell nuclei were stained with Hoechst 33342 dye (1 µg/mL; Sigma-Aldrich). The analysis was performed using confocal microscope Zeiss LSM780 (Carl Zeiss).

Table 2. List of primary antibodies used for immunocytochemistry.

Antigen	Source	Isotype	Dilution	Company	Catalogue Number
Nestin	Mouse monoclonal	IgG1	1:500	Merck Millipore	MAB5326
β-III-Tubulin	Mouse monoclonal	IgG2B	1:500	Sigma-Aldrich	T8660
Neurofilament 200 (NF-200)	Mouse monoclonal	IgG1	1:400	Merck Millipore	N042
NeuN	Mouse monoclonal	IgG1	1:100	Merck Millipore	MAB377
A2B5	Mouse monoclonal	IgM	1:700	Merck Millipore	MAB312R
Ki67	Rabbit polyclonal	IgG(L+H)	1:1000	Abcam	AB15580
SSEA4	Mouse monoclonal	IgG3	1:400	Merck Millipore	MAB4304

2.10. Real Time-Quantitative Polymerase Chain Reaction (RT-qPCR)

Total RNA was isolated from initial, 2D populations, and 3D cultured population using the following kits: Total RNA Mini Plus kit (A&A Biotechnology, Gdynia, Poland) and Total RNA Mini Plus Concentrator (A&A Biotechnology) according to the manufacturer's protocols.

RNA was eluted with 20 µL of RNase-free H_2O (Sigma Aldrich). The quantity and the quality of RNA were assessed using a NanoDrop 2000 spectrophotometer (Thermo Scientific). The elimination of genomic DNA (gDNA) contamination in all RNA samples was performed using a Clean up RNA Concentrator (A&A Biotechnology).

RNA samples were stored at −80 °C until were further used. A complementary strand of DNA (cDNA) from RNA was generated using a High-Capacity RNA-to-cDNA™ Kit (Applied Biosystems, Thermo Fischer Scientific, Waltham, MA, USA) according to the manufacturer's instructions. Following the reverse transcription, samples were diluted in RNase-free water and stored at −20 °C until subsequent testing.

Quantitative polymerase chain reactions were performed using SYBR green Master Mix (Applied Biosystems) and specific primers (Table 3) with the 7500 Real Time PCR System (Applied Biosystems). The relative amount of RNA was calculated with the comparative delta-delta Ct method ($2^{-\Delta\Delta Ct}$) and gene expression was normalized using β-actin (ACTB). Gene expression was compared with the mean level of the corresponding gene expression in cells from initial population (3rd passage of WJ-MSC culture) and expressed as n-fold ratio.

Table 3. List of primers used for Real Time-Quantitative Polymerase Chain Reaction (RT-qPCR).

Gene	NCBI Reference Sequence	Product Size	Primer Sequence (5′ -> 3′)
β-Actin	NM_001101.5	250 bp	F: CATGTACGTTGCTATCCAGGC R: CTCCTTAATGTCACGCACGAT
Nestin1	NM_006617.2	64 bp	F: GGGAAGAGGTGATGGAACCA R: AAGCCCTGAACCCTCTTTGC
β-Tubulin III	NM_001197181.2	126 bp	F: GGAAGAGGGCGAGATGTACG R: GGGTTTAGACACTGCTGGCT
MAP-2	NM_001375545.1	99 bp	F: TTGGTGCCGAGTGAGAAGA R: GTCTGGCAGTGGTTGGTTAA
GFAP	NM_001363846.2	100 bp	F: CCGACAGCAGGTCCATGT R: GTTGCTGGACGCCATTG
Sox2	NM_003106.4	93 bp	F: GTGGAAACTTTTGTCGGAGA R: TTATAATCCGGGTGCTCCTT
Rex1	NM_001304358.2	107 bp	F: GCTCCCTTGAATGTTCTTTG R: GCCTGTCATGTACTCAGAAT
Nanog	NM_024865.4	103 bp	F: GAACCTCAGCTACAAACAGG R: CGTCACACCATTGCTATTCT
Oct3/4 (Pou5F1)	NM_001285986.2	331 bp	F: CCTGAAGCAGAAGAGGATCACC R: AAAGCGGCAGATGGTCGTTTGG

2.11. Statistics

Two-group comparisons were performed with Student's test, whereas multiple groups used one-way analysis of variance (ANOVA). The results are presented as mean values of 3 independent experiments ± SD (* < 0.05, ** < 0.01, *** < 0.001, **** < 0.0001), each experiment was performed with cells obtained from one donor. Statistical analysis was conducted with GraphPad Prism v. 7.00 software.

3. Results

3.1. Characteristics of WJ-MSC Cultured in 2D and 3D Conditions

WJ-MSC chosen for experiments exhibited characteristic features of MSC such as morphology and expression of markers. Cells were adherent to the surface and presented spindle, fibroblast-like morphology. Flow cytometry analysis revealed that cells expressed specific mesenchymal markers (CD73, CD90, and CD105). Less than 1% of WJ-MSC expressed negative markers for MSC (CD34, CD11b, CD19, CD45, and HLA-DR) (Figure 1A). Above described features are accordant with the minimal criteria for MSC established by The International Society for Cellular Therapy.

WJ-MSC were cultured until the third passage in the monolayer with standardly used culture medium. Then, collected cells were divided into two groups: first group was cultured constantly as monolayer (2D cultured), while the second—as spheroids (3D culture) (Figure 1B). Both cultures were conducted for time required for three passages (about 20 div). Then spheroids were dissociated and seeded again to 2D conditions—to observe changes of WJ-MSC. In further analysis, following populations were compared: initial population—third passage of standard WJ-MSC culture; 2D culture—seventh passage of standard WJ-MSC culture; and 3D culture—WJ-MSC cultured as spheroids for 20 div and then reseeded to standard conditions.

Applied 3D method, slightly modified in our laboratory succeeded in sphere formation by WJ-MSC. WJ-MSC were cultured on anti-adhesive surface with the presence of supplements and mitogens (EGF and bFGF)—similarly to the neurospheres formed by NSC. Diameter of WJ-MSC spheroids varied from 20 µm to even 500 µm; however, average size oscillated between 40 and 120 µm. Spheres were cultured up to 20 div; however, the best morphology was observed in the first five div of 3D culture—in later stages spheres spontaneously disintegrated.

Flow cytometry analysis revealed the change in size of single cells—differences were noticed during the cultivation time, as well as between 2D and 3D culture models (Figure 1C). In young spheroids (three div) small, round cells predominated, compared to parallelly cultured 2D cells, whereas in old spheroids (10 div) we could distinguish an increase in the number of large cells. During 2D culture, the ratio of both cells subpopulations (small and large) did not change remarkably.

Adherent properties in WJ-MSC were still detectable after long-term 3D. Although cell morphology was similar to those acquired during standard culture, some differences occurred more frequently (Figure 1D). After 3D culture, we distinguished three subpopulation of cells—standard fibroblast-like cells similar to standard 2D culture; narrow and spindle cells with improved neural potential; and very broad cells indicating senescent morphology.

Live-dead staining using EthD-1 and Cal AM revealed that most of the cells cultured in monolayer were alive (98.91% ± 0.26) (Figure 2A), whereas spheroids contained significantly more dead cells inside (Figure 2B,C). Number of alive cells was reduced to 55.9% ± 6.18 and 72.87% ± 5.29 in 3 and 10 div respectively (Figure 2C). The increase in viability between early and late stage of spheroid culture was also significant.

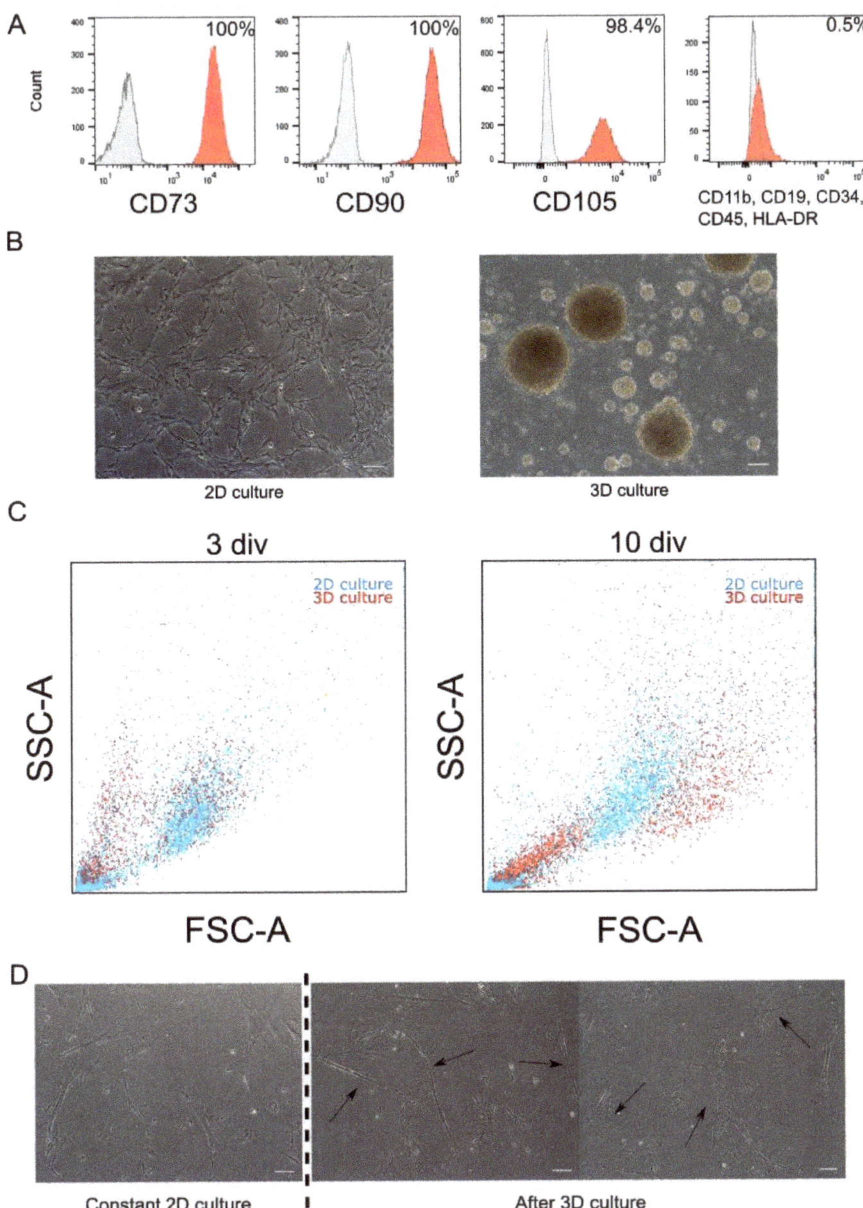

Figure 1. Morphology and phenotype of WJ-MSC cultured in monolayer and as spheroids. (**A**). Flow cytometry analysis. Initial population of WJ-MSC used to experiments expressed specific MSC markers (CD73, CD90, and CD105) and less than 1% of WJ-MSC expressed negative markers (CD11b, CD19, CD34, CD45, and HLA-DR). Red histogram—analyzed marker, grey histogram—isotype control. (**B**). Morphology of 2D and 3D cultured WJ-MSC—monolayer culture of WJ-MSC from 4th passage (up) and WJ-MSC 4 div after formed spheroids (down). (**C**). Flow cytometry—Forward scatter (FSC) and Side scatter (SSC) analysis for 3D culture (spheroids) and 2D (monolayer). Three days in vitro (div) and 10 div after sphere induction, there were differences in size of cells between 2D and 3D culture. (**D**). Different morphology of cells cultured as spheroids after reseeding into 2D conditions. Right: WJ-MSC cultured constantly in 2D for 8 passages. Center and left: WJ-MSC from 3D reseeded to 2D condition—visible two different types of morphology: narrow and small (center) and flat and broad (left) cells. White scalebars represent 100 μm.

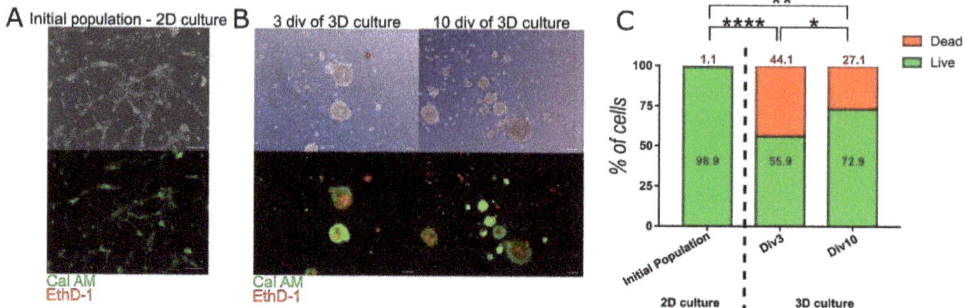

Figure 2. Viability of 3D cultured WJ-MSC—Calcein AM (Cal AM) and Ethidium homodimer-1 (EthD-1) staining. Cal AM stains live cells in green, while EthD-1 stains dead cells in red. Double stained cells are early apoptotic cells. (**A**) Initial population of WJ-MSC contained live cells with almost no dead cells. Scale bars: 100 μm. (**B**) Viability of cells in spheroids—after 3 and 10 div (day in vitro) of 3D culture. Observed darker shade in the contrast phase corresponds to dead cells visible. Scale bars: 100 μm. (**C**) Analysis of live, dead, and early apoptotic cells in initial population and 3D cultured population for 3rd and 10th (div). The results are presented as mean values of 3 experiments ± SD, for * < 0.05, ** <0.01, **** < 0.0001.

3.2. *Physiological Properties of WJ-MSC Cultured in 3D Conditions*

Spheroids culture conditions change not only the morphology of WJ-MSC, but also influence on the physiological features of the cells. We compared culture features such as: doubling ratio of cells, induction of senescence process and content of stem fraction in population. For this purpose, WJ-MSC spheroids were dissociated to the single cells after 20 div of culture and reseeded to monolayer culture conditions. Results of assays obtained from 3D cultured WJ-MSC were compared to those observed in the initial population of WJ-MSC (from passage 3) and WJ-MSC continuously cultured in 2D conditions (from passage 7).

WJ-MSC were monitored after reseeding to monolayer conditions to calculate Population Doubling Time (PDT) required for each of next four passages. During first passage after 3D cultured, cell proliferation decreased compared to cells constantly cultured in monolayer (Figure 3A). In further passages, 3D cultured WJ-MSC restored the doubling rate and divided in the same ratio as 2D cultured WJ-MSC (Supplementary Table S2). There is a shift between passages of 2D and 3D cultured populations when cumulative value is analyzed (Figure 3B).

To evaluate the effect of the culture method on cell senescence, the β-galactosidase activity was measured. Spheroid-cultured WJ-MSC exhibited significantly higher activity of β-galactosidase (49.04 ± 10.72) than initial population of WJ-MSC (1.24 ± 0.59) and WJ-MSC continuously cultured as monolayer (1.84 ± 0.93) (Figure 3C).

ISCT recommends the Colony forming unit frequency assay (CFU-F) to confirm the stemness of MSC and estimate the fraction of stem/progenitor cells in the population. The values of CFU-F from initial population of WJ-MSC (19.63± 9.5) and 2D cultured WJ-MSC (12.78 ± 5.02) does not differ significantly. However, 3D cultured WJ-MSC indicated significantly reduced CFU-F than initial population (5.11 ± 6.95) (Figure 3D).

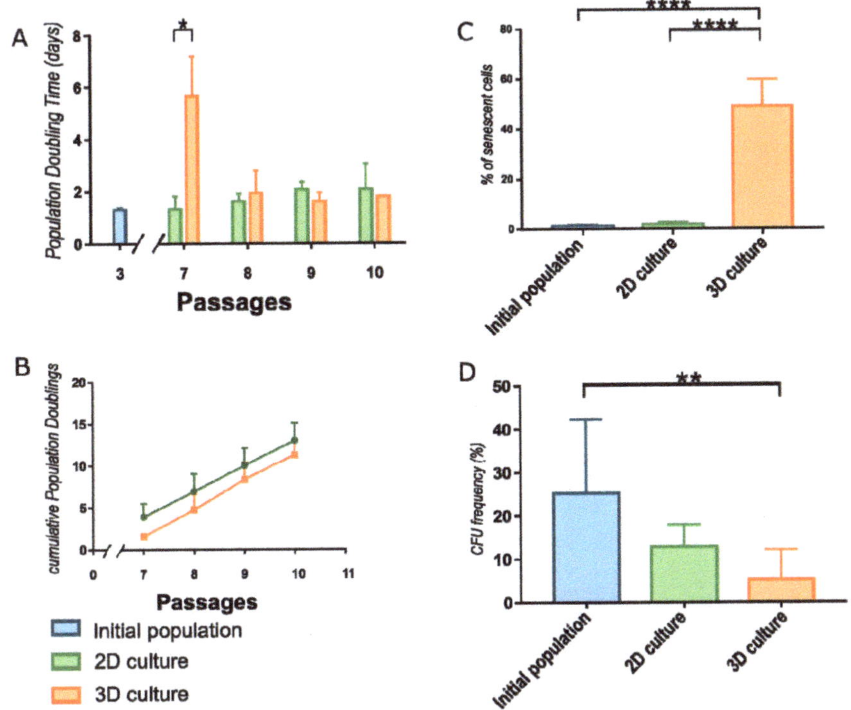

Figure 3. Characteristic of WJ-MSC cultured in different conditions. (**A**) Population doubling time (PDT) for WJ-MSC cultured in 2D and 3D. 3D cultured WJ-MSC divided more slowly than 2D cultured WJ-MSC during the first passage. However, there are observed no differences in next 3 passages. (**B**) Cumulative Population Doublings calculation revealed shift between 2D and 3D cultured populations. (**C**) Senescence process analysis. Number of cells expressing β-galactosidase was significantly higher in 3D cultured WJ-MSC than in initial and 2D cultured populations. (**D**) Colony forming unit (CFU) frequency reduced significantly during 3D culture of WJ-MSC. Following populations of WJ-MSC were used: from 3rd passage (initial populations), from 7th passage (2D culture) and cultured as spheroids (3D culture). The results are presented as mean values of 3 experiments ± SD. For * < 0.05, ** <0.01, **** < 0.0001.

3.3. Neural and Pluripotent Markers Expression in 3D Cultured WJ-MSC

To find out whether 3D culture condition predisposes to increase expression of early neural as well as pluripotent markers, immunocytochemical analysis was performed. Staining of sectioned spheroids revealed the presence in the sphere the early neural markers: Nestin and β-III-Tubulin, early oligodendrocyte marker A2B5 as well as pluripotent surface marker SSEA4. However, we did not observe more mature markers as NF200 or NeuN inside the sphere (Figure 4). Double staining for SSEA4 and Caspase-3 showed that most of SSEA-4 positive cells remained alive (Supplementary Figure S2).

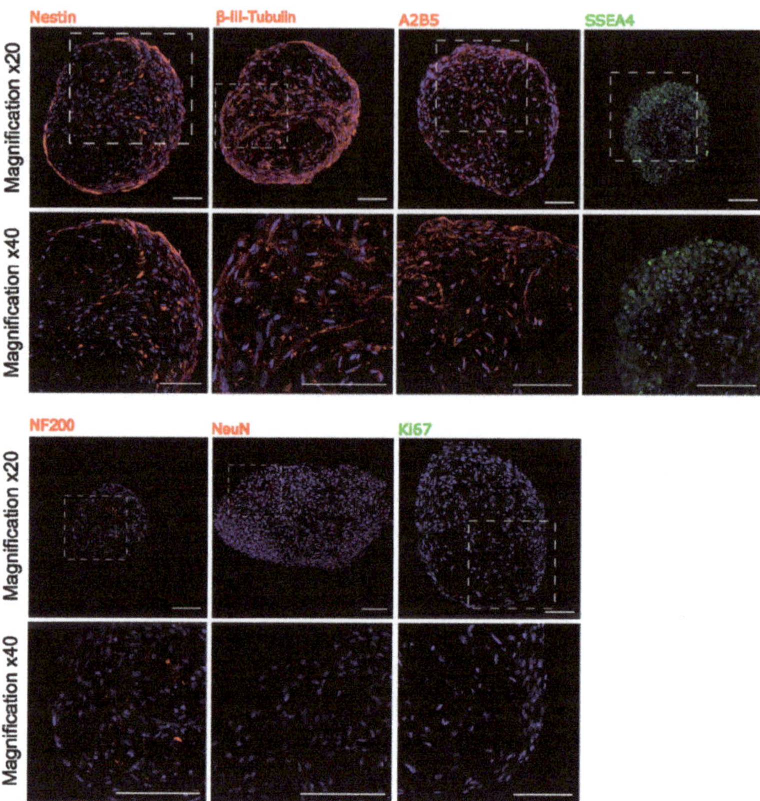

Figure 4. Immunocytochemistry staining of neural and pluripotent markers in sectioned WJ-MSC spheroids in two different magnifications. Early neural markers are also expressed inside the sphere, whereas more mature neural markers were not detected. Spheroids were collected between 10 and 20 day in vitro of 3D culture. Scale bars: 100 μm.

3D population seeded again on the monolayer also displayed early neural and oligodendrocyte markers—the expression was compared to initial population and 2D constantly cultured WJ-MSC (Figure 5, Supplementary Table S3). Expression of Ki67—proliferation marker—was decreased in 3D cultured WJ-MSC compared to initial and 2D cultures, what confirms previously described PDT and CFU-F observations. Early neural markers—Nestin and β-III-Tubulin—were presented more frequently in 3D cultured WJ-MSC than in 2D cultured WJ-MSC. β-III-Tubulin expression in 3D WJ-MSC was even higher than in initial population. However, 3D cell culture conditions did not increase the expression of more mature markers NF-200 and NeuN. Changed culture condition did not affect the expression of A2B5. Expression of SSEA4—pluripotent surface glycosphingolipid—was increased in 3D cultured WJ-MSC, what may indicate that SSEA4-positive cells more favorably survives during long-termed 3D culture.

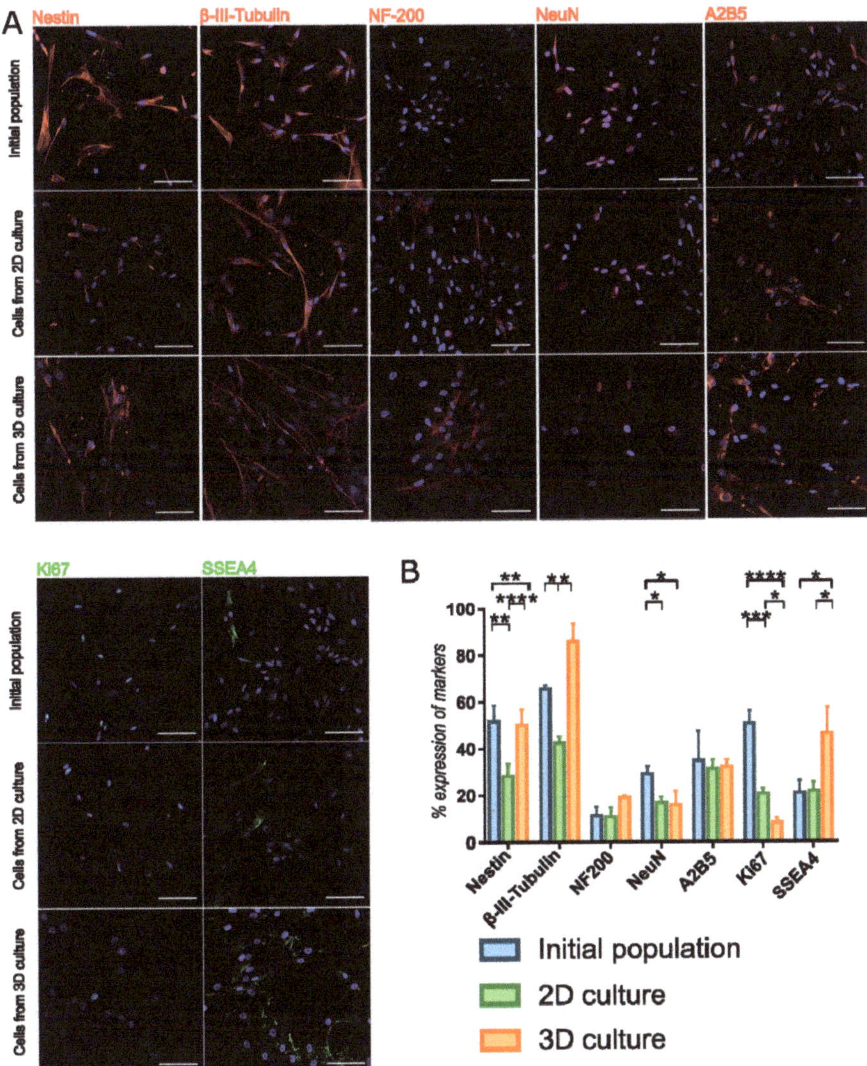

Figure 5. Immunocytochemistry staining of neural and pluripotent markers in WJ-MSC cultured in different conditions. (**A**) 3D cultures changes the expression of some neural and pluripotent markers what is observed even after reseeding to 2D culture. Scale bars: 100 µm. (**B**) Analysis of marker expression in WJ-MSC populations. Following populations of WJ-MSC were used: cells from 3rd passage (initial populations), cells from 7th passage (2D culture) and cells cultured as spheroids after reseeding to 2D (3D culture). The results are presented as mean values of 3 experiments ± SD. For * < 0.05, ** <0.01, *** < 0.001, **** < 0.0001.

3D culture influence was also assessed by the expression changes in genes characteristic for neural or pluripotent phenotype (Figure 6, Supplementary Table S4). Due to the huge fluctuations between genetic material from different isolations we did not reported significant changes upon a long-term 3D cells culture when it comes to early neural marker Nestin and pluripotency genes Nanog, Sox2, Oct3/4, and Rex1. Tendency of increased expression of Nestin and Nanog genes may indicate the maintenance of early undifferentiated state of cells during 3D cell culture. Other analyzed markers—H3Tubulin (early neural marker), MAP2 (mature neural marker) and GFAP (astroglial marker) revealed some changes between initial population and two methods of further culturing. 2D conditions was more favorable for expression of MAP2 and GFAP. Results obtained from gene expression and immunocytochemistry analysis are not fully consistence, however it may be connected with differences between transcriptional and protein level of analyzed markers. We also performed gene expression analysis for RNA isolated directly from 3D spheroids—without reseeding to 2D (Figure S3). We observed the changes between two timepoints of analysis—Nestin was expressed on higher level in 3D spheroids than 48 h after reseeding to 2D culture. We observed this tendency also for other genes; however, differences were not significantly important.

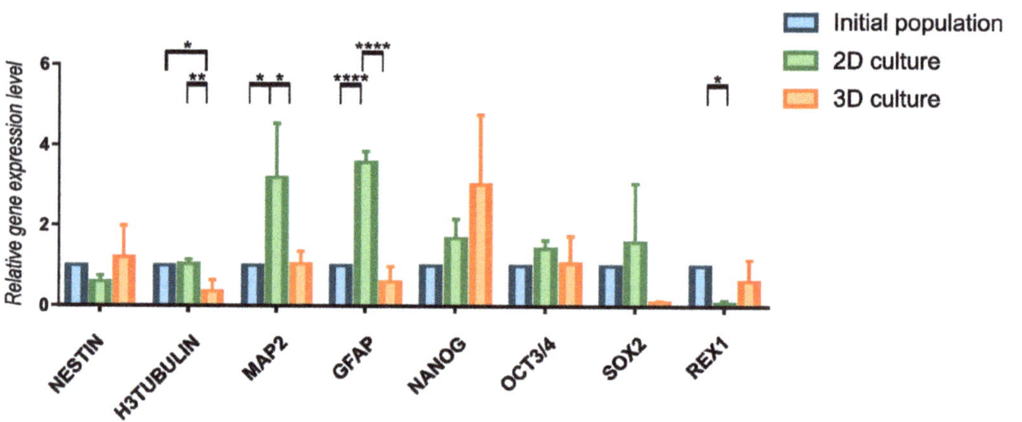

Figure 6. Relative gene expression level (fold change, mean ± SD) of neural and pluripotency phenotype in 2D and 3D WJ-MSC cultures. Quantitation of these genes was determined relative to ACTB as a housekeeping gene by quantitative real-time PCR. Changes in gene expression in WJ-MSC cells cultured in 2D and 3D conditions are shown relative to that in cells grown in the initial population. Following populations of WJ-MSC were used: from 3rd passage (initial populations), from 7th passage (2D culture), cultured as spheroids and transferred to 2D (3D culture). Results shown are the mean of 3 independent RNA isolations, for * < 0.05, ** <0.01, **** < 0.0001.

4. Discussion

Umbilical cord contains Wharton Jelly, which is commonly used source of MSC in stem cell research [14,15]. Different factors and culture conditions are proposed to improve either stemness properties or differentiation potential of WJ-MSC. Our group is especially interested in establishment of protocol inducing differentiation of WJ-MSC toward neural lineage—due to limited alternatives for regeneration of neural system. Although neural differentiation of MSC is controversial, due to the lack functional evidence in vivo, our group confirmed the presence of voltage-sensitive and ligand-gated channels in differentiating neural stem-like cells derived from the nonhematopoietic fraction of human umbilical cord blood [16]. As MSC populations is highly heterogenous [17], we assume initial selection of pro-neural or undifferentiated cells would improve the effect of further neural differentiation. We continued in targeting our research on WJ-MSC and looked for the manner of increasing their neural/undifferentiated potential as we had reported in WJ-MSC spontaneous expression of neuroglial markers such as β-III-Tubulin, GFAP, NF-200 probably due to their premature origin [18]. Change of culture spatial structure appeared to be promising factor that we would like to confront.

The present studies were focused on the effect of prolonged 3D culture system of MSC. We tried to find out if applied culture conditions would result in obtaining population with better proliferation potential, younger phenotype or increased clonogenic abilities like in the cell niche. What is more, we examined the influence on gene expression and presence of markers responsible for neural or pluripotent phenotype—which direction cells would follow after 3D culture. WJ-MSC were cultured as spheroids in long termed culture for about 20 div to examine physiological changes. We compared features of 3D cultured cells with initial population derived from early passage (third passage) as well as population cultured constantly in monolayer up to the seventh passage. Clonogenicity, proliferation potential, senescence, and expression of neural and pluripotency markers were analyzed to confirm whether long-term 3D conditions are favorable for MSC.

We confirmed that despite their adherent properties, WJ-MSC can be directed to form the spheres in culture medium dedicated for NSC formed neurospheres. Spheroids can be generated also from other tissues containing MSC such as bone marrow [19], adipose tissue [20], or dental pulp [21] as well as other [21–24]. Main purpose of 3D neurosphere culture is to selectively isolate and culture the subpopulation of cells with higher neural differentiation capacities.

WJ-MSC in spheroids still exhibited strong adhesive properties—even undissociated spheres in standard culture medium settled down and attached to surface. 3D culture also did not change expression of typical MSC markers (CD73, CD90, and CD105) and multipotent differentiation into mesodermal lines [25,26]. Some research groups observed even enhanced mesodermal differentiation after 3D culture [7,26]. Spheroid culture led to morphology changes, visible especially during the first passage after reseeding on 2D [6,8,11,27]. Except WJ-MSC with characteristic fibroblast-like morphology, we observed more small cells with round or spindle shape. Acquisition of round shape and then elongation is typical for neural cell maturation from progenitors to immature neurons [28]. Morphological changes were also observed in non-neural cells undergoing neuronal differentiation [29]. Small size of cells after 3D culture could be also the effect of cell reorganization in spheroids and decreased packing density [11,27]. Unfortunately, we also noticed more large, flat cells connected with the loss of stemness state and higher senescence process [30,31]. We speculate that 3D culture could promote the survival of quiescent cell subpopulation.

3D cultured WJ-MSC significantly differed in cell division ratio than initial WJ-MSC population and 2D cultured WJ-MSC. What is more, immunochemistry staining for Ki67—proliferation marker—confirmed smaller fraction of dividing cells after 3D culture. Ki67 staining in AD-MSC neurospheres revealed that more than 80% of cells remains in quiescent phase [32]. Reduced proliferation of cells in spheroids suggested arrest in Go/G1 phase [7,8,26,33–35]. This observation could be connected with reduced CFU frequency indicating smaller number of clonogenic cells presented in 3D cultured population. However, many groups noticed better CFU frequency after transferring cells from spheres to 2D conditions [6,36]. Bartosh and colleagues published observations similar to ours, that CFU and population doubling values decreased directly after seeding cells derived from spheres. However, in further passages those features were similar or even higher than in a standard adherent culture [8]. Additionally, AD-MSC cultured as spheres expanded more rapidly than 2D cultured cells, but the difference was visible after 42 div of 2D culture [10]. Some research group did not observe the differences in CFU frequency between 2D and 3D cell culture [7].

Senescence was more pronounced in 3D cultured WJ-MSC. Increased β-galactoside activity was detected in almost a half of population just after changing the environment from 2D to 3D. To the contrast, other research groups reported that population after 3D culture contains less amount of senescent cells [6]. AD-MSC from spheroids contained more senescent cells in longer 3D culture, but still less than cells cultured on monolayer [10]. Differences in senescent processes between the results may arise from the source of MSC. In our experiments MSC were isolated from afterbirth tissue—part of umbilical cord. The initial population of WJ-MSC contained small number of senescent cells (less than 1%) and senescence process did not increase with further passages (up to 5% of senescent cells) [37], whereas for tissues from adult patients percentage of senescent cells was higher:

for BM-MSC about 22,5% [6], for AD-MSC about 10% [10]. In our opinion, changing the culture conditions to less physiological for adherent cells (stress factor) accelerates senescence. Despite that, the other subpopulation that remains in the culture over time—those cells either have greater stress resistance or are developmentally younger and do not necessarily require adhesion for proper function.

These observations are consistent with changes in cell survival. In the majority of published experiments, 3D culture is a transient stage which last usually up to three div, no longer than seven div—especially when it comes to the use of low-attachment surface [4] among others because of poor MSC survival in 3D conditions. In our experiments 3D culture also reduced cell viability. Calcein AM and Ethidium Homodimer-1 staining pointed huge proportion of dead cells inside spheroid, which usually was placed in dark core. We noticed huge fraction of double stained cells, which was counted as dead—those cells were permeabilized with residual activity of esterases. Such situation was reported in other publication, using propidium iodide which has the same properties and emission spectrum as EthD-1 [38]. According to Peng, apoptosis rate in sphere reaches up to 20% of cells [39]. The most visible shift we noticed between third and fourth div of 3D culture. This change in viability was reported also by other research groups [8,32]. Even 3D cultures performed with other method, such as hanging drop, underwent the reduction in viability [8]. However, the spheroid culture contained more alive cells in 10 div of culture than in 3 div. With the duration of 3D cell culture, cellular composition of spheres had to change as the number of dead cells declined. This confirms our assumptions that during 3D culture, spheroid is decomposed, while population is segregated spontaneously—survived cells are more resistant to stress conditions, with slightly different properties than typical adult mesenchymal cells. Decreased viability might be also connected with size and insufficient transportation of nutrients and oxygen to all cells in bigger spheroids. For ESC embryoid bodies, oxygen concentration in the center can vary depends on the structure's radius. In larger aggregates (400 μm radius) was lower than in smaller aggregates (200 μm radius) [40]. For 100 μm sphere diameter, the concentration of glucose and glutamine is 36,38 and 1,33 mmol/L—those values are insufficient, and cells begin to die in spheres core [41]. However, MSC spheroids are usually smaller. Hypoxic core, which could be responsible for increased cell deaths, was observed only in large spheroids consisting of 250,000 cells or more [27]. In smaller spheres oxygen gradient varied less than 10% across the aggregate layers. Decreased packing density might be the other mechanism that enable penetration of nutrients and oxygen to sphere core [27]. Observed population of smaller cells in 3D culture supported the evidence for change in cell densities and morphology in spheroids. Despite that, caspase activity confirmed that viability of cells decrease with sphere size [27]. Other used technique—thermal lifting in 3D culture was reported not only to lower apoptosis processes but also ischemic stress which is another relevant factor [36]. Cells in 3D culture were characterized by metabolism changes such as lower glucose consumption, lower L-lactase production [6,27], as well as changes in mitochondria [6]. Although lowered metabolism rate, there were observed higher level of mitochondrial and total reactive oxide species [6], what might be responsible for lowered viability and increased senescence during long-term 3D culture of MSC.

Except the changes of cell morphology, physiology and viability, 3D culture stimulates the acquisition of unique phenotype by cells. RNA-sequencing revealed the unique transcriptional profile of UC-MSC neurospheres, containing the features of NSC and MSC [42]. In our experiments, immunocytochemistry staining confirmed increased presence of early neural markers such as Nestin and β-III-Tubulin. 3D culture probably improve only the early stages of acquiring neural phenotype—NF-200 and NeuN expression was weak and did not changed between 2D and 3D culture. Yang and colleagues reported similar effect of 3D culture—cells from neurospheres expressed more Nestin after exposition to neural differentiation conditions, but any mature markers such as NeuN, MAP2, or glial marker GFAP [32]. However, 3D culture stage effected on future neural differentiation. Neurosphere culture of UC-MSC increased expression of neural [43] or both neural and glial [44] markers during neural differentiation. According to some reports, neurosphere culture of MSC improved neural differentiation potential [44]. Especially, Feng and colleagues claimed to have obtained even nerve-like cells with properties of astrocytes, neurons, and oligodendrocytes [45]. 3D culture could be an initial step preceding the neural differentiation under the specific differentiation conditions.

Stemness-related transcription factors (SRTF)—Sox2, Nanog, Rex1, Oct3/4, Klf4, are involved in cell divisions during embryonic development. Pluripotency and differentiation of MSC into all germ layers is broadly discussed. Oct3/4 was observed in WJ-MSC cultured in 5% oxygen concentration [46]. Nanog was detected in AD-MSC and BM-MSC, but not Sox-2 and Oct3/4 [47]. Although the discoveries of pluripotency genes in MSC, their expression is much weaker than in ESC [47] or iPSC [48]. 3D culture is proposed as a method to increase the expression of pluripotent genes by MSC. However, we did not observe the change of SRTF genes expression between 2D and 3D culture conditions what is not consistent with observations from other research groups. Pluripotent marker expression in MSC-neurospheres was observed by others in neurosphere-forming media [39,42,44] as well as in other sphere-inducing conditions [6,9,10,33]. Except the increased expression of SRTF, there were also observed epigenetic changes gene promotors and miRNAs responsible for pluripotency [9]. However, some research also did not report increased expression of stem cell markers in MSC-neurospheres—Bonilla-Porras and colleagues observed lower levels of Nanog, Oct4 and Sox2 in cells from WJ-MSC neurospheres than from 2D cultured WJ-MSC [43]. Dromard and colleagues did not notice the significant differences in expression of Nanog and Oct3/4 between spheres and monolayer [7]. The differences in above discussed results may be explained with the time of material collection. Cells in sphere did not exhibit pluripotency during whole period of 3D culture—expression was the highest between 3 and 5 div, after six div it drops [34]. We observed that even short time of adhesion change the RNA expression (Supplementary materials). Other authors reported that during 48 h after spheres return to 2D conditions, pluripotent expression decreases to the level presented in cultured monolayer cells [34]. Unfortunately, expression of pluripotent genes in MSC 3D still was significantly lower than in ESC [34].

It is discussed whether described effects are exaggerated by 3D conditions or could be influence by source of MSC. The comparative analysis are limited. WJ-MSC and BM-MSC generated secondary spheres after dissociation, while AD-MSC did not. What is more, secretomes profile of spheres were different—WJ-MSC secreted more neurotrophic factors. Protein level of neural markers inside sphere and under neural differentiation conditions also varied between different tissues. According to these observations, authors strongly suggested that WJ has better neurogenic potential than other sources [49]. In addition, Peng and colleagues performed identical analysis on MSC-spheres derived from two tissues—adipose and umbilical cord. The relative gene expression was similar, although the time of increase in gene expression differed. For some genes, such as Sox2 or Olig2 for UC-MSC spheres expression reached the maximum in second div, while for AD-MSC the peak was in seventh div [39,42]. It indicates that choice of MSC source be relevant.

Immunostaining revealed increased presence of surface marker SSEA4—glycosphingolipid which is included as a one of pluripotency markers. SSEA4 and SSEA3—other molecule from the same family—are characteristic for ESC cells [50]. SSEA3 is earlier in development—SSEA3 expression is the highest between four and eight cell stages of embryo, whereas SSEA4—morula and early blastocyst stages [51,52]. SSEA4 is also detected in early neuroepithelium [53,54] SSEA4-positive NSC can form neurospheres and SSEA4 expression remains sustained during 3 passages of NSC cultures [55]. Role of SSEA4 in NSC cells still remains unknown [51]. SSEA4 is also detected on MSC surface [18,56,57]. SSEA4 is presented during the MSC culture up to the ninth passage of WJ-MSC culture, while expression of SSEA3 is the highest directly after isolation and is detected only up to fifth passage [56]. SSEA4-positive BM-MSC population characterizes with better clonogenity [57]. SSEA4 together with small cell size are proposed as prognostic markers to distinguish young and old cells, which can be isolated with FACS [58]. SSEA4 expression may depend on cell culture conditions—it was correlated with medium serum concentration [56]. SSEA4, even counted as pluripotency marker, does not always coexist with other pluripotency genes what was shown on population of limbal stem cells (which resembled MSC) [59]. Similar situation was observed in MSC—SSEA4-positive WJ-MSC expressed SRTF genes on the same level as SSEA-4-negative cells [56]. Increased detection of SSEA4 in 3D cultured population might suggest acquiring this marker during spheroids culture or better survival of SSEA4-positive cells. Double staining for SSEA4 and Caspase-3 confirmed that apoptosis did occur only in a small fraction of SSEA-4-positive cells (Supplementary Materials). We rather assume that SSEA4-positve cells are more favorable to endure harsh conditions of neurosphere formation. The presence of SSEA4 positive cells could correspond with their undifferentiated state. According to Arrora and colleagues, undifferentiated state of cells in unfavorable conditions could be connected with quiescence [60]. Stress conditions such as loss of adhesion and high cell density lead to the quiescent state of MSC. Described earlier hypoxia and oxidative stress linked with sphere size [27] seems to be also responsible for cell persistent in a quiescent state in order to be available for further tissue repair and regeneration in beneficial [61].

In accordance with suggested neuroectodermal origin of distinct subpopulation of MSC [62], existing subpopulation of small cells and increased containment of early neural markers Nestin and β-III-Tubulin indicate that prolonged 3D culture enable to select cells with higher neural potential. Especially, very interesting for us was expression of Nestin, which may indicate the other explanation. Even though, quiescent NSC are negative for Nestin, active NSC express Nestin—importantly to mention, active NSC is the population which forms neurospheres, whereas quiescent NSC do not [63]. When it comes to MSC, Nestin may be the key explaining neurosphere formation—Nestin-negative cells do not form spheroids [64,65]. Nestin expression and sphere-forming capacities suggest neuroectodermal origin of this population [64]. Nestin-positive population derived from rat bone marrow contains MSC and Neural Crest Stem Cells as well [66]. During development, cells from neural crest which are not committed to glial lineage yet may migrate along nerves to bone marrow and there reside for the rest of life [67]. Those cells create a niche for hematopoietic stem cells (HSC) and remains quiescent in adulthood [64,65]. Under spheroids culture formed by magnetic levitation, increased Nestin expression is explained as quiescent state of such cells [67]. Neural crest-derived cells occur not only in bone marrow but also in other tissues containing MSC such as adipose [68,69]. Human umbilical cord blood contains neural crest-like progenitor cells which express Nestin, form neurospheres and can differentiate into neuronal and glial lineages [70]. Interestingly, we also reported more SSEA4-positive cells after long 3D cell culture. High content of SSEA3 and SSEA4 was observed during cell progeny activated by antibody mimicking Interferon-I [71]. Taking together, increased levels of Nestin and SSEA4 suggest quiescence of cells which remained over 20 div of 3D culture. Quiescent state can be the factor that enables survival of MSC during long-termed non-adherent conditions.

5. Conclusions

We confirmed that WJ-MSC can be cultured as spheroids on low-attachment surface with culture medium dedicated for neurospheres for at least 20 div. This type of culture characterizes with increased cell death after first three div, but then, viability stabilizes in later stage. Long-termed 3D culture of WJ-MSC as spheroids did not improved the cell condition. In fact, it reduced stemness and increased senescence process. However, it improved the occurrence of early neural markers what might indicate the survival of cell subpopulation with unique features, such as SSEA4 expression and possible quiescent state.

Supplementary Materials: The following are available online at https://www.mdpi.com/2073-4409/10/4/719/s1, Table S1: List of secondary antibodies used for immunocytochemistry, Figure S1: Secondary antibody staining controls for 2D and 3D conditions, Table S2: Population doubling time values (days) for 2D and 3D cultured WJ-MSC, Table S3: Quantitative data of immunocytochemistry staining analysis, Figure S2: Immunocytochemistry double staining for SSEA4 and Cleaved caspase-3 in WJ-MSC spheroid, Table S4: Relative gene expression level quantitative data. Figure S3: Extended analysis for relative gene expression level (fold change, mean ± SD) of neural and pluripotency phenotype in 2D and 3D WJ-MSC. Table S5: Relative gene expression level quantitative data for RNA collected directly from free floating spheroids.

Author Contributions: Conceptualization, A.K. and A.S.; methodology, A.K., M.K., and A.S.; software, A.K.; validation, A.K, A.W., and M.K.; formal analysis, A.K., A.W., and M.K.; investigation, A.K.; resources, A.K.; data curation, A.K. and A.S.; writing—original draft preparation, A.K.; writing—review and editing, A.S.; visualization, A.K.; supervision, M.K. and A.S.; project administration, A.S.; funding acquisition, A.S. All authors have read and agreed to the published version of the manuscript.

Funding: This research was funded by The National Center for Research Grant NCN 2018/31/B/NZ4/03172, and ESF, POWR.03.02.00-00-I028/17-00.

Institutional Review Board Statement: The study was conducted according to the guidelines of the Declaration of Helsinki, and approved by the Ethics Committee of Warsaw Medical University guideline (KB/213/2016).

Informed Consent Statement: Informed consent was obtained from all subjects involved in the study.

Data Availability Statement: The data presented in this study are available on request from the corresponding author.

Acknowledgments: We are thankful to Laboratory of Advanced Microscopy Techniques for assistance with Confocal Microscopy imaging, and colleagues from Translational Platform for Regenerative Medicine and Department of Stem Cell Bioengineering for supporting during research.

Conflicts of Interest: The authors declare no conflict of interest.

References

1. Friedenstein, A.J.; Piatetzky-Shapiro, I.I.; Petrakova, K.V. Osteogenesis in transplants of bone marrow cells. *Development* **1966**, *16*, 381–390.
2. Dominici, M.; Le Blanc, K.; Mueller, I.; Slaper-Cortenbach, I.; Marini, F.C.; Krause, D.S.; Deans, R.J.; Keating, A.; Prockop, D.J.; Horwitz, E.M. Minimal criteria for defining multipotent mesenchymal stromal cells. The International Society for Cellular Therapy position statement. *Cytotherapy* **2006**, *8*, 315–317. [CrossRef] [PubMed]
3. Sart, S.; Tsai, A.C.; Li, Y.; Ma, T. Three-dimensional aggregates of mesenchymal stem cells: Cellular mechanisms, biological properties, and applications. *Tissue Eng. Part B Rev.* **2014**, *20*, 365–380. [CrossRef] [PubMed]
4. Jauković, A.; Abadjieva, D.; Trivanović, D.; Stoyanova, E.; Kostadinova, M.; Pashova, S.; Kestendjieva, S.; Kukolj, T.; Jeseta, M.; Kistanova, E.; et al. Specificity of 3D MSC Spheroids Microenvironment: Impact on MSC Behavior and Properties. *Stem Cell Rev. Rep.* **2020**, *16*, 853–875. [CrossRef]
5. Knight, E.; Przyborski, S. Advances in 3D cell culture technologies enabling tissue-like structures to be created in vitro. *J. Anat.* **2015**, *227*, 746–756. [CrossRef] [PubMed]
6. Liu, Y.; Muñoz, N.; Tsai, A.C.; Logan, T.M.; Ma, T. Metabolic Reconfiguration Supports Reacquisition of Primitive Phenotype in Human Mesenchymal Stem Cell Aggregates. *Stem Cells* **2017**, *35*, 398–410. [CrossRef]
7. Dromard, C.; Bourin, P.; André, M.; de Barros, S.; Casteilla, L.; Planat-Benard, V. Human adipose derived stroma/stem cells grow in serum-free medium as floating spheres. *Exp. Cell Res.* **2011**, *317*, 770–780. [CrossRef]
8. Bartosh, T.J.; Ylöstalo, J.H.; Mohammadipoor, A.; Bazhanov, N.; Coble, K.; Claypool, K.; Lee, R.H.; Choi, H.; Prockop, D.J. Aggregation of human mesenchymal stromal cells (MSCs) into 3D spheroids enhances their antiinflammatory properties. *Proc. Natl. Acad. Sci. USA* **2010**, *107*, 13724–13729. [CrossRef]
9. Guo, L.; Zhou, Y.; Wang, S.; Wu, Y. Epigenetic changes of mesenchymal stem cells in three-dimensional (3D) spheroids. *J. Cell. Mol. Med.* **2014**, *18*, 2009–2019. [CrossRef]
10. Cheng, N.-C.; Chen, S.-Y.; Li, J.-R.; Young, T.-H. Short-Term Spheroid Formation Enhances the Regenerative Capacity of Adipose-Derived Stem Cells by Promoting Stemness, Angiogenesis, and Chemotaxis. *Stem Cells Transl. Med.* **2013**, *2*, 584–594. [CrossRef]
11. Frith, J.E.; Thomson, B.; Genever, P.G. Dynamic three-dimensional culture methods enhance mesenchymal stem cell properties and increase therapeutic potential. *Tissue Eng. Part C Methods* **2010**, *16*, 735–749. [CrossRef]
12. Vescovi, A.L.; Parati, E.A.; Gritti, A.; Poulin, P.; Ferrario, M.; Wanke, E.; Frölichsthal-Schoeller, P.; Cova, L.; Arcellana-Panlilio, M.; Colombo, A.; et al. Isolation and Cloning of Multipotential Stem Cells from the Embryonic Human CNS and Establishment of Transplantable Human Neural Stem Cell Lines by Epigenetic Stimulation. *Exp. Neurol.* **1999**, *156*, 71–83. [CrossRef]
13. Reynolds, B.A.; Weiss, S. Generation of neurons and astrocytes from isolated cells of the adult mammalian central nervous system. *Science* **1992**, *255*, 1707–1710. [CrossRef] [PubMed]

14. Forraz, N.; Mcguckin, C.P. The umbilical cord: A rich and ethical stem cell source to advance regenerative medicine. *Cell Prolif.* **2011**, *44*, 60–69. [CrossRef]
15. Pirjali, T.; Azarpira, N.; Ayatollahi, M.; Aghdaie, M.H.; Geramizadeh, B.; Talai, T. Isolation and Characterization of Human Mesenchymal Stem Cells Derived from Human Umbilical Cord Wharton's Jelly and Amniotic Membrane. *Int. J. Organ Transplant. Med.* **2013**, *4*, 111–116. [PubMed]
16. Sun, W.; Buzanska, L.; Domanska-Janik, K.; Salvi, R.J.; Stachowiak, M.K. Voltage-Sensitive and Ligand-Gated Channels in Differentiating Neural Stem-Like Cells Derived from the Nonhematopoietic Fraction of Human Umbilical Cord Blood. *Stem Cells* **2005**, *23*, 931–945. [CrossRef]
17. Li, Z.; Zhang, C.; Weiner, L.P.; Zhang, Y.; Zhong, J.F. Molecular characterization of heterogeneous mesenchymal stem cells with single-cell transcriptomes. *Biotechnol. Adv.* **2013**, *31*, 312–317. [CrossRef]
18. Drela, K.; Lech, W.; Figiel-Dabrowska, A.; Zychowicz, M.; Mikula, M.; Sarnowska, A.; Domanska-Janik, K. Enhanced neuro-therapeutic potential of Wharton's Jelly-derived mesenchymal stem cells in comparison with bone marrow mesenchymal stem cells culture. *Cytotherapy* **2016**, *18*, 497–509. [CrossRef] [PubMed]
19. Somoza, R.; Conget, P.; Rubio, F.J. Neuropotency of Human Mesenchymal Stem Cell Cultures: Clonal Studies Reveal the Contribution of Cell Plasticity and Cell Contamination. *Biol. Blood Marrow Transplant.* **2008**, *14*, 546–555. [CrossRef] [PubMed]
20. Choi, S.A.; Lee, J.Y.; Wang, K.-C.; Phi, J.H.; Song, S.H.; Song, J.; Kim, S.-K. Human adipose tissue-derived mesenchymal stem cells: Characteristics and therapeutic potential as cellular vehicles for prodrug gene therapy against brainstem gliomas. *Eur. J. Cancer* **2012**, *48*, 129–137. [CrossRef]
21. Pisciotta, A.; Bertoni, L.; Riccio, M.; Mapelli, J.; Bigiani, A.; La Noce, M.; Orciani, M.; de Pol, A.; Carnevale, G. Use of a 3D Floating Sphere Culture System to Maintain the Neural Crest-Related Properties of Human Dental Pulp Stem Cells. *Front. Physiol.* **2018**, *9*, 547. [CrossRef] [PubMed]
22. Abe, S.; Yamaguchi, S.; Sato, Y.; Harada, K. Sphere-Derived Multipotent Progenitor Cells Obtained From Human Oral Mucosa Are Enriched in Neural Crest Cells. *Stem Cells Transl. Med.* **2016**, *5*, 117–128. [CrossRef]
23. Lee, S.-T.; Chu, K.; Jung, K.-H.; Song, Y.-M.; Jeon, D.; Kim, S.U.; Kim, M.; Lee, S.K.; Roh, J.-K. Direct Generation of Neurosphere-Like Cells from Human Dermal Fibroblasts. *PLoS ONE* **2011**, *6*, e21801. [CrossRef]
24. Yang, R.; Xu, X. Isolation and culture of neural crest stem cells from human hair follicles. *J. Vis. Exp.* **2013**, *6*, e3194. [CrossRef]
25. Fu, L.; Zhu, L.; Huang, Y.; Lee, T.D.; Forman, S.J.; Shih, C.C. Derivation of neural stem cells from mesenchymal stem cells: Evidence for a bipotential stem cell population. *Stem Cells Dev.* **2008**, *17*, 1109–1121. [CrossRef] [PubMed]
26. Baraniak, P.R.; McDevitt, T.C. Scaffold-free culture of mesenchymal stem cell spheroids in suspension preserves multilineage potential. *Cell Tissue Res.* **2012**, *347*, 701–711. [CrossRef] [PubMed]
27. Murphy, K.C.; Hung, B.P.; Browne-Bourne, S.; Zhou, D.; Yeung, J.; Genetos, D.C.; Leach, J.K. Measurement of oxygen tension within mesenchymal stem cell spheroids. *J. R. Soc. Interface* **2017**, *14*, 20160851. [CrossRef]
28. Vieira, M.S.; Santos, A.K.; Vasconcellos, R.; Goulart, V.A.M.; Parreira, R.C.; Kihara, A.H.; Ulrich, H.; Resende, R.R. Neural stem cell differentiation into mature neurons: Mechanisms of regulation and biotechnological applications. *Biotechnol. Adv.* **2018**, *36*, 1946–1970. [CrossRef] [PubMed]
29. Jurga, M.; Forraz, N.; Basford, C.; Atzeni, G.; Trevelyan, A.J.; Habibollah, S.; Ali, H.; Zwolinski, S.A.; McGuckin, C.P. Neurogenic properties and a clinical relevance of multipotent stem cells derived from cord blood samples stored in the biobanks. *Stem Cells Dev.* **2012**, *21*, 923–936. [CrossRef]
30. Wagner, W.; Ho, A.D.; Zenke, M. Different facets of aging in human mesenchymal stem cells. *Tissue Eng. Part B Rev.* **2010**, *16*, 445–453. [CrossRef] [PubMed]
31. Gu, Y.; Li, T.; Ding, Y.; Sun, L.; Tu, T.; Zhu, W.; Hu, J.; Sun, X. Changes in mesenchymal stem cells following long-term culture in vitro. *Mol. Med. Rep.* **2016**, *13*, 5207–5215. [CrossRef] [PubMed]
32. Yang, E.; Liu, N.; Tang, Y.; Hu, Y.; Zhang, P.; Pan, C.; Dong, S.; Zhang, Y.; Tang, Z. Generation of neurospheres from human adipose-derived stem cells. *BioMed Res. Int.* **2015**, *2015*, 743714. [CrossRef] [PubMed]
33. Zhang, Q.; Nguyen, A.L.; Shi, S.; Hill, C.; Wilder-Smith, P.; Krasieva, T.B.; Le, A.D. Three-dimensional spheroid culture of human gingiva-derived mesenchymal stem cells enhances mitigation of chemotherapy-induced oral mucositis. *Stem Cells Dev.* **2012**, *21*, 937–947. [CrossRef] [PubMed]
34. Pennock, R.; Bray, E.; Pryor, P.; James, S.; McKeegan, P.; Sturmey, R.; Genever, P. Human cell dedifferentiation in mesenchymal condensates through controlled autophagy. *Sci. Rep.* **2015**, *5*, 13113. [CrossRef]
35. Santos, J.M.; Camões, S.P.; Filipe, E.; Cipriano, M.; Barcia, R.N.; Filipe, M.; Teixeira, M.; Simões, S.; Gaspar, M.; Mosqueira, D.; et al. Three-dimensional spheroid cell culture of umbilical cord tissue-derived mesenchymal stromal cells leads to enhanced paracrine induction of wound healing. *Stem Cell Res. Ther.* **2015**, *6*, 1–19. [CrossRef] [PubMed]
36. Kim, J.; Ma, T. Endogenous extracellular matrices enhance human mesenchymal stem cell aggregate formation and survival. *Biotechnol. Prog.* **2013**, *29*, 441–451. [CrossRef]
37. Lech, W.; Figiel-Dabrowska, A.; Sarnowska, A.; Drela, K.; Obtulowicz, P.; Noszczyk, B.H.; Buzanska, L.; Domanska-Janik, K. Phenotypic, Functional, and Safety Control at Preimplantation Phase of MSC-Based Therapy. *Stem Cells Int.* **2016**, *2016*, 2514917. [CrossRef] [PubMed]
38. Hiraoka, Y.; Kimbara, K. Rapid assessment of the physiological status of the polychlorinated biphenyl degrader Comamonas testosteroni TK102 by flow cytometry. *Appl. Environ. Microbiol.* **2002**, *68*, 2031–2035. [CrossRef] [PubMed]

39. Peng, C.; Lu, L.; Li, Y.; Hu, J. Neurospheres Induced from Human Adipose-Derived Stem Cells as a New Source of Neural Progenitor Cells. *Cell Transplant.* **2019**, *28*, 66S–75S. [CrossRef]
40. van Winkle, A.P.; Gates, I.D.; Kallos, M.S. Mass transfer limitations in embryoid bodies during human embryonic stem cell differentiation. *Cells Tissues Organs* **2012**, *196*, 34–47. [CrossRef] [PubMed]
41. Ge, D. Effect of the neurosphere size on the viability and metabolism of neural stem/progenitor cells. *Afr. J. Biotechnol.* **2012**, *11*, 3976–3985. [CrossRef]
42. Peng, C.; Li, Y.; Lu, L.; Zhu, J.; Li, H.; Hu, J. Efficient One-Step Induction of Human Umbilical Cord-Derived Mesenchymal Stem Cells (UC-MSCs) Produces MSC-Derived Neurospheres (MSC-NS) with Unique Transcriptional Profile and Enhanced Neurogenic and Angiogenic Secretomes. *Stem Cells Int.* **2019**, *2019*, 9208173. [CrossRef]
43. Bonilla-Porras, A.R.; Velez-Pardo, C.; Jimenez-Del-Rio, M. Fast transdifferentiation of human Wharton's jelly mesenchymal stem cells into neurospheres and nerve-like cells. *J. Neurosci. Methods* **2017**, *282*, 52–60. [CrossRef] [PubMed]
44. Mukai, T.; Nagamura-Inoue, T.; Shimazu, T.; Mori, Y.; Takahashi, A.; Tsunoda, H.; Yamaguchi, S.; Tojo, A. Neurosphere formation enhances the neurogenic differentiation potential and migratory ability of umbilical cord-mesenchymal stromal cells. *Cytotherapy* **2016**, *18*, 229–241. [CrossRef]
45. Feng, N.; Han, Q.; Li, J.; Wang, S.; Li, H.; Yao, X.; Zhao, R.C. Generation of highly purified neural stem cells from human adipose-derived mesenchymal stem cells by Sox1 activation. *Stem Cells Dev.* **2014**, *23*, 515–529. [CrossRef]
46. Drela, K.; Sarnowska, A.; Siedlecka, P.; Szablowska-Gadomska, I.; Wielgos, M.; Jurga, M.; Lukomska, B.; Domanska-Janik, K. Low oxygen atmosphere facilitates proliferation and maintains undifferentiated state of umbilical cord mesenchymal stem cells in an hypoxia inducible factor-dependent manner. *Cytotherapy* **2014**, *16*, 881–892. [CrossRef]
47. Pierantozzi, E.; Gava, B.; Manini, I.; Roviello, F.; Marotta, G.; Chiavarelli, M.; Sorrentino, V. Pluripotency regulators in human mesenchymal stem cells: Expression of NANOG but not of OCT-4 and SOX-2. *Stem Cells Dev.* **2011**, *20*, 915–923. [CrossRef]
48. Musiał-Wysocka, A.; Kot, M.; Sułkowski, M.; Badyra, B.; Majka, M. Molecular and Functional Verification of Wharton's Jelly Mesenchymal Stem Cells (WJ-MSCs) Pluripotency. *Int. J. Mol. Sci.* **2019**, *20*, 1807. [CrossRef]
49. Balasubramanian, S.; Thej, C.; Venugopal, P.; Priya, N.; Zakaria, Z.; SundarRaj, S.; Majumdar, A. Sen Higher propensity of Wharton's jelly derived mesenchymal stromal cells towards neuronal lineage in comparison to those derived from adipose and bone marrow. *Cell Biol. Int.* **2013**, *37*, 507–515. [CrossRef]
50. Kallas, A.; Pook, M.; Maimets, M.; Zimmermann, K.; Maimets, T. Nocodazole Treatment Decreases Expression of Pluripotency Markers Nanog and Oct4 in Human Embryonic Stem Cells. *PLoS ONE* **2011**, *6*, e19114. [CrossRef]
51. Itokazu, Y.; Yu, R.K. Glycolipid Antigens in Neural Stem Cells. In *Neural Surface Antigens: From Basic Biology Towards Biomedical Applications*; Elsevier Inc.: Amsterdam, The Netherlands, 2015; pp. 91–102. ISBN 9780128011263.
52. Fenderson, B.A.; Eddy, E.M.; Hakomori, S.-I. Glycoconjugate expression during embryogenesis and its biological significance. *BioEssays* **1990**, *12*, 173–179. [CrossRef]
53. Barraud, P.; He, X.; Caldwell, M.A.; Franklin, R.J. Secreted factors from olfactory mucosa cells expanded as free-floating spheres increase neurogenesis in olfactory bulb neurosphere cultures. *BMC Neurosci.* **2008**, *9*, 24. [CrossRef] [PubMed]
54. Piao, J.-H.; Odeberg, J.; Samuelsson, E.-B.; Kjældgaard, A.; Falci, S.; Seiger, Å.; Sundström, E.; Åkesson, E. Cellular composition of long-term human spinal cord- and forebrain-derived neurosphere cultures. *J. Neurosci. Res.* **2006**, *84*, 471–482. [CrossRef] [PubMed]
55. Barraud, P.; Stott, S.; Møllgård, K.; Parmar, M.; Björklund, A. In vitro characterization of a human neural progenitor cell coexpressing SSEA4 and CD133. *J. Neurosci. Res.* **2007**, *85*, 250–259. [CrossRef]
56. He, H.; Nagamura-Inoue, T.; Tsunoda, H.; Yuzawa, M.; Yamamoto, Y.; Yorozu, P.; Agata, H.; Tojo, A. Stage-Specific Embryonic Antigen 4 in Wharton's Jelly—Derived Mesenchymal Stem Cells Is Not a Marker for Proliferation and Multipotency. *Tissue Eng. Part A* **2014**, *20*, 1314–1324. [CrossRef] [PubMed]
57. Lai, Y.; Sun, Y.; Skinner, C.M.; Son, E.L.; Lu, Z.; Tuan, R.S.; Jilka, R.L.; Ling, J.; Chen, X.-D. Reconstitution of Marrow-Derived Extracellular Matrix Ex Vivo: A Robust Culture System for Expanding Large-Scale Highly Functional Human Mesenchymal Stem Cells. *Stem Cells Dev.* **2010**, *19*, 1095–1107. [CrossRef] [PubMed]
58. Block, T.J.; Marinkovic, M.; Tran, O.N.; Gonzalez, A.O.; Marshall, A.; Dean, D.D.; Chen, X.D. Restoring the quantity and quality of elderly human mesenchymal stem cells for autologous cell-based therapies. *Stem Cell Res. Ther.* **2017**, *8*, 239. [CrossRef]
59. Lim, M.N.; Hussin, N.H.; Othman, A.; Umapathy, T.; Baharuddin, P.; Jamal, R.; Zakaria, Z. Ex vivo expanded SSEA-4+ human limbal stromal cells are multipotent and do not express other embryonic stem cell markers. *Mol. Vis.* **2012**, *18*, 1289–1300. [PubMed]
60. Arora, R.; Rumman, M.; Venugopal, N.; Gala, H.; Dhawan, J. Mimicking muscle stem cell quiescence in culture: Methods for synchronization in reversible arrest. In *Methods in Molecular Biology*; Humana Press Inc.: Totowa, NJ, USA, 2017; Volume 1556, pp. 283–302.
61. Demaria, M.; Campisi, J. Matters of life and breath: A role for hypoxia in determining cell state. *Aging* **2012**, *4*, 523–524. [CrossRef]
62. Niibe, K.; Zhang, M.; Nakazawa, K.; Morikawa, S.; Nakagawa, T.; Matsuzaki, Y.; Egusa, H. The potential of enriched mesenchymal stem cells with neural crest cell phenotypes as a cell source for regenerative dentistry. *Jpn. Dent. Sci. Rev.* **2017**, *53*, 25–33. [CrossRef] [PubMed]
63. Codega, P.; Silva-Vargas, V.; Paul, A.; Maldonado-Soto, A.R.; DeLeo, A.M.; Pastrana, E.; Doetsch, F. Prospective Identification and Purification of Quiescent Adult Neural Stem Cells from Their In Vivo Niche. *Neuron* **2014**, *82*, 545–559. [CrossRef]

64. Méndez-Ferrer, S.; Michurina, T.V.; Ferraro, F.; Mazloom, A.R.; MacArthur, B.D.; Lira, S.A.; Scadden, D.T.; Ma'ayan, A.; Enikolopov, G.N.; Frenette, P.S. Mesenchymal and haematopoietic stem cells form a unique bone marrow niche. *Nature* **2010**, *466*, 829–834. [CrossRef]
65. Isern, J.; García-García, A.; Martín, A.M.; Arranz, L.; Martín-Pérez, D.; Torroja, C.; Sánchez-Cabo, F.; Méndez-Ferrer, S. The neural crest is a source of mesenchymal stem cells with specialized hematopoietic stem cell niche function. *eLife* **2014**, *3*, e03696. [CrossRef]
66. Wislet-Gendebien, S.; Laudet, E.; Neirinckx, V.; Alix, P.; Leprince, P.; Glejzer, A.; Poulet, C.; Hennuy, B.; Sommer, L.; Shakhova, O.; et al. Mesenchymal stem cells and neural crest stem cells from adult bone marrow: Characterization of their surprising similarities and differences. *Cell. Mol. Life Sci.* **2012**, *69*, 2593–2608. [CrossRef]
67. Lewis, E.E.L.; Wheadon, H.; Lewis, N.; Yang, J.; Mullin, M.; Hursthouse, A.; Stirling, D.; Dalby, M.J.; Berry, C.C. A Quiescent, Regeneration-Responsive Tissue Engineered Mesenchymal Stem Cell Bone Marrow Niche Model via Magnetic Levitation. *ACS Nano* **2016**, *10*, 8346–8354. [CrossRef]
68. Coste, C.; Neirinckx, V.; Sharma, A.; Agirman, G.; Rogister, B.; Foguenne, J.; Lallemend, F.; Gothot, A.; Wislet, S. Human bone marrow harbors cells with neural crest-associated characteristics like human adipose and dermis tissues. *PLoS ONE* **2017**, *12*, e0177962. [CrossRef]
69. Sowa, Y.; Imura, T.; Numajiri, T.; Takeda, K.; Mabuchi, Y.; Matsuzaki, Y.; Nishino, K. Adipose stromal cells contain phenotypically distinct adipogenic progenitors derived from neural crest. *PLoS ONE* **2013**, *8*, e84206. [CrossRef]
70. Al-Bakri, Z.; Ishige-Wada, M.; Fukuda, N.; Yoshida-Noro, C.; Nagoshi, N.; Okano, H.; Mugishima, H.; Matsumoto, T. Isolation and characterization of neural crest-like progenitor cells in human umbilical cord blood. *Regen. Ther.* **2020**, *15*, 53–63. [CrossRef]
71. Hatzfeld, A.; Eid, P.; Peiffer, I.; Li, M.L.; Barbet, R.; Oostendorp, R.A.J.; Haydont, V.; Monier, M.N.; Milon, L.; Fortunel, N.; et al. A sub-population of high proliferative potential-quiescent human mesenchymal stem cells is under the reversible control of interferon α/β. *Leukemia* **2007**, *21*, 714–724. [CrossRef]

cells

MDPI

Article

The Effects of the Combination of Mesenchymal Stromal Cells and Nanofiber-Hydrogel Composite on Repair of the Contused Spinal Cord

Agnes E. Haggerty [1], Ines Maldonado-Lasunción [1,2,3,4], Yohshiro Nitobe [1,5], Kentaro Yamane [1,6], Megan M. Marlow [1], Hua You [7], Chi Zhang [8,9,10], Brian Cho [8,11], Xiaowei Li [8,9,10,12], Sashank Reddy [10,11,13], Hai-Quan Mao [8,9,10,13,*] and Martin Oudega [3,4,14,15,*]

1 The Miami Project to Cure Paralysis, University of Miami, Miami, FL 33136, USA; ahaggerty@akoyabio.com (A.E.H.); imaldola@gmail.com (I.M.-L.); n1t0bey0sh1r0@gmail.com (Y.N.); woodblocks0311@gmail.com (K.Y.); mmarlow135@yahoo.com (M.M.M.)
2 Department of Regeneration of Sensorimotor Systems, Netherlands Institute for Neuroscience, Institute of the Royal Netherlands Academy of Arts and Sciences, 1105 BA Amsterdam, The Netherlands
3 Shirley Ryan AbilityLab, Chicago, IL 60611, USA
4 Department of Physical Therapy and Human Movements Sciences, Northwestern University, Chicago, IL 60611, USA
5 Department of Orthopedic Surgery, Hirosaki University Graduate School of Medicine, Hirosaki 036-8562, Japan
6 Department of Orthopedic Surgery, Okayama University Graduate School of Medicine, Dentistry, and Pharmaceutical Science, Kitaku, Okayama 700-8558, Japan
7 Department of Oncology and Hematology, Affiliated Cancer Hospital & Institute of Guangzhou Medical University, Guangzhou 510095, China; youhua307@163.com
8 Translational Tissue Engineering Center, Johns Hopkins School of Medicine, Baltimore, MD 21231, USA; czhang96@jhu.edu (C.Z.); bcho8@uw.edu (B.C.); xiaoweili@wustl.edu (X.L.)
9 Department of Materials Science and Engineering, Johns Hopkins University, Baltimore, MD 21218, USA
10 Institute for NanoBioTechnology, Johns Hopkins University, Baltimore, MD 21218, USA; sreddy6@jhmi.edu
11 Department of Plastic and Reconstructive Surgery, Johns Hopkins School of Medicine, Baltimore, MD 21287, USA
12 Division of Plastic and Reconstructive Surgery, Department of Surgery, Washington University School of Medicine, St. Louis, MO 63110, USA
13 Department of Biomedical Engineering, Johns Hopkins University, Baltimore, MD 21205, USA
14 Department of Neuroscience, Northwestern University, Chicago, IL 60611, USA
15 Edward Hines Jr. VA Hospital, Hines, IL 60141, USA
* Correspondence: hmao@jhu.edu (H.-Q.M.); moudega@sralab.org (M.O.)

Citation: Haggerty, A.E.; Maldonado-Lasunción, I.; Nitobe, Y.; Yamane, K.; Marlow, M.M.; You, H.; Zhang, C.; Cho, B.; Li, X.; Reddy, S.; et al. The Effects of the Combination of Mesenchymal Stromal Cells and Nanofiber-Hydrogel Composite on Repair of the Contused Spinal Cord. *Cells* **2022**, *11*, 1137. https://doi.org/10.3390/cells11071137

Academic Editors: Joni H. Ylostalo, Nisha C. Dur and Eva Sykova

Received: 17 February 2022
Accepted: 23 March 2022
Published: 28 March 2022

Abstract: A bone marrow-derived mesenchymal stromal cell (MSC) transplant and a bioengineered nanofiber-hydrogel composite (NHC) have been shown to stimulate nervous tissue repair in the contused spinal cord in rodent models. Here, these two modalities were combined to assess their repair effects in the contused spinal cord in adult rats. Cohorts of contused rats were treated with MSC in NHC (MSC-NHC), MSC in phosphate-buffered saline (MSC-PBS), NHC, or PBS injected into the contusion site at 3 days post-injury. One week after injury, there were significantly fewer CD68+ cells in the contusion with MSC-NHC and NHC, but not MSC-PBS. The reduction in CD86+ cells in the injury site with MSC-NHC was mainly attributed to NHC. One and eight weeks after injury, we found a greater CD206+/CD86+ cell ratio with MSC-NHC or NHC, but not MSC-PBS, indicating a shift from a pro-inflammatory towards an anti-inflammatory milieu in the injury site. Eight weeks after injury, the injury size was significantly reduced with MSC-NHC, NHC, and MSC-PBS. At this time, astrocyte, and axon presence in the injury site was greater with MSC-NHC compared with MSC-PBS. We did not find a significant effect of NHC on MSC transplant survival, and hind limb function was similar across all groups. However, we did find fewer macrophages at 1 week post-injury, more macrophages polarized towards a pro-regenerative phenotype at 1 and 8 weeks after injury, and reduced injury volume, more astrocytes, and more axons at 8 weeks after injury in rats with MSC-NHC and NHC alone compared with MSC-PBS; these findings were especially significant between rats with MSC-NHC and MSC-PBS. The data support further study in the use of an NHC-MSC combination transplant in the contused spinal cord.

Keywords: nanofiber-hydrogel composite; spinal cord injury; inflammation; macrophages; secondary injury; astrocytes; axon growth

---

1. Introduction

The prevalent mechanism of spinal cord injury (SCI) in humans is a contusion, which typically leads to nervous tissue damage and sensory and motor function loss [1,2]. The limited endogenous repair of central nervous tissue and typically poor recovery of function after a spinal cord contusion in humans motivates the ongoing search for reparative treatments [1,3,4].

Studies in animal models of contusive SCI have shown that a transplant of bone marrow-derived mesenchymal stromal cells (MSC) in the injury site elicited nervous tissue repair and, albeit not in all cases, function improvements [5–11]. MSC secrete paracrine factors [12–14] that can direct immunomodulatory effects [15–18], promote neuroprotection [12,13], and increase axon growth [19,20]. Transplanting MSC is considered a promising strategy to treat SCI, but the overall effects of an MSC transplant on nervous tissue repair in the injured spinal cord remain limited [6,9,21].

We demonstrated that an in situ forming, injectable nanofiber-hydrogel composite (NHC) elicited nervous tissue repair in an adult rat model of contusive SCI [22]. The NHC consists of poly(ε-caprolactone) (PCL) nanofibers interfacially bound to a hydrogel network formed from thiolated hyaluronic acid (HA-SH) and poly(ethylene glycol) diacrylate (PEGDA). The composite has a shear storage modulus similar to that of the nervous tissue in the spinal cord [23,24] and provides a sufficiently high porosity to support host cell infiltration and migration [22,25]. Importantly, we showed that NHC modulated the inflammatory response towards anti-inflammatory, pro-regenerative, macrophage phenotypes, and facilitated tissue formation in the injury site [22].

In the present study, we combined two repair modalities by using the NHC as a transplant matrix for MSC. We investigated the combination's therapeutic potential in the contused adult rat thoracic spinal cord. The repair was assessed by evaluating immunomodulation, neuroprotection, and hindlimb locomotor function.

2. Materials and Methods

2.1. Animals

Adult female Sprague Dawley rats (n = 68, 200–220 g, Charles Rivers Laboratory; Wilmington, MA, USA) were used in this study. All animal procedures were performed according to the guidelines of the National Institutes of Health and the United States Department of Agriculture at the Miami Project to Cure Paralysis at the University of Miami, Miami, FL, and approved by the local Institutional Animal Care and Use Committee. The Assessment and Accreditation of Laboratory Animal Care accredited the animal facility where the procedures were performed. Pairs of rats were housed under a 12-h light/dark cycle with ad libitum access to food and water.

2.2. NHC Preparation

NHC was made following the protocol previously described [22]. Briefly, nanofibers were electrospun from a PCL solution (16% *w/w*; Sigma-Aldrich, St. Louis, MO, USA) in a mixture of dichloromethane (Sigma-Aldrich) and dimethylformamide (9/1, *v/v*; Sigma-Aldrich). Poly(9,9-dioctylfluorene-alt-benzothiadiazole) (F8BT; Sigma-Aldrich), a green fluorescent dye, was added to the PCL solution to enable fiber identification after injection [22]. Carboxyl groups were introduced to the surface of the fibers in a plasma cleaner (expanded plasma cleaner; PDC-001; Harrick Plasma; Ithaca, NY, USA). These carboxyl groups were initiated by ethyl dimethylaminopropyl carbodiimide (Sigma-Aldrich) and N-hydroxysuccinimide (Sigma-Aldrich) and then converted to maleimide (MAL) groups by N-(2-aminoethyl) maleimide (Sigma-Aldrich). The MAL-modified fibers were cryogenically milled, sterilized with 70% ethanol. All three components of NHC were stored individually at −20 °C, and thawed 30 min prior to use. NHC was prepared by mixing MAL-modified fibers (10 mg/mL) into a mix of HA-SH (4 mg/mL; ESI BIO, Alameda, CA) and PEGDA (2 mg/mL; ESI BIO) in sterile phosphate-buffered saline (PBS) [22,25] on ice. We mixed and injected NHC or MSC-NHC within 15 min after exposing the contused spinal cord (see below Section 2.4). Once mixed, NHC can be kept on ice for 6 h for injections.

2.3. MSC Preparation

MSCs were harvested from bone marrow from femurs and tibias of adult female Sprague Dawley rats (n = 6) following a previously described protocol [9,26]. MSCs were cultured on poly-D-lysine-coated dishes in Dulbecco's Modified Eagle Medium (DMEM; Sigma-Aldrich) with 10% fetal bovine serum (Mediatech; Herndon, VA, USA), 0.03% L-glutamine (Sigma-Aldrich), and 0.1% gentamycin (VWR; Radnor, PA, USA). The cells were transduced with lentiviral vectors encoding for green fluorescent protein (GFP) at a multiplicity of infection of 100, passaged 4 times, and then harvested for transplantation [9]. MSC remained on ice until mixing with freshly prepared NHC or sterile PBS prior to injection into the epicenter of the spinal cord contusion.

2.4. Spinal Cord Contusion and Injection

Rats were anesthetized using an intraperitoneal injection of ketamine (50 mg/kg; Zoetis; Parsippany, NJ, USA) and dexmedetomidine (0.5 mg/kg; Zoetis). In the absence of corneal, hindlimb, and tail pinch reflexes, the rats were shaved and cleaned with Nolvasan® skin and wound cleaner (Zoetis). Refresh® Lacri-Lube® eye ointment (Allergan; Madison, NJ, USA) was applied to the eyes to prevent drying during surgery. The 9th thoracic spinal cord segment was exposed and impacted onto its dorsal midline with a force of 175 kDyne using the Infinite Horizon IH-0400 impactor (Precision Systems and Instrumentation LLC; Versailles, KY, USA) [27] resulting in a bilateral injury, as previously described [9,10,28]. After rinsing the injury site with sterile lactated Ringers' solution containing 0.1% gentamicin (VWR), the muscles were sutured, and the skin closed with Michel wound clips (Fine Science Tools; Foster City, CA, USA). The rats received a subcutaneous injection of the α2-adrenergic receptor antagonist, atipamezole (antisedan; 1.5 mg/kg; Zoetis) to reverse the sedative and analgesic effects of dexmedetomidine. Laboratory personnel monitored the rats until fully awake and applied after-surgery treatments as previously described [9,10].

At 3 days after injury, rats were anesthetized using an intraperitoneal injection of ketamine (50 mg/kg; Zoetis) and dexmedetomidine (0.5 mg/kg; Zoetis). The contused spinal cord was exposed and the injury epicenter was injected with a total volume of 5 μL with 500,000 MSC in NHC or PBS, or with NHC or PBS only at a rate of 1 μL/min using a 10 μL Hamilton syringe, fitted with a pulled glass needle, fixed to a stereotaxic device [6,9]. The internal diameter of the pulled glass needles was, on average, 120 μm; no clogging of the needle was observed during the injections. The needle was kept in place for 5 min after the injection was completed to prevent backflow. Possible leakage was verified using a UV flashlight to detect green fluorescence on the spinal cord. Note that for NHC, the individual components were mixed and injected within 15 min after exposing the contused spinal cord. For MSC-NHC, the NHC was mixed first and then the MSC was mixed into the NHC and injected within 15 min after exposing the contusion site. After the injection was completed, the muscles were sutured, and the skin was closed with Michel wound clips (Fine Science Tools). The rats received an intraperitoneal injection of antisedan (1.5 mg/kg; Zoetis) to reverse the anesthesia. Laboratory personnel monitored the rats until fully awake and applied after-surgery treatments as previously described [9,10]. Each of the four experimental groups contained 15 rats, which survived for 1 week (n = 5) or 8 weeks (n = 10) after injury. The rats that survived for 1 week were fixed (see histological preparations) and their spinal cord was used for histological and anatomical assessments. The rats that survived for 8 weeks were used to evaluate hind limb motor function during survival (n = 10) and, after fixation (see histological preparations), for histological and anatomical assessments (n = 5; randomly selected). Two rats died during surgery and were replaced.

2.5. Assessment of Hind Limb Function

Hind limb overground walking was assessed weekly using the Basso, Bresnahan, Beattie (BBB) open-field walking test [9,29]. Individual values were averaged per experimental group per time point. In addition, we assessed hind limb sensorimotor function at 4 and 8 weeks after injury using the horizontal ladder test [9,30,31]. For this, personnel blinded to the treatment used high-definition video recordings of the ladder crossings to enable accurate quantification of foot and leg slips, which were qualified as medium or large slips, respectively. The total sum of medium and large slips was expressed as a percentage of the number of steps taken to cross the ladder and averaged per rat and per group for each time point.

2.6. Histological Preparations

At 1 and 8 weeks after injury, rats were deeply anesthetized and transcardially perfused with 300 mL of PBS followed by 400 mL of 4% paraformaldehyde in PBS [9]. We dissected the spinal cord from the rats and kept them in the same fixative overnight, followed by 30% sucrose in PBS for 2 days. We then removed a 12 mm long segment centered on the contusion of each spinal cord and embedded these in frozen section medium (NEG 50; Richard-Allan Scientific, Thermo Fisher Scientific, Kalamazoo, MI, USA). The embedded frozen segments were cut into twelve series of 20 μm thick horizontal sections on a cryostat (CM 1950; Leica Biosystems; Buffalo Grove, IL, USA). Each series represented the width of the spinal cord with sections at 240 μm intervals. The sections were collected on glass slides and stored at −20 °C until staining.

2.7. Immunohistochemistry

For immunostaining, 20 μm thick tissue sections were rinsed for 5 min in PBS and incubated in PBS with 5% normal goat serum and 0.3% Triton X-100 for 1 h at room temperature to block non-specific antibody binding and permeabilize the tissue. Next, the sections were incubated with primary antibodies (Table 1) mixed in PBS with 5% normal goat serum and 0.3% Triton X-100 for 2 h at room temperature, followed by overnight at 4 °C. Subsequently, the sections were rinsed 3 times for 5 min in PBS and then incubated with goat secondary antibodies (Alexa Fluor 488, Alexa Fluor 555, or Alexa Fluor 647; 1:500; Invitrogen; Carlsbad, CA, USA) against the host of the primary antibody, mixed in PBS, for 2 h at room temperature. After this incubation, the sections were rinsed 3 times for 5 min in PBS, counterstained for 3 min with the nuclear marker, 4′,6-diamidino-2-phenylindole (DAPI; ThermoFisher Scientific; Waltham, MA, USA), rinsed 2 times for 5 min in PBS, and then covered with a glass slip with fluorescence mounting medium (DAKO; Agilent; Santa Clara, CA, USA). The 3 middle sections of the contusion site in a series were identified and imaged using the Olympus VS120 slide scanner with 10× (UPISAPO, 0.4NA, Air) or 20× (UPISAPO, 0.75NA, Air) objectives and fitted with a Hamamatsu ORCA-Flash 4.0 camera (Hamamatsu; Bridgewater, NJ, USA). Autofocus was set on the DAPI channel for cell quantification, and on the axon marker channel for axon quantification.

2.8. Injury Volume Assessment

For measuring the injury volume, we used the ImageJ measure function on sections stained for glial fibrillary acidic protein (GFAP) at 1 and 8 weeks after injury. The injury site was defined by the inner border of the surrounding GFAP+ scar. The injury site volume was determined by taking the surface area of the injury site in the middle section, multiplied by the known distance between sections, and adding up the surface area of the edge sections multiplied by 1/2 of the known distance between sections [22]. The individual injury volumes were averaged per experimental groups.

Table 1. Source, catalog number, and dilution of the used primary antibodies.

Primary Antibody *	Source	Catalog Number	Dilution **
Rabbit anti-GFP	Millipore	AB3080P	1:1000
Mouse anti-GFAP	Sigma	G3893	1:500
Rabbit anti-GFAP	PhosphoSolutions	620-GFAP	1:1000
Mouse anti-ED1 (CD68)	Millipore	MAB1435	1:200
Rabbit anti-NeuN	Millipore	ABN78	1:500
Rabbit anti-CD206	Abcam	Ab64693	1:500
Rabbit anti-CD86	Abcam	Ab53004	1:500
Rabbit anti-NFh	Millipore	AB1991	1:500

* All antibodies are commercially available. Their source and catalog number are listed. ** Dilutions in PBS with 5% normal goat serum and 0.3% Triton X-100.

2.9. Automated Quantification

An ImageJ Find Maxima plugin (modified from [32]) was used for the automated quantification of cells and axons. Images were corrected for the background by subtracting a Gaussian filtered image, converted to 8-bit, and thresholded to create a count mask. For CD86+ or CD206+ macrophages, two thresholded images from the general macrophage marker, CD68, and from each of the polarization markers were used to create double count masks, each overlaying the thresholded DAPI filter image. For GFP+ MSC, a thresholded image from GFP was used to create a count mask to overlay the thresholded DAPI filter image. Particles that were positive for both the count mask and DAPI were summed to determine the total count of positive hits. The average particle size of ten randomly selected nuclei for the specific cell of interest was used to create a multiplication factor to account for multiple overlaying nuclei. Total counts were multiplied by the multiplication factor to acquire standardized counts, which were averaged per rat and per experimental group.

2.10. Macrophage Quantification

We used automated quantification to evaluate CD68+, CD86+, and CD206+ macrophages in the spinal cord at 1 and 8 weeks after injury. Quantification was performed in each of the middle three sections of the contusion, in a 1.5 $\times$ 2 mm region of interest centered on the midpoint of the injury. This selected 2 mm long and 1.5 mm wide region of interest covered the injury site for 1 mm caudal and rostral from the midpoint of the contusion. CD68 is a pan-macrophage marker, CD86 is a pro-inflammatory macrophage marker, and CD206 is an anti-inflammatory macrophage marker. While recognizing the range of phenotypes among macrophages within an injury site, we used CD86 to indicate a pro-inflammatory (M1-like) macrophage phenotype and CD206 to indicate an anti-inflammatory (M2-like) macrophage phenotype. We determined cell segmentation counts (DAPI+) residing within double-positive (CD86+/CD68+ and CD206+/CD68+) masks. We standardized the counts by dividing the average particle size by the average size of 10 nuclei of each of the macrophage populations to create a multiplication factor to account for multiple overlaying nuclei. We reported our particle counts multiplied by the multiplication factor as the standardized average count, which were averaged per rat and then per group. We determined the total number of macrophages and the CD206/CD86 ratio. Counts were averaged per rat and per experimental group.

2.11. Astrocyte Presence Examination

Automated quantification was used to evaluate GFAP+ astrocytes in the injured spinal cord at 1 and 8 weeks after injury. Quantification was performed in each of the middle three sections of the contusion, in a 2.5 × 4 mm region of interest centered on the midpoint of the injury. This selected 4 mm long and 2.5 mm wide region of interest covered the injury site and adjacent glia scar for 2 mm caudal and rostral from the midpoint of the contusion. The total area and the percent area of GFAP were determined using the batch threshold and measure function of ImageJ. Measurements were averaged per rat and per experimental group.

2.12. Axon Presence

We used automated quantification to evaluate axons, recognized with antibodies against neurofilament high molecular weight (NFh), in the contusion at 8 weeks after injury. Quantification was performed in each of the middle three sections of the contusion, in a 1.5 × 2 mm region of interest centered on the midpoint of the injury. This selected 2 mm long and 1.5 mm wide region of interest covered the injury site for 1 mm caudal and rostral from the midpoint of the contusion. The total area and the percent area positive for NFh were determined at 8 weeks after injury. Standardized ImageJ algorithms were used for batch thresholding of all images, and all measurements were averaged per rat and per experimental group.

2.13. Transplant Survival Assessment

Automated quantification was used to quantify GFP+ MSC in the contusion at 1 and 8 weeks after injury. Quantification was performed in each of the middle three sections of the contusion, in a 2 × 2 mm region of interest centered on the midpoint of the injury. This selected 2 mm long and 2 mm wide region of interest covered the injury site for 1 mm caudal and rostral from the midpoint of the contusion. Positive particles for the GFP count mask and DAPI were quantified to determine the total number of positive hits. We standardized the counts by dividing the average particle size by the average size of 10 nuclei of GFP+ MSC to create a multiplication factor to account for multiple overlaying nuclei. We reported our particle counts multiplied by the multiplication factor as the standardized average count, which were averaged per rat and per experimental group.

2.14. Statistical Analyses

Data were shown as means ± SEM (standard error of the mean). For statistical analyses, we used IBM SPSS Statistics for Windows, version 24 (IBM; Armonk, NY, USA). Data were analyzed by nonparametric ANOVA (independent-samples Kruskal–Wallis test) or parametric (or nonparametric) repeated-measures ANOVA (Friedman's test), followed by Bonferroni post hoc correction, unless otherwise stated. Pearson's correlation analysis was used to determine the relationship between variables expressed by the coefficient of determination (r^2) and considered strong when $r^2 > 0.5$. Differences were accepted as statistically significant if $p < 0.05$.

3. Results

3.1. Inflammatory Response in the Injury Site

We assessed the number of macrophages in the injury site at 1 and 8 weeks after injury. Macrophages were present in the injury site at both time points, irrespective of the treatment. Figure 1 shows CD68+ macrophages in the injury site with MSC-NHC (Figure 1A), MSC-PBS (Figure 1B), NHC (Figure 1C), or PBS (Figure 1D) at 1 week after injury. The areas outlined in Figure 1A–D are enlarged in Figure 1A′–D′, respectively. We found 458 ± 75 (mean ± SEM) macrophages in the injury site with MSC-NHC, 1997 ± 174 with MSC-PBS, 565 ± 94 with NHC, and 1699 ± 151 with PBS (Figure 1E). The 4.4-fold decrease in total macrophage count with MSC-NHC compared with MSC-PBS ($p < 0.05$) and the 3.7-fold decrease in macrophages between MSC-NHC and PBS ($p < 0.05$) were significantly different. The 3-fold decrease in macrophages with NHC compared with PBS ($p < 0.05$) and the 3-fold decrease with NHC compared with MSC-PBS ($p < 0.05$) were significantly different. At 8 weeks after injury, there were 833 ± 128 (average ± SEM) macrophages in the injury site with MSC-NHC, 684 ± 135 with MSC-PBS, 649 ± 60 with NHC, and 914 ± 114 with PBS (Figure 1F). The differences in CD68+ macrophage counts at 8 weeks after injury were not significant. The data indicated that early macrophage infiltration in the contusion was attenuated with MSC-NHC or NHC compared with MSC-PBS.

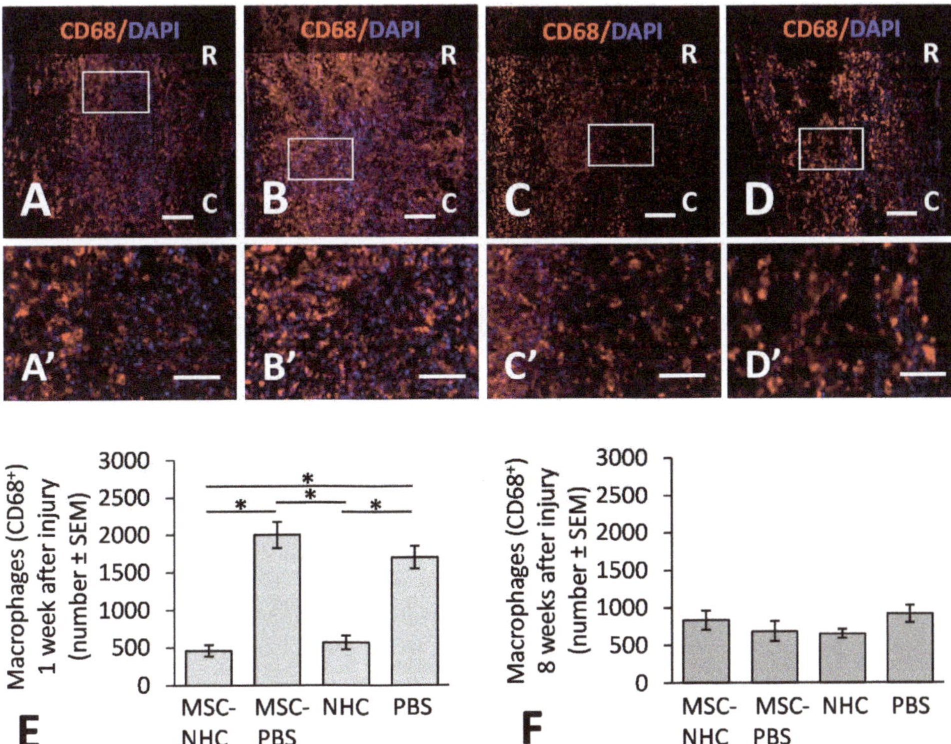

Figure 1. CD68 inflammatory response in the injury site. Photomicrographs showing macrophages stained for CD68 (red), a pan-macrophage marker, and the nucleus marker, DAPI (blue), in the injury site with MSC-NHC (**A**), MSC-PBS (**B**), NHC (**C**), or PBS (**D**) at 1 week after injury. Images in panels (**A′–D′**) are enlargements of the outlined area in panels (**A–D**), respectively. Scale bar in (**A**,**C**,**D**) is 250 μm, and 200 μm in (**B**). The scale bar in (**A′–D′**) is 100 μm. In (**A–D**), the rostral (R) and caudal (C) orientation of the horizontal sections are indicated. (**E**) Bar graph showing the average number of CD68+ macrophages in the injury site of each group at 1 week after injury. There are significantly less macrophages in the injury site with MSC-NHC and NHC only compared with MSC-PBS and PBS only. (**F**) Bar graph showing the average number of CD68+ macrophages in the injury site of each group at 8 weeks after injury. The differences in numbers of CD68+ macrophages in the injury site of the four groups were not statistically significant. In both graphs, the asterisks indicate $p < 0.05$, and the bars represent SEM. Abbreviations: DAPI = 4′,6-diamidino-2-phenylindole; MSC = mesenchymal stromal cells; NHC = nanofiber-hydrogel composite; PBS = phosphate-buffered saline; SEM = standard error of the mean.

To assess the relative amount of CD206+ anti-inflammatory, pro-regenerative (M2-like) macrophages vs. CD86+ pro-inflammatory (M1-like) macrophages, we quantified the CD206+/CD86+ ratios at 1 and 8 weeks after injury by analyzing tissue sections stained with antibodies against CD206/CD68 and CD86/CD68. CD206 and CD86 are markers for M2-like and M1-like macrophages, respectively. Figure 2 shows CD206+/CD68+ macrophages in the injury site with MSC-NHC (Figure 2A), MSC-PBS (Figure 2B), NHC

(Figure 2C), or PBS (Figure 2D) at 8 weeks after injury. The areas outlined in Figure 2A–D are enlarged in Figure 2A′–D′ and show CD206+/CD68+ macrophages, and Figure 2A″–D″ shows the same macrophages stained only for CD68. Moreso, shown are the CD86+/CD68+ macrophages in the injury site with MSC-NHC (Figure 2E), MSC-PBS (Figure 2F), NHC (Figure 2G), or PBS (Figure 2H) at 8 weeks after injury. The areas outlined in Figure 2A–D are enlarged in Figure 2A′–D′ and show CD206+/CD68+ macrophages, and Figure 2A″–D″ shows the same macrophages stained only for CD68. At 1 week after injury, we found that the CD206+/CD86+ ratio was 1.72 ± 0.35 in the injury site with MSC-NHC, 0.57 ± 0.09 with MSC-PBS, 1.29 ± 0.33 with NHC, and 0.24 ± 0.11 with PBS (Figure 2I). The CD206+/CD86+ ratio was significantly higher ($p < 0.05$) with MSC-NHC than with MSC-PBS and with PBS. The CD206+/CD86+ ratio was significantly higher with NHC than with PBS ($p < 0.05$). At 8 weeks after injury, the CD206+/CD86+ ratio was 4.96 ± 1.32 in the injury site with MSC-MHC, 1.41 ± 0.97 with MSC-PBS, 3.34 ± 0.81 with NHC, and 0.79 ± 0.38 with PBS (Figure 2J). At this time point, the CD206+/CD86+ ratio was significantly higher ($p < 0.05$) with MSC-NHC than with MSC-PBS or PBS ($p < 0.05$). The results showed that treatment with MSC-NHC compared with MSC-PBS facilitated the polarization of macrophages in the injury site towards the pro-regenerative phenotype.

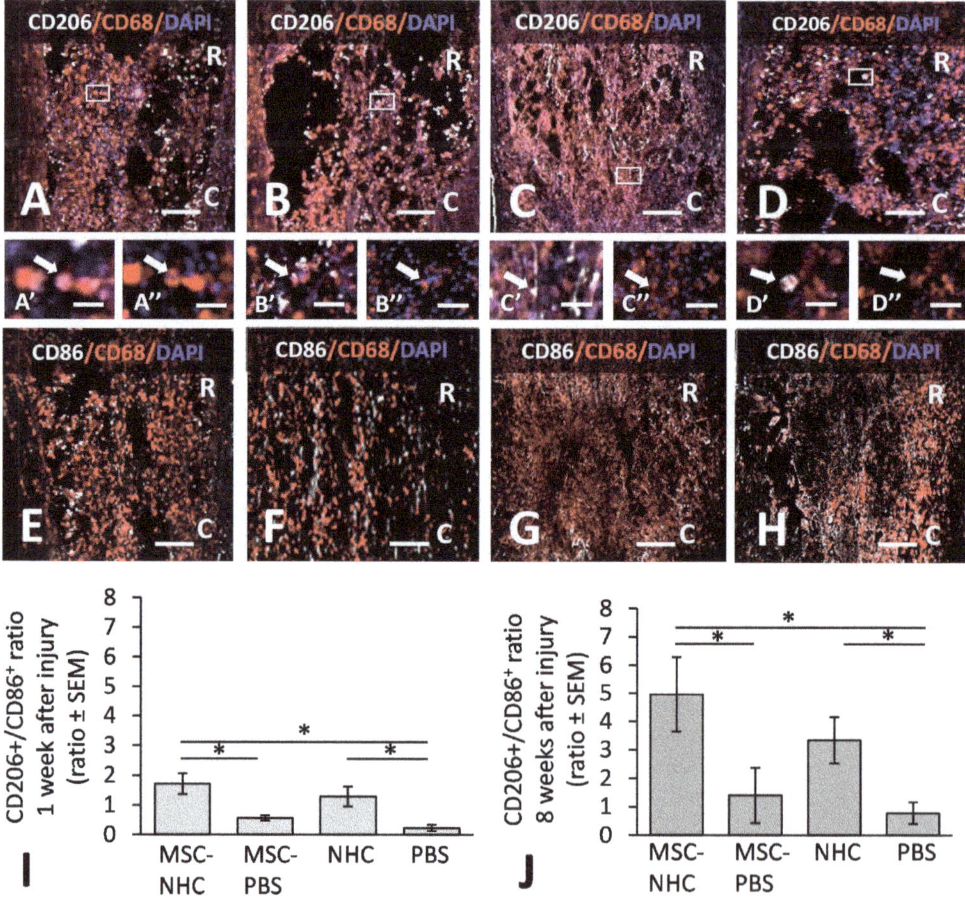

Figure 2. Inflammation polarization ratios in the injury site. Photomicrographs showing macrophages stained for CD206 (white), an anti-inflammatory macrophage marker, CD68 (red), a pan-macrophage marker, and the nucleus marker, DAPI (blue), in the injury site with MSC-NHC (**A**), MSC-PBS (**B**) NHC

(**C**), or PBS (**D**) at 8 weeks after injury. Images in panels (**A′–D′**) (showing CD206, CD68, and DAPI) and (**A″–D″**) (showing CD68 and DAPI) are enlargements of the outlined area in panels (**A–D**), respectively. Scale bar in (**A**,**C**) is 250 μm, and 200 μm in (**B**,**D**). The scale bar in (**A′/A″**) and (**D′/D″**) is 15 μm, and 25 μm in (**B′/B″**) and (**C′/C″**). In (**A–D**), the rostral (R) and caudal (C) orientation of the horizontal sections are indicated. Photomicrographs showing macrophages stained for CD86 (white), a pro-inflammatory macrophage marker, CD68 (red), a pan-macrophage marker, and the nucleus marker, DAPI (blue), in the injury with MSC-NHC (**E**), MSC-PBS (**F**) NHC (**G**), or PBS (**H**) at 8 weeks after injury. The scale bar in (**E–H**) is 250 μm. (**I**) Bar graph of the M2/M1 macrophage ratio in the injury site of each group at 1 week after injury. The M2/M1 ratio was significantly higher with MSC-NHC and NHC only than with MSC-PBS and PBS only. There was no statistically significant difference in the M2/M1 ratio of MSC-PBS and NHC only. (**J**) Bar graph of the M2/M1 macrophage ratio in the injury site of each group at 8 weeks after injury. The M2/M1 ratio was significantly higher with MSC-NHC and NHC only than with MSC-PBS and PBS only. There was no statistically significant difference in the M2/M1 ratio of MSC-PBS and NHC only. In both graphs, the asterisks indicate $p < 0.05$, and the bars represent SEM. Abbreviations: DAPI = 4′,6-diamidino-2-phenylindole; MSC = mesenchymal stromal cells; NHC = nanofiber-hydrogel composite; PBS = phosphate-buffered saline; SEM = standard error of the mean.

3.2. Injury Site Volume

We measured the average volume of the injury site at 1 and 8 weeks after injury. In all animals, an injury site surrounded by a GFAP+ astrocyte scar was discernable at both times after injury. Figure 3 shows the injury site in rats treated with MSC-NHC (Figure 3A), MSC-PBS (Figure 3B), NHC (Figure 3C), or PBS (Figure 3D) at 8 weeks after injury. The volume of the injury site at 1 week after injury was 2.48 ± 0.40 mm^3 (average ± SEM) with MSC-NHC, 2.72 ± 0.34 mm^3 with MSC-PBS, 2.88 ± 0.24 mm^3 with NHC, and 2.44 ± 0.31 mm^3 with PBS (Figure 3E). There were no statistical differences among these volumes. At 8 weeks after injury, the volume of the injury site was 1.77 ± 0.28 mm^3 (average ± SEM) with MSC-NHC, 2.58 ± 0.35 mm^3 with MSC-PBS, 1.91 ± 0.28 mm^3 with NHC, and 3.95 ± 0.30 mm^3 with PBS (Figure 3F). Thus, we found a 31 % decrease in injury size with MSC-NHC compared with MSC-PBS ($p < 0.05$), a 55 % decrease in injury size with MSC-NHC compared with PBS ($p < 0.05$), and a 51 % decrease in injury size with NHC compared with PBS ($p < 0.05$). These results showed that MSC-NHC delivery did not result in a significant change in lesion size at 1 week after injury (i.e., 4 days after treatment) over MSC-PBS, NHC, or PBS; however, MSC-NHC and NHC only both limited the secondary injury and yielded a significantly smaller injury compared with MSC-PBS and PBS at week 8.

3.3. Astrocytes in the Injury Site

The presence of GFAP+ astrocytes in the injury site was evaluated at 1 and 8 weeks after injury. Figure 3 shows the GFAP-surrounded injury site in rats treated with MSC-NHC (Figure 3A), MSC-PBS (Figure 3B), NHC (Figure 3C), or PBS (Figure 3D) at 8 weeks after injury. At 1 week after injury, GFAP staining occupied 3.43 ± 0.66% (average ± SEM) of the contused segment with MSC-NHC, 3.17 ± 0.73% with MSC-PBS, 4.21 ± 0.68% with NHC, and 4.24 ± 0.75% with PBS (Figure 3G). There were no statistical differences among these measurements. At 8 weeks after injury, GFAP staining occupied 10.46 ± 1.83% (average ± SEM) of the contused segment with MSC-NHC, 5.08 ± 0.4 % with MSC-PBS, 8.59 ± 0.88% with NHC, and 5.36 ± 0.72% with PBS (Figure 3H). The 2.1-fold increase in astrocytes within the injury site with MSC-NHC compared with MSC-PBS ($p < 0.05$) and the 1.9-fold increase in astrocytes within the injury site with MSC-NHC compared with PBS ($p < 0.05$) were statistically significant. The results indicated that treatment with MSC-NHC compared with MSC-PBS resulted in a higher degree of astrocyte presence in the injury site.

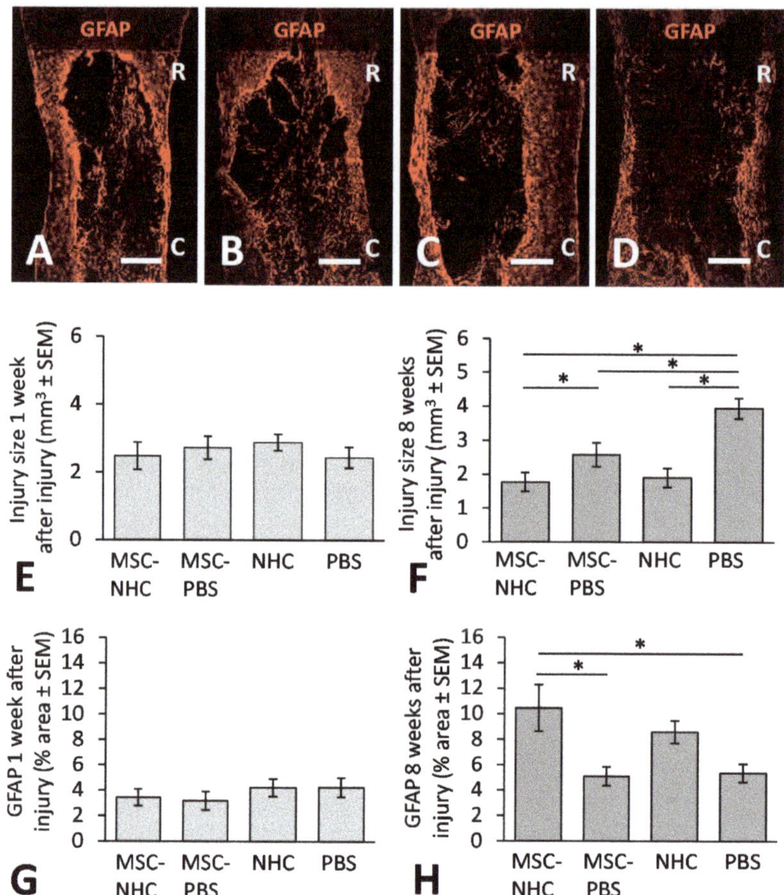

Figure 3. Injury size volume and astrocyte presence. Photomicrographs showing astrocytes stained for GFAP (red) in the injury site with MSC-NHC (**A**), MSC-PBS (**B**), NHC (**C**), or PBS (**D**) at 8 weeks after injury. The injury site was defined by the inner border of the GFAP+ scar. The scale bar in A is 600 µm, and 550 µm in (**B–D**). In (**A–D**), the rostral (R) and caudal (C) orientation of the horizontal sections are indicated. (**E**) Bar graph showing the volume of the injury site at 1 week after injury. There were no statistically significant differences between groups. (**F**) Bar graph showing the volume of the injury site at 8 weeks after injury. The injury site was significantly smaller with MSC-NHC compared with MSC-PBS or PBS only, with NHC compared with PBS, and with MSC-PBS compared with PBS only. (**G**) Bar graph of the percentage area of the injury site positive for GFAP at 1 week after injury. There were no statistically significant differences in astrocyte presence in the injury site between the groups. (**H**) Bar graph of the percentage area of the injury site positive for GFAP at 8 weeks after injury. There were significantly more astrocytes in the injury site with MSC-NHC compared with MSC-PBS and PBS only. In all graphs, the asterisks indicate $p < 0.05$, and the bars represent SEM. Abbreviations: GFAP = glial fibrillary acidic protein; MSC = mesenchymal stromal cells; NHC = nanofiber-hydrogel composite; PBS = phosphate-buffered saline; SEM = standard error of the mean.

3.4. Axons in the Injury Site

We quantified the presence of axons in the injury site at 8 weeks after injury. Figure 4 shows axons stained for NFh in the injury site treated with MSC-NHC (Figure 4A), MSC-

PBS (Figure 4B), NHC (Figure 4C), or PBS (Figure 4D). The areas outlined in Figure 4A–D are enlarged in Figure 4A′–D′, respectively. The NFh+ staining occupied 4.05 ± 0.29% (average ± SEM) of the injury site area with MSC-NHC, 2.62 ± 0.28% with MSC-PBS, 3.07 ± 0.33% with NHC, and 1.28 ± 0.43% with PBS (Figure 4E). The 55% increase in NFh+ axons within the injury site with MSC-NHC compared with MSC-PBS was statistically significant ($p < 0.05$). The 140% increase in NFh+ axons within the injury site with NHC compared with PBS was statistically significant ($p < 0.05$). These results showed that treatment with MSC-NHC compared with MSC-PBS resulted in an increased axon presence in the injury site. Pearson's correlation analysis showed a strong positive association between the axons and astrocytes ($r^2 = 0.56$) in the injury site.

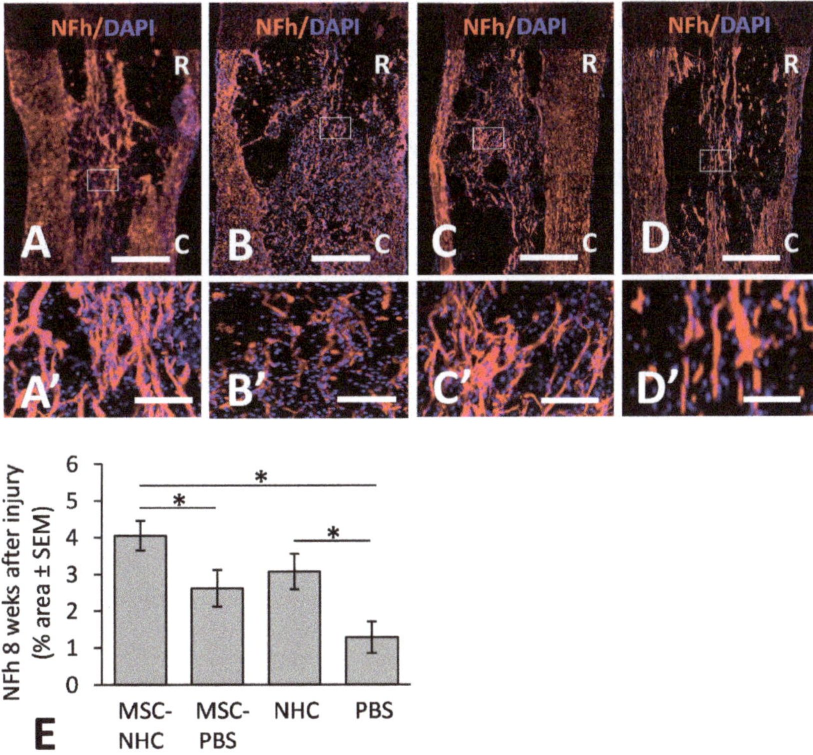

Figure 4. Axon presence in the injury site. Photomicrographs showing axons stained for NFh (red) in the injury site with MSC-NHC (**A**), MSC-PBS (**B**), NHC (**C**), or PBS (**D**) at 8 weeks after injury. Sections were counterstained with the nucleus marker, DAPI (blue). Images in panels (**A′–D′**) are enlargements of the outlined area in panels (**A–D**), respectively. The scale bar in (**A–D**) is 450 µm. Scale bar in (**A′,C′**) is 125 µm, and 115 µm in (**B′,D′**). In (**A–D**), the rostral (R) and caudal (C) orientation of the horizontal sections are indicated. (**E**) Bar graph of the percentage area of injury site positive for NFh at 8 weeks after injury. There were significantly more NFh+ axons in the injury site with MSC-NHC compared with MSC-PBS and PBS only. Asterisks indicates $p < 0.05$. Bars in the graph represent SEM. Abbreviations: DAPI = 4′,6-diamidino-2-phenylindole; MSC = mesenchymal stromal cells; NFh = neurofilament high molecular weight; NHC = nanofiber-hydrogel composite; PBS = phosphate-buffered saline; SEM = standard error of the mean.

3.5. MSC Transplant Survival

We examined transplanted MSC survival in the injury site at 1 and 8 weeks after injury. GFP+ MSC were identified in the injury site at 1 week after injury (Figure 5A–D), but not at 8 weeks after injury. At 1 week after injury, the relative MSC survival, in reference to the number of MSCs injected, was not significantly different between the group with NHC (MSC-NHC; 41 ± 11%; average ± SEM) and the group without NHC (MSC-PBS; 29 ± 13%) (Figure 5E). Our data showed that mixing MSC into NHC compared with PBS prior to injection into damaged nervous tissue in an adult rat contused spinal cord does not significantly affect their survival.

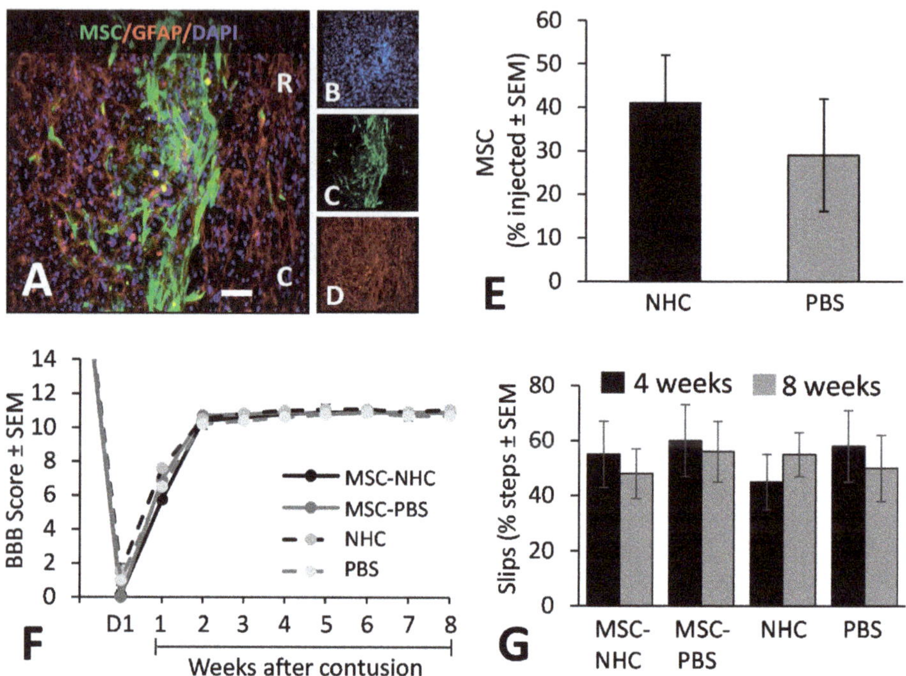

Figure 5. MSC transplant survival and hindlimb function. (**A**) Photomicrograph showing MSC (green), astrocytes (red, anti-GFAP), and cell nuclei (blue, DAPI) in the injury site with MSC-NHC at 1 week after injury. The scale bar in (**A**) is 180 μm. Scale bar in (**B**–**D**) is 550 μm. In (**A**), the rostral (R) and caudal (C) orientation of the horizontal section are indicated. (**E**) Bar graph showing the number of MSCs present in the injury site 1 week after injury. The difference in number between groups was not significant. (**F**) Line graph showing the BBB scores of the experimental groups on day 1 (d1) after injury and weekly thereafter. Differences between groups per time point were not statistically significant. In both graphs, the bars represent SEM. (**G**) Bar graph showing the total slips as a percentage of the total steps made by the rats to cross the horizontal ladder at 4 and 8 weeks after injury. Differences between groups per time point were not statistically significant. In all graphs, the bars represent SEM. Abbreviations: BBB = Basso, Beattie, Bresnahan; d = day; DAPI = 4′,6-diamidino-2-phenylindole; GFAP = glial fibrillary acidic protein; MSC = mesenchymal stromal cells; NHC = nanofiber-hydrogel composite; PBS = phosphate-buffered saline; SEM = standard error of the mean.

3.6. Hindlimb Locomotor Function

We assessed hindlimb motor function using an open-field (BBB) test [29] and hind limb sensorimotor function using a horizontal ladder test [30]. Hindlimb function, as determined by the BBB score, was similar across the groups after injury and after the subsequent treatment. The average BBB score per experimental group at 1 day after injury was 0.1 ± 0.1 (average ± SEM) for rats with MSC-NHC, 0.0 ± 0.0 with MSC-PBS, 1.5 ± 0.5 with NHC, and 1.0 ± 0.5 with PBS (Figure 5F). The average BBB scores gradually increased to reach 10.8 ± 0.1 in rats with MSC-NHC, 10.9 ± 0.1 with MSC-PBS, 11.0 ± 0.1 with NHC, and

10.6 ± 0.2 with PBS (Figure 5F) at 4 weeks after injury. These scores were maintained for the following weeks until the 8-week endpoint. Hindlimb sensorimotor function as assessed on the horizontal ladder was similar among all experimental groups at 4 and 8 weeks after injury (Figure 5G).

4. Discussion

The repair potential of MSC-NHC was evaluated in a model of spinal cord contusion. A contusion is the prevalent mechanism of SCI in the clinic. We found that the early inflammatory response in the contused spinal cord of adult rats was substantially attenuated by treatment with MSC-NHC or NHC only, but not with MSC-PBS. Macrophages are necessary in the damaged spinal cord nervous tissue for clearance of cellular/tissue debris and the restoration of homeostasis [33]. During the process, macrophages secrete cytotoxic molecules that may contribute to the propagation of nervous tissue damage in the spinal cord [34,35]. Attenuation of the number of macrophages early after SCI is considered to support nervous tissue repair [36], and this notion was substantiated in previous studies (e.g., [33,34]). The observation that treatment with NHC resulted in a reduction in the macrophage number in the contusion site was consistent with previous findings [22]. We anticipated that treatment with MSC-PBS would not cause a decrease in the number of macrophages in the injury site. While studies have reported anti-inflammatory effects of an MSC transplant [15–18], this particular MSC-mediated effect has not been associated so far with fewer total macrophages in the injured spinal cord.

The absence of a reduction in the macrophage number in the contusion site with MSC-PBS treatment combined with the decrease in the macrophage number with MSC-NHC and NHC only treatment, points at NHC as the chief mediator of the observed decline in macrophages elicited by MSC-NHC treatment. This proposition would also explain the similar reduction in macrophage numbers by MSC-NHC and NHC only. The ability of NHC to lower the presence of macrophages in an injury site in the spinal cord supports its use as a matrix for cell transplants. Moreover, this finding warrants further investigation of possible mechanisms that underlie the anti-inflammatory effects of NHC.

The inflammatory response in the contused spinal cord of adult rats shifted from a pro-inflammatory-dominant response (low CD206+/CD86+ ratio) towards an anti-inflammatory, pro-regenerative-dominant response (high CD206+/CD86+ ratio) by treatment with MSC-NHC or NHC only, but not with MSC-PBS. A shift in macrophage polarization in the injury site changes their secretome, and thus the molecular environment, in support of tissue remodeling and repair [34,36]. In the injured spinal cord, the chronic pro-inflammatory environment is considered a key contributing factor to continuing nervous tissue damage [33,36]. Our earlier published report on NHC effects on spinal cord repair following SCI supports the finding that NHC modulated macrophages towards a pro-regenerative phenotype [22]. The mechanisms of NHC-mediated macrophage modulation are unknown, but the specific structural design of NHC, including its stiffness and the presence of electrospun fibers, has been proposed to be involved in its immunomodulatory effects [22]. The absent immunomodulatory effect in the contusion site with MSC-PBS treatment was unexpected because previous studies reported that a transplant of MSC exerts such an effect among macrophages present in damaged nervous tissue [7,15–17]. It has been suggested that transplanted MSC mediate a change in macrophages towards a pro-regenerative phenotype through paracrine signaling by molecules, such as indoleamine [37] and interleukin-10 [38]. The absence of this effect in our study may be related to factors, such as the severity of the injury or the dose of MSC. In addition, because of the dynamic nature of the inflammatory response, the time of treatment may also play a role in the degree of MSC-mediated modulation of the immune response in the injury site.

The macrophage polarizing effect was greater with MSC-NHC than with MSC-PBS. It is possible that MSC within NHC is less susceptible to injury-derived factors that would decrease their production of molecules directing a macrophage phenotype shift, or alternatively, are more susceptible to factors that would support the secretion of molecules that promote a macrophage phenotype shift. Another possibility is that immunomodulation of the macrophages is promoted by mechanical/physical cues from the NHC besides the soluble factors [39,40]. We can rule out the possibility that MSC survival was improved—which could also result in more molecules that lead to macrophage polarization—because we found similar degrees of survival of MSC in NHC or PBS. The observed interaction between MSC and NHC to trigger the crucial macrophage phenotype shift is a compelling advantage for their combined use for SCI treatment. The critical role of inflammation in SCI [33,41] warrants a future study of mechanisms underlying NHC-mediated modulation of the inflammatory response.

Secondary loss of nervous tissue in the contused spinal cord segment was limited by treatment with MSC-NHC, NHC only, and MSC-PBS. Loss of nervous tissue after the primary damage is a hallmark of SCI. Mechanistically, the primary insult triggers a series of molecular and cellular cytotoxic events that, in concert, cause further nervous tissue damage and degeneration [1,2]. Protecting nervous tissue from secondary damage is an important early line of defense to limit the devastating consequences of SCI [42]. We found that MSC and NHC both have neuroprotective effects that limit injury expansion in the spinal cord. An MSC transplant may exert neuroprotective effects through the secretion of neuroprotective molecules, including neurotrophins (e.g., [12,13,43]. The finding that NHC has neuroprotective effects is in agreement with earlier published observations [22]. The mechanisms of NHC-mediated neuroprotection are under investigation. It is possible that the anti-inflammatory effects of NHC indirectly contribute to its neuroprotective effects [9,34,44].

Neuroprotection in the contused spinal cord by NHC alone and MSC-PBS treatment was similar, while neuroprotection by MSC-NHC treatment was significantly stronger than by MSC-PBS treatment. It is possible that the presence of NHC provided a more favorable environment for the MSC to secrete neuroprotective molecules. There was no significant difference in MSC survival in NHC or PBS, which rules out the possibility that neuroprotective molecules were secreted for a prolonged time due to an extended MSC presence. The significance of neuroprotection in SCI is a compelling argument to use NHC as a transplant matrix for MSC. Future studies will need to focus on elucidating mechanisms underlying the MSC-NHC interactions that support nervous tissue repair.

The presence of axons and astrocytes in the injury site was increased with MSC-NHC. Providing an environment in the injury site that supports the presence of astrocytes and axons is important for the overall repair after spinal cord damage because they may contribute to re-establishing neural tissue and neural connections. Astrocytes and axons are often found closely associated with an injury site in the spinal cord [22]; such a relationship was also observed in the present study. Interestingly, our data suggest that the combined use of MSC and NHC facilitated the establishment of an environment beneficial for astrocytes and axons. Several studies demonstrated that a transplant of MSC supports axon presence in an injury site in the spinal cord [19,20]. The presence of NHC in a spinal cord contusion was shown to result in an increased axon presence [22]. We now show that the MSC and NHC combination resulted in more axons and astrocytes in the injury site compared with either of the single treatments.

Our results showed that an MSC-NHC transplant resulted in an attenuated immune response, macrophage polarization, reduced injury size, and increased axons and astrocytes in the injury site compared with MSC-PBS. These changes did not translate into improved hindlimb function recovery in the used model of contusive SCI, as was shown using the BBB scale and the horizontal ladder walking test. The combinatorial MSC-NHC treatment was investigated for its hypothesized ability to elicit nervous tissue repair in the contused spinal cord. Improved morphological outcomes do not necessarily guarantee improvements in functional outcomes. In this study, the benefit of the observed improvements in morphological outcomes may lie in an increased likelihood and/or efficacy of additional interventions that aim to recover function.

We found that the volume of the injury site was significantly smaller in rats with MSC-PBS compared with PBS, which is in agreement with previous reports [5,6,8,9]. On the other hand, we did not find significant differences in macrophage number, macrophage polarization towards a pro-regenerative phenotype, and astrocyte and axon presence in the injury site between rats with MSC-PBS compared with PBS, which is in discord with prior publications [7,18–20]. The absence of these latter repair-supporting effects in rats with MSC-PBS relative to rats with PBS only could be due to the use of MSC derived from bone marrow, which, relative to, for instance, adipose-derived MSC, are characterized by lower in vivo survival and less axon protection in the injured spinal cord [45,46]. Another possibility is that the number of MSC transplanted into the injury may have been too small, compared with previous studies [6,9]. Moreso, the day of transplantation may have affected the outcomes. We injected MSC three days after injury, which is when the inflammatory response reaches its peak [33,34,36].

Our results show that MSC-NHC treatment elicits greater immunomodulation and neuroprotection compared to MSC-PBS or NHC alone. It is possible that the larger effect of MSC-NHC on nervous tissue protection is, at least in part, due to the stronger modulation of the immune response towards a pro-regenerative phenotype. Importantly, our data showed the strongest shift in the immune response and largest neuroprotection with MSC-NHC than with MSC-PBS. These effects were not the result of improved survival of MSC within the NHC matrix. Future studies will need to investigate mechanisms of the relationship between MSC and NHC and explore if other types of repair-mediating cells would also benefit from NHC as their transplant matrix. The observed interactions between MSC and NHC provide an exciting advantage for their combined use for SCI treatment.

Author Contributions: The study was designed by A.E.H., I.M.-L., H.Y., H.-Q.M. and M.O. Cell culture were carried out by A.E.H. and I.M.-L. In vivo procedures were carried out by A.E.H., I.M.-L., Y.N., K.Y. and M.M.M. Histological procedures, imaging, data collection and data analysis/validation were performed by A.E.H., I.M.-L., M.M.M., C.Z., B.C., X.L., S.R. and X.L. Data interpretation was performed by A.E.H., I.M.-L., H.-Q.M. and M.O. The project was supervised by H.-Q.M. and M.O. The manuscript was reviewed, edited and approved by all co-authors. All authors have read and agreed to the published version of the manuscript.

Funding: This research was funded by NINDS, NS101298 (M.O.); the Craig H. Neilsen Foundation, 460461 (M.O.); the Department of Defense, W81XWH1810245 (M.O.); the Department of Veterans Affairs, RR&D Merit Review I01RX001807 (M.O.); and RR&D Merit Review I01RX002848 (M.O.); NINDS, R21 NS085714 (H.-Q.M.); the Maryland Stem Cell Fund Launch Grant, (S.R., H.-Q.M.); and the AHA Career Development Award, 18CDA34110314 (X.L.).

Institutional Review Board Statement: The study was conducted following the guidelines of the National Institutes of Health and the United States Department of Agriculture for all animal procedures. The University of Miami Institutional Animal Care and Use Committee approved the procedures under protocol #15-231 (approved 2 March 2017).

Informed Consent Statement: Not applicable.

Data Availability Statement: The data presented in this study are available on request from the corresponding author.

Acknowledgments: We thank the Animal Core Facility and Yelena Pressman of the Miami Project to Cure Paralysis, and Maria Boulina of the Imaging Core Facility of the Diabetes Research Institute at the University of Miami, FL for their contributions.

Conflicts of Interest: H.-Q.M. and S.R. are co-founders of LifeSprout LLC, a startup company that has licensed the technology of the NHC that was used in this study. The other authors declare that they have no known competing financial interests or personal relationship that could have appeared to influence the described work.

References

1. Hachem, L.D.; Fehlings, M.G. Pathophysiology of Spinal Cord Injury. *Neurosurg. Clin. N. Am.* **2021**, *32*, 305–313. [CrossRef] [PubMed]
2. Hagg, T.; Oudega, M. Degenerative and spontaneous regenerative processes after spinal cord injury. *J. Neurotrauma* **2006**, *23*, 264–280. [CrossRef]
3. van Niekerk, E.A.; Tuszynski, M.H.; Lu, P.; Dulin, J.N. Molecular and cellular mechanisms of axonal regeneration after spinal cord injury. *Mol. Cell. Proteom.* **2016**, *15*, 394–408. [CrossRef] [PubMed]
4. Walsh, C.M.; Wychowaniec, J.K.; Brougham, D.F.; Dooley, D. Functional hydrogels as therapeutic tools for spinal cord injury: New perspectives on immunopharmacological interventions. *Pharmacol. Ther.* **2021**, *233*, 108043. [CrossRef] [PubMed]
5. Cizkova, D.; Rosocha, J.; Vanicky, I.; Jergova, S.; Cizek, M. Transplants of human mesenchymal stem cells improve functional recovery after spinal cord injury in the rat. *Cell. Mol. Neurobiol.* **2006**, *26*, 1167–1180. [CrossRef] [PubMed]
6. Nandoe Tewarie, R.D.; Hurtado, A.; Ritfeld, G.J.; Rahiem, S.T.; Wendell, D.F.; Barroso, M.M.; Grotenhuis, J.A.; Oudega, M. Bone marrow stromal cells elicit tissue sparing after acute but not delayed transplantation into the contused adult rat thoracic spinal cord. *J. Neurotrauma* **2009**, *26*, 2313–2322. [CrossRef] [PubMed]
7. Nakajima, H.; Uchida, K.; Guerrero, A.R.; Watanabe, S.; Sugita, D.; Takeura, N.; Yoshida, A.; Long, G.; Wright, K.T.; Johnson, W.E.; et al. Transplantation of mesenchymal stem cells promotes an alternative pathway of macrophage activation and functional recovery after spinal cord injury. *J. Neurotrauma* **2012**, *29*, 1614–1625. [CrossRef] [PubMed]
8. Quertainmont, R.; Cantinieaux, D.; Botman, O.; Sid, S.; Schoenen, J.; Franzen, R. Mesenchymal stem cell graft improves recovery after spinal cord injury in adult rats through neurotrophic and pro-angiogenic actions. *PLoS ONE* **2012**, *7*, e39500. [CrossRef] [PubMed]
9. Ritfeld, G.J.; Tewarie, R.N.; Vajn, K.; Rahiem, S.T.; Hurtado, A.; Wendell, D.F.; Roos, R.A.; Oudega, M. Bone marrow stromal cell-mediated tissue sparing enhances functional repair after spinal cord contusion in adult rats. *Cell Transpl.* **2012**, *21*, 1561–1575. [CrossRef] [PubMed]
10. Ritfeld, G.J.; Rauck, B.M.; Novosat, T.L.; Park, D.; Patel, P.; Roos, R.A.; Wang, Y.D.; Oudega, M. The effect of a polyurethane-based reverse thermal gel on bone marrow stromal cell transplant survival and spinal cord repair. *Biomaterials* **2014**, *35*, 1924–1931. [CrossRef] [PubMed]
11. Morita, T.; Sasaki, M.; Kataoka-Sasaki, Y.; Nakazaki, M.; Nagahama, H.; Oka, S.; Oshigiri, T.; Takebayashi, T.; Yamashita, T.; Kocsis, J.D.; et al. Intravenous infusion of mesenchymal stem cells promotes functional recovery in a model of chronic spinal cord injury. *Neurosci* **2016**, *335*, 221–231. [CrossRef]
12. Liu, Y.; Dulchavsky, D.S.; Gao, X.; Kwon, D.; Chopp, M.; Dulchavsky, S.; Gautam, S.C. Wound repair by bone marrow stromal cells through growth factor production. *J. Surg. Res.* **2006**, *136*, 336–341. [CrossRef] [PubMed]
13. Nakano, N.; Nakai, Y.; Seo, T.B.; Yamada, Y.; Ohno, T.; Yamanaka, A.; Nagai, Y.; Fukushima, M.; Suzuki, Y.; Nakatani, T.; et al. Characterization of conditioned medium of cultured bone marrow stromal cells. *Neurosci. Lett.* **2010**, *483*, 57–61. [CrossRef] [PubMed]
14. Ritfeld, G.J.; Patel, A.; Chou, A.; Novosat, T.L.; Castillo, D.G.; Roos, R.A.; Oudega, M. The role of brain-derived neurotrophic factor in bone marrow stromal cell-mediated spinal cord repair. *Cell Transpl.* **2015**, *24*, 2209–2220. [CrossRef] [PubMed]
15. Tomchuck, S.L.; Zwezdaryk, K.J.; Coffelt, S.B.; Waterman, R.S.; Danka, E.S.; Scandurro, A.B. Toll-like receptors on human mesenchymal stem cells drive their migration and immunomodulating responses. *Stem. Cells* **2008**, *26*, 99–107. [CrossRef] [PubMed]
16. Shin, T.; Ahn, M.; Moon, C.; Kim, S.; Sim, K.B. Alternatively activated macrophages in spinal cord injury and remission: Another mechanism for repair? *Mol. Neurobiol.* **2013**, *47*, 1011–1019. [CrossRef] [PubMed]
17. He, X.; Wang, H.; Jin, T.; Xu, Y.; Mei, L.; Yang, J. TLR4 activation promotes bone marrow MSC proliferation and osteogenic differentiation via wnt3a and wnt5a signaling. *PLoS ONE* **2016**, *11*, e0149876. [CrossRef] [PubMed]
18. Cofano, F.; Boido, M.; Monticelli, M.; Zenga, F.; Ducati, A.; Vercelli, A.; Garbossa, D. Mesenchymal stem cells for spinal cord injury: Current options, limitations, and future of cell therapy. *Int. J. Mol. Sci.* **2019**, *20*, 2698. [CrossRef]
19. Okuda, A.; Horii-Hayashi, N.; Sasagawa, T.; Shimizu, T.; Shigematsu, H.; Iwata, E.; Morimoto, Y.; Masuda, K.; Koizumi, M.; Akahane, M.; et al. Bone marrow stromal cell sheets may promote axonal regeneration and functional recovery with suppression of glial scar formation after spinal cord transection injury in rats. *J. Neurosurg. Spine* **2017**, *26*, 388–395. [CrossRef] [PubMed]
20. Lin, L.; Lin, H.; Bai, S.; Zheng, L.; Zhang, X. Bone marrow mesenchymal stem cells (BMSCs) improved functional recovery of spinal cord injury partly by promoting axonal regeneration. *Neurochem. Int.* **2018**, *115*, 80–84. [CrossRef]
21. Nandoe Tewarie, R.D.; Hurtado, A.; Levi, A.D.; Grotenhuis, J.A.; Oudega, M. Bone marrow stromal cells for repair of the spinal cord: Towards clinical application. *Cell Transplant* **2006**, *15*, 563–577. [CrossRef] [PubMed]
22. Li, X.; Zhang, C.; Haggerty, A.E.; Yan, J.; Lan, M.; Seu, M.; Yang, M.; Marlow, M.M.; Maldonado-Lasunción, I.; Cho, B.; et al. The effect of a nanofiber-hydrogel composite on neural tissue repair and regeneration in the contused spinal cord. *Biomaterials* **2020**, *245*, 119978. [CrossRef]
23. Ozawa, H.; Matsumoto, T.; Ohashi, T.; Sato, M.; Kokubun, S. Comparison of spinal cord gray matter and white matter softness: Measurement by pipette aspiration method. *J. Neurosurg.* **2001**, *95* (Suppl. S2), 221–224. [CrossRef] [PubMed]
24. Ware, T.; Simon, D.; Arreaga-Salas, D.E.; Reeder, J.; Rennaker, R.; Keefer, E.W.; Voit, W. Fabrication of responsive, softening neural interfaces. *Adv. Funct. Mater.* **2012**, *22*, 3470–3479. [CrossRef]

25. Li, X.; Cho, B.; Martin, R.; Seu, M.; Zhang, C.; Zhou, Z.; Choi, J.S.; Jiang, X.; Chen, L.; Walia, G.; et al. Nanofiber-hydrogel composite mediated angiogenesis for soft tissue reconstruction. *Sci. Transl. Med.* **2019**, *11*, eaau6210. [CrossRef]
26. Azizi, S.A.; Stokes, D.; Augelli, B.J.; DiGirolamo, C.; Prockop, D.J. Engraftment and migration of human bone marrow stromal cells implanted in the brains of albino rats—Similarities to astrocyte grafts. *Proc. Natl. Acad. Sci. USA* **1998**, *95*, 908–913. [CrossRef] [PubMed]
27. Scheff, S.W.; Rabchevsky, A.G.; Fugaccia, I.; Main, J.A.; Lumpp, J.E., Jr. Experimental modeling of spinal cord injury: Characterization of a force-defined injury device. *J. Neurotrauma* **2003**, *20*, 179–193. [CrossRef]
28. Rauck, B.M.; Novosat, T.L.; Oudega, M.; Wang, Y. Biocompatibility of a coacervate-based controlled release system for protein delivery to the injured spinal cord. *Acta Biomater.* **2015**, *11*, 204–211. [CrossRef]
29. Basso, M.; Beattie, M.S.; Bresnahan, J.C. A sensitive and reliable locomotor rating scale for open field testing in rats. *J. Neurotrauma* **1995**, *12*, 1–21. [CrossRef]
30. Kunkel-Bagden, E.; Dai, H.N.; Bregman, B.S. Methods to assess the development and recovery of locomotor function after spinal cord injury in rats. *Exp. Neurol.* **1993**, *119*, 153–164. [CrossRef]
31. Metz, G.A.; Whishaw, I.Q. Drug-induced rotation intensity in unilateral dopamine-depleted rats is not correlated with end point or qualitative measures of forelimb or hindlimb motor performance. *Neuroscience* **2002**, *111*, 325–336. [CrossRef]
32. Noller, C.M.; Boulina, M.; McNamara, G.; Szeto, A.; McCabe, P.M.; Mendez, A.J. A practical approach to quantitative processing and analysis of small biological structures by fluorescent imaging. *J. Biomol. Tech.* **2016**, *27*, 90–97. [CrossRef] [PubMed]
33. Gensel, J.C.; Zhang, B. Macrophage activation and its role in repair and pathology after spinal cord injury. *Brain Res.* **2015**, *1619*, 1–11. [CrossRef] [PubMed]
34. Kigerl, K.A.; Gensel, J.C.; Ankeny, D.P.; Alexander, J.K.; Donnelly, D.J.; Popovich, P.G. Identification of two distinct macrophage subsets with divergent effects causing either neurotoxicity or regeneration in the injured mouse spinal cord. *J. Neurosci.* **2009**, *29*, 13435–13444. [CrossRef] [PubMed]
35. Kroner, A.; Greenhalgh, A.D.; Zarruk, J.G.; Passos Dos Santos, R.; Gaestel, M.; David, S. TNF and increased intracellular iron alter macrophage polarization to a detrimental M1 phenotype in the injured spinal cord. *Neuron* **2014**, *83*, 1098–1116. [CrossRef] [PubMed]
36. Maldonado-Lasunción, I.; Verhaagen, J.; Oudega, M. Mesenchymal stem cell-macrophage choreography supporting spinal cord repair. *Neurotherapeutics* **2018**, *15*, 578–587. [CrossRef]
37. Ge, W.; Jiang, J.; Arp, J.; Liu, W.; Garcia, B.; Wang, H. Regulatory T-cell generation and kidney allograft tolerance induced by mesenchymal stem cells associated with indoleamine 2,3-dioxygenase expression. *Transplantation* **2010**, *90*, 1312–1320. [CrossRef] [PubMed]
38. Wheeler, K.C.; Jena, M.K.; Pradhan, B.S.; Nayak, N.; Das, S.; Hsu, C.D.; Wheeler, D.S.; Chen, K.; Nayak, N.R. VEGF may contribute to macrophage recruitment and M2 polarization in the decidua. *PLoS ONE* **2018**, *13*, e0191040. [CrossRef] [PubMed]
39. Garg, K.; Pullen, N.A.; Oskeritzian, C.A.; Ryan, J.J.; Bowlin, G.L. Macrophage functional polarization (M1/M2) in response to varying fiber and pore dimensions of electrospun scaffolds. *Biomaterials* **2013**, *34*, 4439–4451. [CrossRef] [PubMed]
40. McWhorter, F.Y.; Wang, T.; Nguyen, P.; Chung, T.; Liu, W.F. Modulation of macrophage phenotype by cell shape. *Proc. Natl. Acad. Sci. USA* **2013**, *110*, 17253–17258. [CrossRef] [PubMed]
41. Popovich, P.G.; Guan, Z.; McGaughy, V.; Fisher, L.; Hickey, W.F.; Basso, D.M. The neuropathological and behavioral consequences of intraspinal microglial/macrophage activation. *J. Neuropathol. Exp. Neurol.* **2002**, *61*, 623–633. [CrossRef]
42. Fiani, B.; Kondilis, A.; Soula, M.; Tao, A.; Alvi, M.A. Novel methods of necroptosis inhibition for spinal cord injury using translational research to limit secondary injury and enhance endogenous repair and regeneration. *Neurospine* **2021**, *18*, 261–270. [CrossRef] [PubMed]
43. Labrador-Velandia, S.; Alonso-Alonso, M.L.; Di Lauro, S.; García-Gutierrez, M.T.; Srivastava, G.K.; Pastor, J.C.; Fernandez-Bueno, I. Mesenchymal stem cells provide paracrine neuroprotective resources that delay degeneration of co-cultured organotypic neuroretinal cultures. *Exp. Eye Res.* **2019**, *185*, 107671. [CrossRef] [PubMed]
44. Krupa, P.; Vackova, I.; Ruzicka, J.; Zaviskova, K.; Dubisova, J.; Koci, Z.; Turnovcova, K.; Urdzikova, L.M.; Kubinova, S.; Rehak, S.; et al. The effect of human mesenchymal stem cells derived from Wharton's Jelly in spinal cord injury: Treatment is dose-dependent and can be facilitated by repeated application. *Int. J. Mol. Sci.* **2018**, *19*, 1503. [CrossRef] [PubMed]
45. Zhou, Z.; Chen, Y.; Zhang, H.; Min, S.; Yu, B.; He, B.; Jin, A. Comparison of mesenchymal stromal cells from human bone marrow and adipose tissue for the treatment of spinal cord injury. *Cytotherapy* **2013**, *15*, 434–448. [CrossRef] [PubMed]
46. Takahashi, A.; Nakajima, H.; Uchida, K.; Takeura, N.; Honjoh, K.; Watanabe, S.; Kitade, M.; Kokubo, Y.; Johnson, W.E.B.; Matsumine, A. Comparison of Mesenchymal Stromal Cells Isolated from Murine Adipose Tissue and Bone Marrow in the Treatment of Spinal Cord Injury. *Cell Transpl.* **2018**, *27*, 1126–1139. [CrossRef] [PubMed]

Article

Inhibition of Human Malignant Pleural Mesothelioma Growth by Mesenchymal Stromal Cells

Valentina Coccè [1,†], Silvia La Monica [2,*,†], Mara Bonelli [2], Giulio Alessandri [3], Roberta Alfieri [2], Costanza Annamaria Lagrasta [2], Denise Madeddu [2], Caterina Frati [2], Lisa Flammini [4], Daniela Lisini [5], Angela Marcianti [5], Eugenio Parati [6], Francesca Paino [1], Aldo Giannì [1,7], Giampietro Farronato [7,8], Angela Falco [2], Lorenzo Spaggiari [9,10], Francesco Petrella [1,9,10,‡] and Augusto Pessina [1,‡]

1 CRC StaMeTec, Department of Biomedical, Surgical and Dental Sciences, University of Milan, 20122 Milan, Italy; valentina.cocce@unimi.it (V.C.); francesca.paino@unimi.it (F.P.); aldo.gianni@unimi.it (A.G.); francesco.petrella@unimi.it (F.P.); augusto.pessina@unimi.it (A.P.)
2 Department of Medicine and Surgery, University of Parma, 43126 Parma, Italy; mara.bonelli@unipr.it (M.B.); roberta.alfieri@unipr.it (R.A.); costanzaannamaria.lagrasta@unipr.it (C.A.L.); denise.madeddu@unipr.it (D.M.); caterina.frati@unipr.it (C.F.); angela.falco@unipr.it (A.F.)
3 Department of Molecular and Translational Medicine, Section of Microbiology and Virology, Medical School, University of Brescia, 25100 Brescia, Italy; cisiamo2@yahoo.com
4 Food and Drug Department, University of Parma, 43124 Parma, Italy; lisa.flammini@unipr.it
5 Cell Therapy Production Unit-UPTC, Fondazione IRCCS Istituto Neurologico Carlo Besta, 20133 Milan, Italy; daniela.lisini@istituto-besta.it (D.L.); angela.marcianti@istituto-besta.it (A.M.)
6 IRCCS Istituti Clinici Scientifici Maugeri, 20138 Milan, Italy; eugenio.parati@icsmaugeri.it
7 Maxillo-Facial and Dental Unit, Fondazione Ca' Granda IRCCS Ospedale Maggiore Policlinico, 20122 Milan, Italy; giampietro.farronato@unimi.it
8 Unit of Orthodontics and Paediatric Dentistry, Fondazione Ca' Granda IRCCS Ospedale Maggiore Policlinico, 20122 Milan, Italy
9 Department of Thoracic Surgery, IRCCS European Institute of Oncology, 20139 Milan, Italy; lorenzo.spaggiari@ieo.it
10 Department of Oncology and Hemato-Oncology, University of Milan, 20122 Milan, Italy
* Correspondence: silvia.lamonica@unipr.it
† These authors share equal contribution.
‡ These authors share equal senior authorship.

Citation: Coccè, V.; La Monica, S.; Bonelli, M.; Alessandri, G.; Alfieri, R.; Lagrasta, C.A.; Madeddu, D.; Frati, C.; Flammini, L.; Lisini, D.; et al. Inhibition of Human Malignant Pleural Mesothelioma Growth by Mesenchymal Stromal Cells. *Cells* **2021**, *10*, 1427. https://doi.org/10.3390/cells10061427

Academic Editors: Joni H. Ylostalo and Nisha C. Durand

Received: 26 March 2021
Accepted: 4 June 2021
Published: 8 June 2021

Abstract: Background: Malignant Pleural Mesothelioma (MPM) is an aggressive tumor that has a significant incidence related to asbestos exposure with no effective therapy and poor prognosis. The role of mesenchymal stromal cells (MSCs) in cancer is controversial due to their opposite effects on tumor growth and in particular, only a few data are reported on MSCs and MPM. Methods: We investigated the in vitro efficacy of adipose tissue-derived MSCs, their lysates and secretome against different MPM cell lines. After large-scale production of MSCs in a bioreactor, their efficacy was also evaluated on a human MPM xenograft in mice. Results: MSCs, their lysate and secretome inhibited MPM cell proliferation in vitro with S or G0/G1 arrest of the cell cycle, respectively. MSC lysate induced cell death by apoptosis. The efficacy of MSC was confirmed in vivo by a significant inhibition of tumor growth, similar to that produced by systemic administration of paclitaxel. Interestingly, no tumor progression was observed after the last MSC treatment, while tumors started to grow again after stopping chemotherapeutic treatment. Conclusions: These data demonstrated for the first time that MSCs, both through paracrine and cell-to-cell interaction mechanisms, induced a significant inhibition of human mesothelioma growth. Since the prognosis for MPM patients is poor and the options of care are limited to chemotherapy, MSCs could provide a potential new therapeutic approach for this malignancy.

Keywords: mesenchymal stromal cells; mesothelioma; malignant pleural mesothelioma (MPM); cell therapy

1. Introduction

Malignant Pleural Mesothelioma (MPM) is an aggressive tumor that has a significant incidence related to widespread asbestos exposure [1]. There is still no effective therapy for MPM patients, the prognosis is poor and, furthermore, conventional chemotherapy entails remarkable toxic side effects without a clear clinical benefit [2–5].

As discussed in many reviews, the relationships between cancer and mesenchymal stromal cells (MSCs) are a controversial matter because MSCs can exert opposite effects on tumor growth. A tumor-promoting capacity of MSCs has been described as the result of paracrine activity of growth factors, exosomes and anti-apoptotic molecules secreted by MSCs [6,7]. Furthermore, the ability of MSCs to differentiate into tumor-associated fibroblasts has been suggested to be a mechanism able to stimulate tumor growth, metastasis formation and increase drug-resistance [8].

On the other hand, many reports suggest that MSCs and/or MSCs-derived secretome can exert antitumor activity against many different types of cancers such as leukemia, prostate carcinoma, colon carcinoma, and breast cancer, both in vitro and in vivo models [9–12]. To explain these anticancer effects, different mechanisms have been investigated such as apoptosis induction due to Tumor Necrosis Factor-Related Apoptosis-Inducing Ligand (TRAIL) up-regulation, cell cycle arrest, over-expression of tumor suppressor genes, and/or cytokine-mediated process [13]. By homing to the tumor site, MSCs can integrate into the tumor mass and exert suppression of growth, as reported for SKMES1 and A549 lung adenocarcinoma cells via the production of some soluble factors [14] or other mechanisms [15]. In addition to the abovementioned mechanisms, MSCs can impact on tumor growth and progression by modulating the innate and adaptive immune response [16]. All the controversial different roles of MSCs have been discussed in many reviews [17,18] and some authors consider that the discrepancies in the ability of MSCs to promote or suppress cancer could be attributable to many factors such as: the remarkable differences among tumor models, the MSCs tissue source, the amount of MSCs used or the mode of cell administration, and sometimes, the different criteria to select the experimental controls.

In the literature, there are only a few reports regarding the role of MSCs in MPM. A significant tumor-inhibiting effect in vitro on MPM cell lines exerted by conditioned medium from human lung MSCs was reported [19] and a reduction in tumor growth in an in vivo MPM model after intravenous delivery of TRAIL-expressing MSCs has been described [20,21]. In this context, the aim of this study was to investigate the in vitro efficacy of human adipose tissue-derived MSCs (AT-MSCs), their cell lysates and secretome on the proliferation of three human mesothelioma cell lines. A further large-scale bioreactor production of MSCs has been set up for treating the human mesothelioma mouse xenograft model. Our in vitro results demonstrated that MSCs, lysates and conditioned media (CM) from AT-MSCs inhibited the proliferation of the mesothelioma cells. The in vivo study confirmed that a loco-regional treatment of a well-established mesothelioma xenograft with AT-MSCs resulted in a substantial inhibition of tumor growth that was comparable with that produced by the chemotherapeutic drug Paclitaxel (PTX), given through a systemic administration.

2. Materials and Methods

2.1. Tumor Cell Lines

The human MPM cell lines MSTO-211H (biphasic histotype), NCI-H2452 (epithelioid histotype) and NCI-H2052 (sarcomatoid histotype) were obtained from ATCC (Manassas, VA, USA), which authenticates the phenotypes of these cell lines on a regular basis. Cells were cultured in RPMI-1640 supplemented with 10% Fetal Bovine Serum (FBS, Euroclone, Milan, Italy) and maintained at 37 °C in a water-saturated atmosphere of 5% CO_2 in air. The cell lines were routinely tested for mycoplasma contamination.

2.2. Mesenchymal Stromal Cells (MSCs) Expansion and Characterization

Adipose tissue lipoaspirates were collected, under general anesthesia, from healthy volunteer donors undergoing plastic surgery for aesthetic purposes (age ranged from 18 to 66 years). Samples were collected after signed informed consent of no objection for the use for research of surgical tissues (otherwise eliminated) in accordance with the Declaration of Helsinki. Informed consent was obtained prior to tissue collection; the Institutional Review Board of the IRCCS Neurological Institute C. Besta Foundation approved the design of the study.

MSC starting batches (MSCs expanded in flasks up to passage 3) were characterized analyzing: (a) the typical spindle-shaped MSC morphology and the adhesion capacity to the plastic support assessed at every culture passage of each MSC line; (b) MSCs' viability assessed at every culture passage of each MSC line (cut-off value $\geq$ 75%); (c) the percentage of expression of typical MSC markers CD90, CD105 and CD73 (cut-off value $\geq$ 80%) and the absence of the hematopoietic/endothelial markers CD34, CD45 and CD31 (cut-off value $\leq$ 15%) by flow cytometry analysis on each culture at passage 3; (d) population doubling time at every culture passage of each line. The osteogenic, adipogenic and chondrogenic differentiation capacities of MSCs were evaluated as previously reported [22]. After expansion in the bioreactor, MSCs were counted and frozen. An aliquot of cells was dedicated to the following controls: (a) viability; (b) flow cytometry analysis (same markers and cut-off as the MSCs starting batch).

2.3. MSCs for In Vitro Studies

For in vitro studies, adipose tissue-derived mesenchymal stromal cells (AT-MSCs) were expanded in 25 cm^2 flasks at a density of 1.2 $\times$ 10^4 cells/cm^2 in 5mL of DMEM low glucose medium (Euroclone) supplemented with 5% platelet lysate Stemulate (Cook Regentec, Indianapolis, IN, USA) and 2 mM L-glutamine (Euroclone) and incubated at 37 °C, 5% CO_2. To study the MSCs' secretome, the conditioned medium of the cells (MSCs CM) obtained from three different donors was collected after 6 days of culture and stored in 1 mL aliquots at -80 °C. To study the cell lysate (MSCs LYS), the cell monolayers were detached with trypsin-EDTA (Euroclone) and after counting, the cells were suspended in 3 mL of complete medium and lysed through a sonication procedure (Labsonic UBraun, Reichertshausen, Germany). The procedure was performed by three cycles of 0.4 s pulse at 30% amplitude each; then, the sample was centrifuged 10 min at 2500$\times$ g to eliminate debris and the supernatant (MSCs LYS) was stored at -80 °C until use.

2.4. Large Scale Expansion of MSCs

Starting from the lipoaspirate of a donor, the AT-MSCs were expanded in flasks until a number of at least 20 $\times$ 10^6 cells was reached for each passage, not exceeding P3. In order to produce a high amount of cells for in vivo experiments, the MSCs were expanded using the bioreactor Quantum Cell Expansion System (Terumo BCT Inc., Lakewood, CO, USA) and GMP-compliant reagents, as previously described [23]. Briefly, after priming of the disposable expansion set (the bioreactor was coated overnight with 5 mg of human fibronectin (Corning Incorporated, Deeside, UK)) to promote cell adhesion, a 4 L media bag was then attached to the appropriate inlet line on the Quantum disposable expansion set. The expanded MSCs were analyzed for the expression of the typical MSCs markers (CD90, CD73, CD105) by using monoclonal antibodies (Becton Dickinson, Franklin Lake, NJ, USA), as previously described [22].

2.5. Analysis of Cell Proliferation, Cell Death and Cell Cycle

Both MSCs LYS and MSCs CM were tested in vitro for their anti-proliferative activity on MSTO-211H, NCI-H2452 and NCI-H2052 cells in 96 multi-well plates (Sarstedt, Nümbrecht, Germany), as previously described [23,24]. Briefly, 1:2 serial dilutions MSCs LYS and MSCs CM were performed in 100 µL of culture medium/well, and then, 10^3 tumor cells were added to each well. Cell growth was evaluated after 7 days of culture by measuring the optical density at 550 nm in an MTT (3-(4,5-dimethyl-2-thiazolyl)-2,5-diphenyl-2-H-tetrazoliumbromide) assay [25]. Cell death was assessed by Hoechst 33342 and propidium iodide dual staining and by using the Apoptosis/Necrosis Detection Kit (Abcam, Cambridge, UK). Caspase-3 activity was measured by the Caspase-3 Assay Kit (Abcam) following the supplier's protocols. The distribution of the cells in the cell cycle (determined by PI staining and flow cytometry analysis) was determined as described elsewhere [26].

2.6. Transwell Assay

The effect of MSCs on MSTO-211H, NCI-H2452 and NCI-H2052 cell proliferation was analyzed using transwell inserts. Aliquots (2×10^4; 4×10^4; 8×10^4) of MSCs were seeded in a 24-well plate (diameter 1.9 cm^2), while 1×10^3 MSTO-211H or NCI-H2452 or NCI-H2052 cells were seeded (ratio 1:6; 1:13; 1:26) onto the insert (0.4 μm pore size; Becton Dickinson). After 5 days of incubation (37 °C, 5% CO_2), the cells in the insert were stained with 0.25% crystal violet (Sigma Aldrich, St. Louis, MO, USA) for 10 min, washed with PBS buffer and eluted with 0.3 mL of 33% glacial acetic acid. The absorbance of the eluted dye was measured at 550 nm.

2.7. Cytokines Measure in MSCs Secretome

The conditioned media of standardized cultures of MSCs (1.2×10^4 cells/cm^2 at 6 days of culture) were analyzed for cytokine content. The qualitative/quantitative analysis was performed by using "multiplex bead-based xMAP technology" (Bio-Plex Human Cytokine 27-Plex Panel, Bio-Plex Human Group II Cytokine 21-Plex Panel, Bio-Rad Laboratories (Hercules, CA, USA)).

2.8. Direct Tumor MSCs Cells Interaction

To study the direct interaction between mesothelioma and mesenchymal stromal cells, MSTO-211H cells were co-cultured with fluorescent MSCs (hASCs-TS/GFP^+) that were previously established in our laboratory [27–29]. The study was performed by using a cyto-inclusion technique and the relationship between MSCs and tumor was analyzed under a confocal microscope [30]. Briefly, MSTO-211H cells (5×10^6) were co-cultured with hASCs-TS/GFP^+ (at ratio 5:1 tumor/MSCs) and after 72 h of incubation, the mixed cell cultures were detached by trypsin and centrifuged. The final pellets were then gently resuspended in 40 μL of Matrigel (Corning, NY, USA) and, after gelification for one hour at 37 °C, were fixed in paraformaldehyde 4% for 15 min at room temperature. Cyto-included were placed onto slides permeabilized with 0.2% TritonX-100 for 5 min at room temperature, washed once in PBS and stained with 1 ug/mL DAPI (PBS) for 3 min to be observed under a confocal microscope. To confirm the direct activity of hASCs-TS/GFP^+ against MSTO-211H, specific co-cultures were set up by seeding $5 \times 10^3/cm^2$ of MSTO-211H cells in 24 well plates in the presence of a different ratio (1:1 and 1:2) of hASCs-TS/GFP^+. The co-cultures were incubated for 48 h under standard culture conditions, then the cells were detached (0.25% trypsin in 0.2 mM EDTA) and counted under fluorescence microscopy.

2.9. In Vivo Experiments

A total of 10^6 human MPM MSTO-211H cells were suspended in 200 μL of Matrigel (Corning) and PBS (1:1) and were subcutaneously injected in the right flank of 6-week-old Balb/c-Nude female mice (Charles River Laboratories, Calco, Italy). The animals were housed in a protected unit for immunodeficient animals with 12 h light–dark cycles and provided with sterilized food and water ad libitum. When tumor volume reached an average size of 100 mm^3, the animals were randomized into three groups: control (CTRL, n = 7), paclitaxel (PTX, n = 8) and mesenchymal cells (MSCs, n = 7). Once a week, paclitaxel (20mg/kg) or vehicle alone (control group) was administered intraperitoneally and a total of 5×10^6 of MSCs in 200 μL of Matrigel and PBS (1:1) were subcutaneously injected very close to the tumor. The treatments were repeated four times, at day 0, 7, 14 and 21 and then, suspended for a further two weeks, to evaluate tumor mass growth after stopping treatments. Tumor xenografts were measured three times per week using a digital caliper and tumor volume was determined using the formula: (length $\times$ width^2)/2 as previously described [31]. At the same time, animal body weight, posture and gait were monitored. At day 35, mice were euthanized by cervical dislocation and the tumor nodules were collected for further analyses. All experiments involving animals and their care were performed with the approval of the Local Ethical Committee of University of Parma (Organismo per la Protezione e il Benessere degli Animali, OPBA) and of the Italian Ministry of Health, in accordance with the institutional guidelines that are in compliance with national (D.Lgs. 26/2014) and international (Directive 2010/63/EU) laws and policies.

2.10. Histological and Morphometric Analysis of Tumor Xenografts

Subcutaneous nodules were excised, formalin fixed, paraffin embedded and processed for histochemical analysis. The morphometric evaluation of xenograft composition was performed on Masson's trichrome-stained sections. In detail, the number of points overlying neoplastic tissue, fibrosis or necrosis was counted and expressed as percentage of the total number of points explored to define the volume fractions of each tissue component. All these morphometric measurements were obtained with the aid of a grid, defining a tissue area of 0.22 mm^2 and containing 42 sampling points, each covering an area of 0.0052 mm^2. These evaluations were performed on the entire section of each tumor sample using an optical microscope (200$\times$ final magnification) [32]. To test angiogenesis in the tumor xenografts, some histological sections were processed for immunohistochemistry after antigen retrieval pretreatment (Proteinase K Working Solution, 20 μg/mL for 10–20 min at 37 °C in humidified chamber). Detection of CD31 antigen was performed using primary anti-CD31 antibody (CD31/PECAM-1 (H-3): sc-376764 Santa Cruz, Dallas TX) and IHC Select® Immunoperoxidase Secondary Detection System (Millipore, Burlington, MA, USA) following the manufacturer's procedures.

2.11. Statistical Analysis

Data are expressed as the mean $\pm$ standard error (SEM). For statistics, one-way analysis of variance (ANOVA), linear regression analysis and Student's t-test were performed by using GraphPad Prism 6.0 (GraphPad Software, La Jolla, CA, USA). For in vivo studies, comparison among groups was made using two-way repeated measures ANOVA followed by Bonferroni's post hoc test (to adjust for multiple comparisons) and/or Student–Newman–Keuls Multiple Comparisons Test. p values of less than 0.05 were considered statistically significant.

3. Results

3.1. Effects of MSCs Lysate and Secretome on Proliferation of Mesothelioma Cell Lines

The lysates (MSCs LYS) and the conditioned medium (MSCs CM) of MSCs obtained from three different donors have been tested concerning the proliferation of MSTO-211H, NCI-H2452 and NCI-H2052 mesothelioma cell lines. Both the lysates and the conditioned medium produced a significant inhibition of the proliferation of mesothelioma cell lines (Figure 1A,B). The dose–response kinetics showed a significant slope ($p < 0.05$) of linear regression with high correlation coefficients (R2 ranged from 0.77 to 0.97). These results demonstrated the paracrine action of MSCs acting without cell-to-cell contact and were also confirmed in a co-culture transwell system. Indeed, as shown in Figure 1C, MCSs produced factors which inhibited mesothelioma cells proliferation and this effect was correlated with the amount of MSCs seeded. The picture of Figure 1D shows representative images of the monolayer of control MSTO-211H in comparison to MSTO-211H cells co-cultured in transwell with 8×10^4 MSCs. We then performed cell cycle analysis and a deeper evaluation of cell viability and cell death in MSTO-211H cells. After 24 h of exposure to MSCs LYS, a block in the S phase of the cell cycle was observed, while the exposure to MSCs CM induced a G0/G1 arrest after 48 h (Figure 1E). Cell death was present in both conditions (Figure 1F); however, apoptosis was documented only with MSCs LYS, as confirmed by morphological analysis (Figure 1G) and caspase 3 activation (Figure 1H).

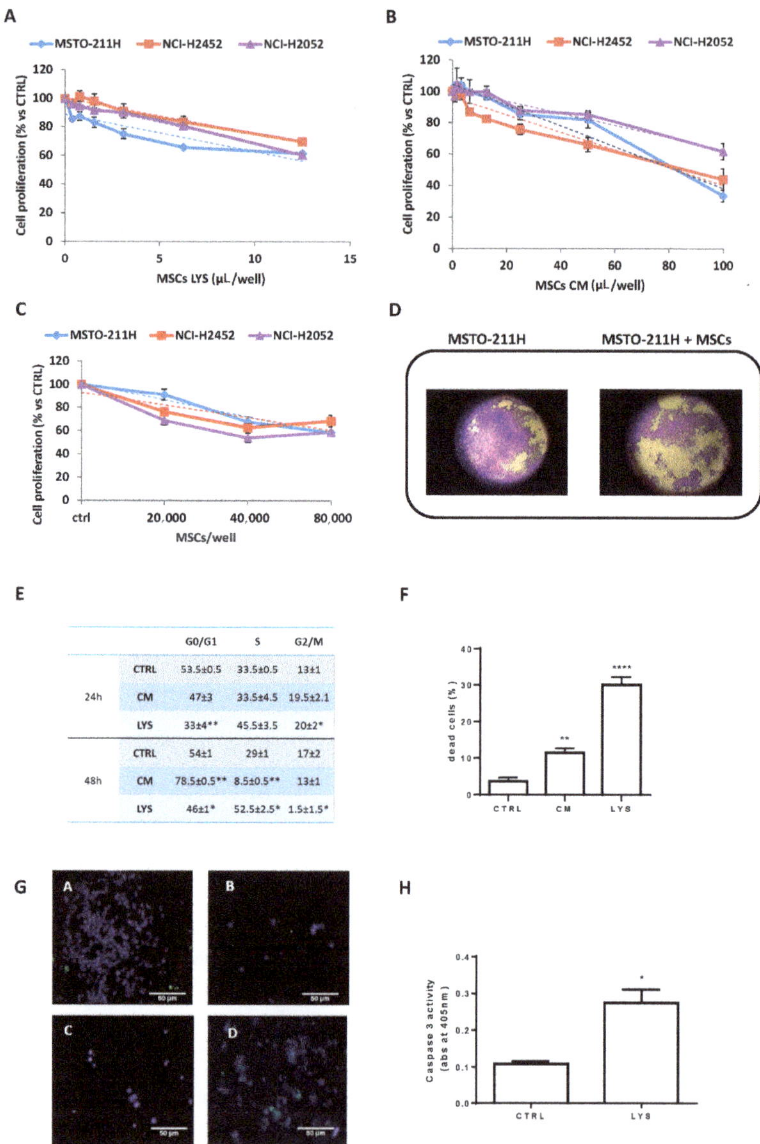

		G0/G1	S	G2/M
24h	CTRL	53.5±0.5	33.5±0.5	13±1
	CM	47±3	33.5±4.5	19.5±2.1
	LYS	33±4**	45.5±3.5	20±2*
48h	CTRL	54±1	29±1	17±2
	CM	78.5±0.5**	8.5±0.5**	13±1
	LYS	46±1*	52.5±2.5*	1.5±1.5*

Figure 1. Inhibitory effect of MSCs LYS and MSCs CM on mesothelioma proliferation and co-culture assay. The activity on cell proliferation of MSCs LYS (**A**) and MSCs CM (**B**), expressed as µL/well from 3 donors of adipose tissue-derived MSCs, were evaluated after 7 days in MSTO-211H, NCI-H2452 and NCI-H2052 cell lines by an MTT assay. Each point represents the mean ± standard error (SEM) of three replicates. Linear regression was reported (dashed lines). (**C**) MSTO-211H, NCI-H2452 and NCI-H2052 cells were co-cultured with MSCs and after 5 days, cell proliferation was evaluated by a crystal violet assay. Data are means ± SEM of six independent replicates. Linear regression was

reported (dashed line). (**D**) Representative images of transwell inserts with MSTO-211H cells co-cultured alone or in the presence of 8 × 10^4 MSCs after staining with crystal violet (400×). (**E**) MSTO-211H cells were treated with MSCs LYS (1:4) and MSCs CM (1:2). After 24 and 48 h, cells were stained with propidium iodide and analyzed by flow cytometry for cell cycle phase distribution. Percentage values ± SEM of two independent experiments are reported in the table (* $p < 0.05$, ** $p < 0.01$ vs. control). (**F**) MSTO-211H cells were treated with MSCs LYS (1:4) and MSCs CM (1:2). After 72 h, cell death was quantified by fluorescence microscopy analysis on Hoechst 33342 and propidium iodide-stained cells. Data are expressed as percentage values ± SEM of three independent experiments (** $p < 0.01$, **** $p < 0.0001$ vs. control). (**G**) Representative confocal images of control cells (**A**,**C**) and cells treated for 48 h with MSCs LYS 1:2 (**B**,**D**). Healthy viable cells were stained with CytoCalcein Violet 450 (blue), necrotic cells with 7-aminoactinomycin D (red), and apoptotic cells with phosphatidylserine (green). (**A**,**B**): adherent cells; (**C**,**D**): detached cells. (Objective 100×). (**H**) Caspase-3 activity measurement after 48 h of treatment. The histogram represents absorbance (abs) at 405 nm. Data are expressed as means ± SEM of three replicates. * $p<0.05$ vs. control.

3.2. *Direct Effect of MSCs on Proliferation of Mesothelioma Cells*

The cell-to-cell interference/interaction between MSCs and tumor cells was studied by mixing fluorescent MSCs (hASCs-TS/GFP$^+$) with MSTO-211H cells and further analyzing the cyto-inclusion under a confocal microscope (Figure 2A). Both MSTO-211H and hASCs-TS/GFP$^+$ cells are present in the cell mass according to the initial seeding ratio and hASCs-TS/GFP$^+$ cells do not appear to be affected by the interaction with cancer cells.

Specific co-culture experiments showed that the proliferation of MSTO-211H, NCI-H2452 and NCI-H2052 cells was significantly impaired by the interaction with hASCs-TS/GFP$^+$ cells, and this anti-proliferative effect was more pronounced in the presence of the higher ratio of hASCs-TS/GFP$^+$: cancer cells (Figure 2B–D).

3.3. *Cytokines Analysis of Secretome*

The secretome analysis of the MSCs derived from nine different donors was performed on the conditioned/culture medium, by a qualitative/quantitative measure of 38 cytokines. Based on the quantitative analysis of cytokines/growth factors secreted, only eight molecules were produced over 2000 pg/mL. As shown in Figure 3, even if an individual variability was detected, a statistically significant difference in cytokines/growth factors production ($p < 0.02$) was found, with the lowest level for IL12p70 (1925.04 ± 298 pg/mL, 2% of the total amount) and the highest release for SCGFb (27.419 ± 10.073 pg/mL, 34% of the total amount).

3.4. *Large-Scale Expansion*

The MSCs were expanded in a closed bioreactor, and after 6 days, we obtained 480 × 10^6 MSCs with a viability of 94%. All the cultured cells obtained displayed the typical MSCs spindle-shaped morphology and a high adhesion capacity to the plastic support (Figure 4A). The MSCs obtained after the expansion also displayed differentiation ability into adipogenic, osteogenic and chondrogenic elements, as reported in Figure 4B–D, respectively. The amount of cumulative population doubling is shown in Figure 4E and it was similar to that of original MSCs. Population doubling time values were of 35.34 ± 9.07 h at P2 and 38.51 ± 11.82 h at P3. Flow cytometry analysis confirmed that MSCs after bioreactor expansion displayed a pattern of CD expression typical of MSCs being CD90$^+$, CD105$^+$ and CD73$^+$ and negative for CD31, CD34 and CD45. In addition, the profile of cytokine production was in the range of the above-described production (Figure 4F). Aliquots of cells, frozen to be used in in vivo experiments, showed a good performance with a high recovery and viability confirmed after 6 months of storing in liquid nitrogen [23].

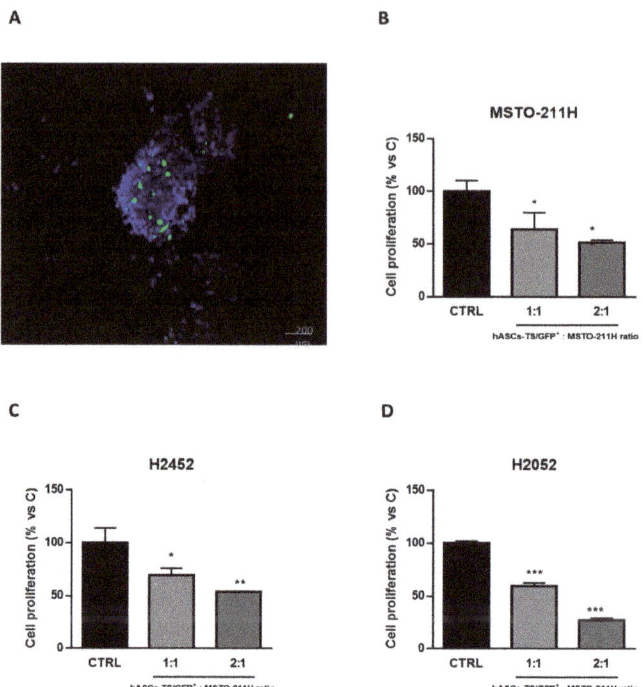

Figure 2. Interaction of MSCs with MSTO-211H cells in vitro. (**A**) The cell-to-cell interference/interaction between MSCs and tumor cells was evaluated by mixing fluorescent MSCs (hASCs-TS/GFP$^+$) with MSTO-211H cells and further analyzing the cyto-inclusion under a confocal microscope. (**B**) MSTO-211H, NCI-H2452 and NCI-H2052 cells were co-cultured in absence (ctrl) or in presence of hASCs-TS/GFP$^+$ cells in the ratio of 1:1 or 1:2. After 2 days for MSTO-211H and after 4 days for NCI-H2452 and NCI-H2052, the cells were detached and counted under a fluorescence microscopy. Data are the means ± SEM of 3 replicates. * $p < 0.05$, ** $p < 0.01$, *** $p < 0.001$ vs. control.

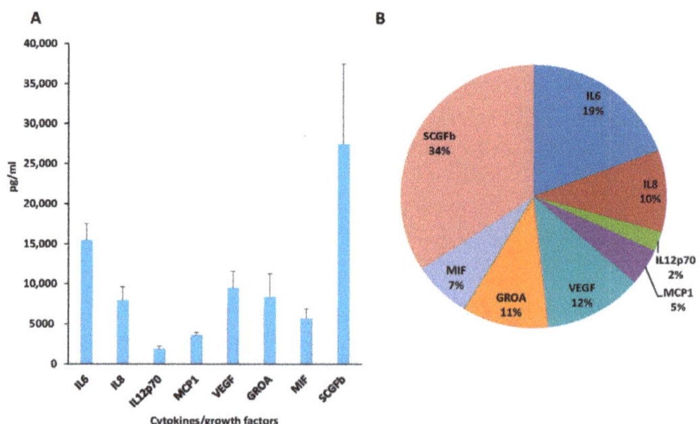

Figure 3. MSCs secretome analysis. (**A**) The histogram reports eight cytokines/growth factors measured in MSC-conditioned media (expressed in pg/mL). Each point represents the mean ± standard error (SEM) of the determinations performed on nine different MSCs donors. (**B**) Cytokines/growth factors expressed as percentage calculated on the total amount of the eight molecules.

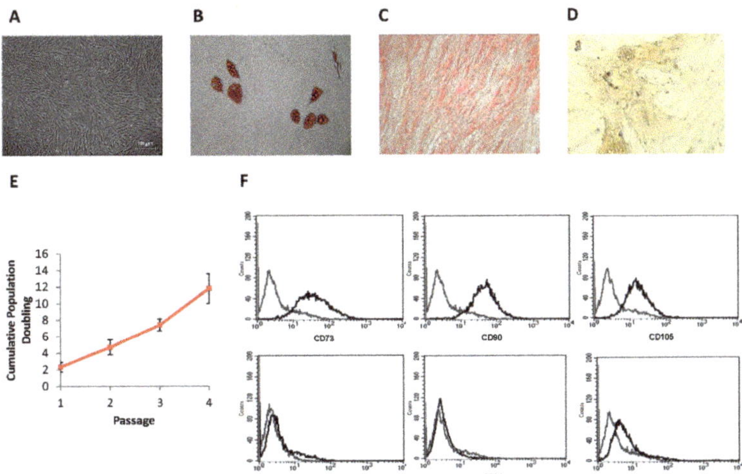

Figure 4. MSC characterization. (**A**): Spindle-shaped MSCs morphology (magnification 5×). (**B**): Adipogenic differentiation evaluated by Oil Red staining (presence of red cytoplasmic inclusions); (**C**): Osteogenic differentiation evaluated by Alizarin Red S staining; (**D**): Chondrogenic differentiation evaluated by micro-mass methods. (All at 200× magnification). (**E**): Number of cumulative population doubling with respect to cellular passage. Each point represents the mean ± SEM of 14 replicates. (**F**): Phenotypic characterization by FACS of a representative sample of MSCs.

3.5. *In Vivo Efficacy of MSCs*

The efficacy of mesenchymal stromal cells on tumor growth was investigated on MSTO-211H xenograft models. MSTO-211H cells were subcutaneously inoculated into Balb/c-Nude female mice and after tumors had reached an average size of about 100 mm^3, the animals were randomized into three different groups: control (CTRL), paclitaxel i.p. (PTX) and MSCs. Tumor growth was monitored for 35 days and during this period, the mice showed no signs of toxicity and regularly gained body weight (Figure 5A); no animal death was observed. As illustrated in Figure 5B, the systemic treatment with PTX produced a reduction in tumor growth kinetics, and 14 days after the last treatment (on day 35), the mean tumor volume was of 817 ± 228 mm^3, which was significantly lower if compared with the mean tumor volume of untreated mice (1893 ± 93 mm^3; $p < 0.0001$). Moreover, the local administration of MSCs showed a significant reduction in tumor volume which was similar to that of mice receiving the systemic administration of PTX. By comparing the mean tumor volumes at day 22 (Figure 5C), it seems that systemic chemotherapy exerted more efficacy than MSCs in controlling the tumor growth (325 ± 78 mm^3 versus 898 ± 90 mm^3; $p < 0.001$). However, after stopping the therapy, in mice treated with PTX the tumors restarted to growth with a significant increase in the mean tumor volume (from 325 ± 78 mm^3 to 817± 228 mm^3; $p < 0.05$) underlined also by the significant R2 (0.94). By contrast, no relapses were observed in mice treated with MSCs (Figure 5C).

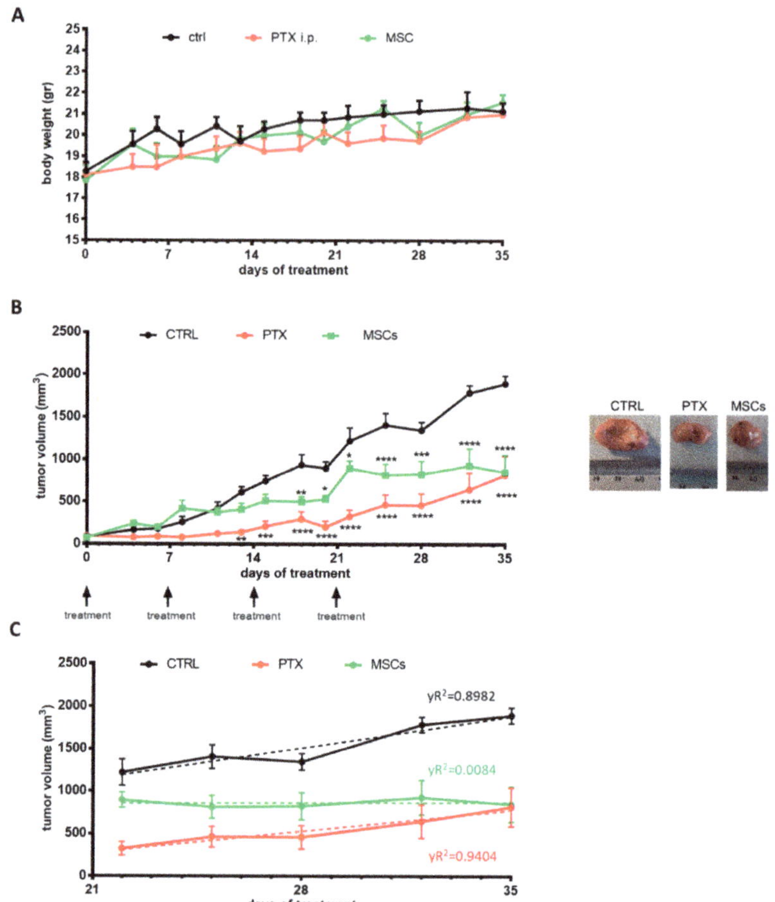

Figure 5. Effects of MSCs in MSTO-211H xenograft model. (**A**) MSTO-211H cells were subcutaneously inoculated into BALB/C nude mice, and after tumors had reached an average size of approximately 100 mm^3, the animals were treated once a week (at days 0, 7, 14 and 21) with vehicle alone (CTRL), paclitaxel (20 mg/kg) or MSCs (5×10^6). (**A**) Mice body weight was monitored for the entire duration of the treatment. (**B**) Tumor volumes were measured twice per week and data are expressed as means of ± SEM. * $p < 0.05$, ** $p < 0.01$, *** $p < 0.001$, **** $p < 0.0001$ vs. CTRL. Representative images of dissected xenograft tumors are shown. (**C**) The graph shows the tumor growth during the 14 days after stopping treatments (from 22 to 35 days). Data are expressed as means ± SEM and for each group, the linear regression (dashed line) with R2 value is reported.

3.6. Histology and Morphometric Analysis of Xenograft

The morphometric evaluation allowed to quantify the percentage of tissue occupied by the neoplastic, fibrotic and necrotic component (Figure 6A–C) on Masson's Trichrome-stained sections of subcutaneous MSTO-211H tumor nodules from each experi-mental group. As shown in Figure 6D, the volume of each component was calculated on the whole masses explanted at day 35 and the volume referred to the neoplastic tissue in untreated mice was compared to that of mice treated with PTX or MSCs. Data analysis by the Student–Newman–Keuls Multiple Comparisons Test indicated a significant decrease ($p < 0.02$) in the fraction of tissue occupied by tumor cells, following administration of PTX i.p. or MSCs. On the contrary, there were no relevant variations in the percentage of necrotic or fibrotic tissues that represent the main tissue component. To test whether angiogenesis in the tumor was affected by MSCs, we evaluated the expression of CD31 in the explanted tumors nodules and we observed that the staining pattern in MSTO-211H treated with MSCs was similar to that of the untreated MSTO-211H (Figures S1 and S2).

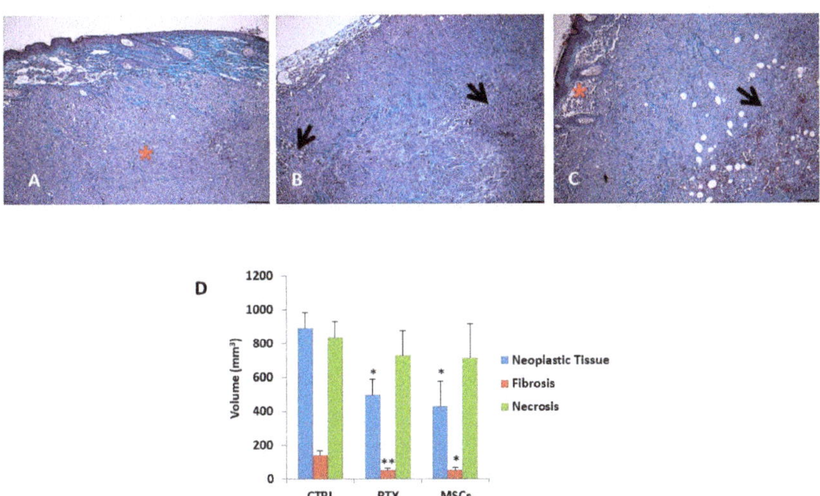

Figure 6. Morphometric analysis of tumor xenografts. Selected sections from MSTO-211H xenografts untreated (CTRL, **A**), treated with paclitaxel (PTX) (**B**) or with MSCs (**C**) stained by Masson's Trichrome to distinguish the fibrotic tissue (greenish) from neoplastic cells (purple). Black arrows in B and C point to necrotic areas, some of which show pigmented (reddish) debris. Red asterisks in A and C indicate skin and adnexa. Scale bars = 100 µm. (**D**) Bar graph showing the quantitative evaluation of tissue composition (neoplastic tissue, fibrosis and necrosis) in tumor xenografts. Data are expressed as means ± SEM. * $p < 0.05$, ** $p < 0.02$ vs. control group.

4. Discussion

The role of MSCs in neoplastic growth is controversial due to their pleiotropic activity and a recent systematic review summarized the application of MSCs of human origin in experimental anticancer therapies [33]. Few data have been reported in the literature on MSCs' role in mesothelioma and these observations regard in vitro studies on the anticancer activity of the secretome of lung-derived human MSCs [19]. Our preliminary experiments investigated if even the secretome from MSCs expanded from human adipose tissue (easier to isolate respect to MSCs from lung tissue) had some inhibitory activity against three MPM cell lines (MSTO-211H, NCI-H2452 and NCI-H2052). We found that both the cell lysates and the MSC-conditioned medium produced a significant inhibition of MPM cell proliferation with cell cycle arrest. In addition, MSC lysate induced apoptosis. The analysis of 38 molecules detected in the secretome of MSCs from nine different donors indicated a strong difference in their relative amount as also previously reported by other authors [34–36]. Even by only considering the molecules produced over 2000 pg/mL, our results did not allow us to reach a clear-cut conclusion to attribute the observed inhibition of mesothelioma cell growth to recognized anticancer molecules secreted by MSCs. In fact, the panel of the factors secreted at the highest level contains molecules that have been described as capable of expressing different and sometimes contradictory/discrepant activities. In this context, even the production of TRAIL (which has been reported as a possible anticancer molecule) was detectable only in two out of nine donor samples at a level (569.3 ± 325 pg/mL) that does not seem sufficient to demonstrate a significant anticancer activity. Regarding interleukin 6 (IL6), it has been reported that this cytokine is present in high concentration in the sera of patients with different cancers including mesothelioma. However, in mesothelioma, the high serum level of IL6 has been indicated as a poor prognostic factor [37]. Many other inflammatory mediators such as chemokines or growth factors produced by MSCs may exert different actions inside the tumor microenvironment involving both antitumor and pro-tumor activity [38]. The conditioned medium of MSCs contains an enormous quantity of molecules (cytokines, chemokines, hormones and many other factors) with different biological activity that can also work together with exosomes/microvesicles produced by MSCs [39,40]. As reported by Mirabdollahi et al. [41], the secretome of umbilical cord-derived mesenchymal cells can have an anticancer effect on MCF-7 tumor cells by inducing apoptosis in a dose-dependent manner. Therefore, although our preliminary study confirmed the anti-proliferative activity of MSCs secretome, it did not help to identify one or more molecules with recognized antitumor activity, suggesting that the observed anticancer activity could be the result of more factors acting together. Moreover, the secretome may change over time when MSCs interact with cancer cells and the secretome of naïve MSCs may be different after co-culturing MSCs with mesothelioma cancer cells. Of course, it is also important to take into account the limitation of in vitro studies performed with MSC secretome because MSCs are physiologically well integrated into very different tissues and their functions are strongly related to the type of tissue in which they are working and to cell-to-cell interaction [42]. In our in vivo model, the cytokines/growth factors could exert their anticancer ability in synergy with the cell-to-cell-interaction, as demonstrated by the close cell interference/interaction observed in the in vitro co-culture model (Figure 2).

On the other hand, the in vivo treatment of tumors with MSCs that secrete VEGF should promote angiogenesis in the xenograft and induce tumor growth. However, as reported by Otsu et al. [43], MSCs, locally injected into tumor tissue, are cytotoxic to newly forming vessels. Therefore, in our model, we tested whether angiogenesis was affected by MSCs by evaluating the expression of vascular marker CD31 in the explanted tumor nodules. The immunohistochemical analysis did not indicate changes in CD31 expression, suggesting that VEGF was not important in inhibiting or stimulating angiogenesis into the tumor mass. Therefore, our findings indicated that tumor regression does not seem to be related to the effects on tumor vascularity.

According to Sage et al. [21], MSCs migrate to pleural mesothelioma tumors in vivo when delivered both intravenously and intrapleurally, but a significant reduction in tumor growth was observed only with intravenous treatment due to the higher number of MSCs capable of tumor engraftment.

As the aim of our study was to evaluate the in vivo effect of MSCs on MPM, we preliminary verified in vitro the ability of MSCs to integrate and interact with mesothelioma cells in a context of cell-to-cell interaction by co-culturing mesothelioma cancer cells with engineered MSCs expressing Green fluorescent protein (GFP). The results confirmed that MSCs were well integrated with cancer cells and that also, in the condition of cell-to-cell contact, MSCs exert a significant inhibitory action on MPM cell proliferation. Based on these data, we evaluated the therapeutic efficacy of MSCs in a model of subcutaneous xenograft of a human MPM in nude mice by a treatment in situ with MSCs that were able to incorporate into tumors. A significant inhibition of tumor growth was observed by MSCs treatment at a level comparable with that observed after systemic treatment with paclitaxel. To better analyze tumor nodule composition, Masson's Trichrome staining of sections was employed to evaluate the percentage of tissue occupied by the neoplastic, fibrotic and necrotic component; the morphometric analysis clearly confirmed that the neoplastic tissue was significantly reduced in mice treated with MSCs. Most relevant, the reduction in tumor growth was comparable among MSCs and paclitaxel-treated animals.

Taken together, our data support the hypothesis that in the MPM environment, the MSCs could have an important role in controlling tumor growth and this could open the way to consider MSCs as a possible therapeutic tool. In this regard, since MSCs have been shown to incorporate and then release different types of anticancer drugs (e.g., paclitaxel, gemcitabine, doxorubicin) [44,45], we suppose that MSCs could improve their basal anticancer efficacy once loaded with chemotherapy molecules. Whether this cell therapy approach could represent a new adjuvant therapy (also associated with the surgery) for human MPM remains to be investigated. However, the demonstration, for the first time, that MSCs per se can act as an "anti-tumor drug" against human MPM has a significant translational value that supports the possibility of future clinical trials with large-scale production in bioreactors, according to the GMP requested by the regulatory agencies.

Supplementary Materials: The following are available online at https://www.mdpi.com/article/10.3390/cells10061427/s1, Figure S1: Immunostaining of the MSTO-211H tumor xenograft sections with anti-mouse CD31 antibody, Figure S2: Immunostaining of the MSTO-211H-MSC tumor xenograft sections with anti-mouse CD31 antibody.

Author Contributions: Conceptualization, G.A., F.P. (Francesco Petrella) and A.P.; Data curation, V.C., S.L.M., M.B. and R.A.; Funding acquisition, F.P. (Francesco Petrella); Investigation, V.C., S.L.M., M.B., C.A.L., D.M., C.F., L.F., F.P. (Francesca Paino) and A.F.; Methodology, D.L., A.M., V.C., E.P. and F.P. (Francesca Paino); Resources, G.F. and F.P. (Francesco Petrella); Supervision, G.A., E.P., A.G. and L.S.; Writing—original draft, A.P.; Writing—review and editing, V.C., S.L.M., M.B., G.A., R.A., C.A.L., D.M., C.F., L.F., E.P., F.P. (Francesca Paino), A.G., G.F., A.F., D.L., A.M., L.S. and F.P. (Francesco Petrella). All authors have read and agreed to the published version of the manuscript.

Funding: This work was partially supported by the IRCCS European Institute Foundation and by the Italian Ministry of Health with Ricerca Corrente and "5 × 1000" funds.

Institutional Review Board Statement: Experiments involving animals and their care were performed with the approval of the Local Ethical Committee of University of Parma (Organismo per la Protezione e il Benessere degli Animali, OPBA) and of the Italian Ministry of Health, in accordance with the institutional guidelines that are in compliance with national (D.Lgs. 26/2014) and in-ternational (Directive 2010/63/EU) laws and policies. Authorization number 325/2018-PR.

Informed Consent Statement: Not applicable.

Data Availability Statement: The data presented in this study are available on request from the corresponding author.

Acknowledgments: The authors are very grateful to Loredana Cavicchini for her excellent technical assistance.

Conflicts of Interest: The authors declare no conflict of interest.

References

1. Carbone, M.; Ly, B.H.; Dodson, R.F.; Pagano, I.; Morris, P.T.; Dogan, U.A.; Gazdar, A.F.; Pass, H.I.; Yang, H. Malignant mesothelioma: Facts, myths, and hypotheses. *J. Cell Physiol.* **2012**, *227*, 44–58. [CrossRef] [PubMed]
2. Peto, J.; Decarli, A.; La Vecchia, C.; Levi, F.; Negri, E. The European mesothelioma epidemic. *Br. J. Cancer* **1999**, *79*, 666–672. [CrossRef] [PubMed]
3. Fennell, D.A.; Gaudino, G.; O'Byrne, K.J.; Mutti, L.; van Meerbeeck, J. Advances in the systemic therapy of malignant pleural mesothelioma. *Nat. Clin. Pract. Oncol.* **2008**, *5*, 136–147. [CrossRef]
4. Krug, L.M.; Pass, H.I.; Rusch, V.W.; Kindler, H.L.; Sugarbaker, D.J.; Rosenzweig, K.E.; Flores, R.; Friedberg, J.S.; Pisters, K.; Monberg, M.; et al. Multicenter phase II trial of neoadjuvant pemetrexed plus cisplatin followed by extrapleural pneumonectomy and radiation for malignant pleural mesothelioma. *J. Clin. Oncol.* **2009**, *27*, 3007–3013. [CrossRef]
5. Mutti, L.; Peikert, T.; Robinson, B.W.S.; Scherpereel, A.; Tsao, A.S.; de Perrot, M.; Woodard, G.A.; Jablons, D.M.; Wiens, J.; Hirsch, F.R.; et al. Scientific Advances and New Frontiers in Mesothelioma Therapeutics. *J. Thorac. Oncol.* **2018**, *13*, 1269–1283. [CrossRef] [PubMed]
6. Li, W.; Zhou, Y.; Yang, J.; Zhang, X.; Zhang, H.; Zhang, T.; Zhao, S.; Zheng, P.; Huo, J.; Wu, H. Gastric cancer-derived mesenchymal stem cells prompt gastric cancer progression through secretion of interleukin-8. *J. Exp. Clin. Cancer Res.* **2015**, *34*, 52. [CrossRef]
7. Li, G.C.; Zhang, H.W.; Zhao, Q.C.; Sun, L.I.; Yang, J.J.; Hong, L.; Feng, F.; Cai, L. Mesenchymal stem cells promote tumor angiogenesis via the action of transforming growth factor β1. *Oncol. Lett.* **2016**, *11*, 1089–1094. [CrossRef] [PubMed]
8. Houthuijzen, J.M.; Daenen, L.G.; Roodhart, J.M.; Voest, E.E. The role of mesenchymal stem cells in anti-cancer drug resistance and tumour progression. *Br. J. Cancer* **2012**, *106*, 1901–1906. [CrossRef]
9. Ramasamy, R.; Lam, E.W.; Soeiro, I.; Tisato, V.; Bonnet, D.; Dazzi, F. Mesenchymal stem cells inhibit proliferation and apoptosis of tumor cells: Impact on in vivo tumor growth. *Leukemia* **2007**, *21*, 304–310. [CrossRef]

10. Chanda, D.; Isayeva, T.; Kumar, S.; Hensel, J.A.; Sawant, A.; Ramaswamy, G.; Siegal, G.P.; Beatty, M.S.; Ponnazhagan, S. Therapeutic potential of adult bone marrow-derived mesenchymal stem cells in prostate cancer bone metastasis. *Clin. Cancer Res.* **2009**, *15*, 7175–7185. [CrossRef]
11. Ohlsson, L.B.; Varas, L.; Kjellman, C.; Edvardsen, K.; Lindvall, M. Mesenchymal progenitor cell-mediated inhibition of tumor growth in vivo and in vitro in gelatin matrix. *Exp. Mol. Pathol.* **2003**, *75*, 248–255. [CrossRef]
12. Ayuzawa, R.; Doi, C.; Rachakatla, R.S.; Pyle, M.M.; Maurya, D.K.; Troyer, D.; Tamura, M. Naïve human umbilical cord matrix derived stem cells significantly attenuate growth of human breast cancer cells in vitro and in vivo. *Cancer Lett.* **2009**, *280*, 31–37. [CrossRef]
13. Yuan, Z.; Kolluri, K.K.; Gowers, K.H.; Janes, S.M. TRAIL delivery by MSC-derived extracellular vesicles is an effective anticancer therapy. *J. Extracell. Vesicles* **2017**, *6*, 1265291. [CrossRef]
14. Li, L.; Tian, H.; Chen, Z.; Yue, W.; Li, S.; Li, W. Inhibition of lung cancer cell proliferation mediated by human mesenchymal stem cells. *Acta Biochim. Biophys. Sin. (Shanghai)* **2011**, *43*, 143–148. [CrossRef] [PubMed]
15. Ramdasi, S.; Sarang, S.; Viswanathan, C. Potential of Mesenchymal Stem Cell based application in Cancer. *Int. J. Hematol. Oncol. Stem Cell Res.* **2015**, *9*, 95–103.
16. Galland, S.; Stamenkovic, I. Mesenchymal stromal cells in cancer: A review of their immunomodulatory functions and dual effects on tumor progression. *J. Pathol.* **2020**, *250*, 555–572. [CrossRef]
17. Hmadcha, A.; Martin-Montalvo, A.; Gauthier, B.R.; Soria, B.; Capilla-Gonzalez, V. Therapeutic Potential of Mesenchymal Stem Cells for Cancer Therapy. *Front. Bioeng. Biotechnol.* **2020**, *8*, 43. [CrossRef] [PubMed]
18. Nwabo Kamdje, A.H.; Kamga, P.T.; Simo, R.T.; Vecchio, L.; Seke Etet, P.F.; Muller, J.M.; Bassi, G.; Lukong, E.; Goel, R.K.; Amvene, J.M.; et al. Mesenchymal stromal cells' role in tumor microenvironment: Involvement of signaling pathways. *Cancer Biol. Med.* **2017**, *14*, 129–141. [CrossRef]
19. Cortes-Dericks, L.; Froment, L.; Kocher, G.; Schmid, R.A. Human lung-derived mesenchymal stem cell-conditioned medium exerts in vitro antitumor effects in malignant pleural mesothelioma cell lines. *Stem Cell Res. Ther.* **2016**, *7*, 25. [CrossRef]
20. Lathrop, M.J.; Sage, E.K.; Macura, S.L.; Brooks, E.M.; Cruz, F.; Bonenfant, N.R.; Sokocevic, D.; MacPherson, M.B.; Beuschel, S.L.; Dunaway, C.W.; et al. Antitumor effects of TRAIL-expressing mesenchymal stromal cells in a mouse xenograft model of human mesothelioma. *Cancer Gene Ther.* **2015**, *22*, 44–54. [CrossRef] [PubMed]
21. Sage, E.K.; Kolluri, K.K.; McNulty, K.; Lourenco Sda, S.; Kalber, T.L.; Ordidge, K.L.; Davies, D.; Gary Lee, Y.C.; Giangreco, A.; Janes, S.M. Systemic but not topical TRAIL-expressing mesenchymal stem cells reduce tumour growth in malignant mesothelioma. *Thorax* **2014**, *69*, 638–647. [CrossRef]
22. Lisini, D.; Nava, S.; Pogliani, S.; Avanzini, M.A.; Lenta, E.; Bedini, G.; Mantelli, M.; Pecciarini, L.; Croce, S.; Boncoraglio, G.; et al. Adipose tissue-derived mesenchymal stromal cells for clinical application: An efficient isolation approach. *Curr. Res. Transl. Med.* **2019**, *67*, 20–27. [CrossRef]
23. Lisini, D.; Nava, S.; Frigerio, S.; Pogliani, S.; Maronati, G.; Marcianti, A.; Coccè, V.; Bondiolotti, G.; Cavicchini, L.; Paino, F.; et al. Automated Large-Scale Production of Paclitaxel Loaded Mesenchymal Stromal Cells for Cell Therapy Applications. *Pharmaceutics* **2020**, *12*, 411. [CrossRef] [PubMed]
24. Petrella, F.; Coccè, V.; Masia, C.; Milani, M.; Salè, E.O.; Alessandri, G.; Parati, E.; Sisto, F.; Pentimalli, F.; Brini, A.T.; et al. Paclitaxel-releasing mesenchymal stromal cells inhibit in vitro proliferation of human mesothelioma cells. *Biomed. Pharmacother.* **2017**, *87*, 755–758. [CrossRef]
25. Mossman, T. Rapid colorimetric assay for cellular growth and survival: Application to proliferation and cytotoxicity assays. *J. Immunol. Methods* **1983**, *65*, 55–63. [CrossRef]
26. Fumarola, C.; La Monica, S.; Alfieri, R.R.; Borra, E.; Guidotti, G.G. Cell size reduction induced by inhibition of the mTOR/S6K-signaling pathway protects Jurkat cells from apoptosis. *Cell Death Differ.* **2005**, *12*, 1344–1357. [CrossRef]
27. Balducci, L.; Blasi, A.; Saldarelli, M.; Soleti, A.; Pessina, A.; Bonomi, A.; Coccè, V.; Dossena, M.; Tosetti, V.; Ceserani, V.; et al. Immortalization of human adipose-derived stromal cells: Production of cell lines with high growth rate, mesenchymal marker expression and capability to secrete high levels of angiogenic factors. *Stem Cell Res. Ther.* **2014**, *5*, 63. [CrossRef]
28. Coccè, V.; Farronato, D.; Brini, A.T.; Masia, C.; Giannì, A.B.; Piovani, G.; Sisto, F.; Alessandri, G.; Angiero, F.; Pessina, A. Drug Loaded Gingival Mesenchymal Stromal Cells (GinPa-MSCs) Inhibit In Vitro Proliferation of Oral Squamous Cell Carcinoma. *Sci. Rep.* **2017**, *7*, 9376. [CrossRef]
29. Coccè, V.; Balducci, L.; Falchetti, M.L.; Pascucci, L.; Ciusani, E.; Brini, A.T.; Sisto, F.; Piovani, G.; Alessandri, G.; Parati, E.; et al. Fluorescent Immortalized Human Adipose Derived Stromal Cells (hASCs-TS/GFP+) for Studying Cell Drug Delivery Mediated by Microvesicles. *Anticancer Agents Med. Chem.* **2017**, *17*, 1578–1585. [CrossRef] [PubMed]
30. Berenzi, A.; Steimberg, N.; Boniotti, J.; Mazzoleni, G. MRT letter: 3D culture of isolated cells: A fast and efficient method for optimizing their histochemical and immunocytochemical analyses. *Microsc. Res. Tech.* **2015**, *78*, 249–254. [CrossRef]
31. La Monica, S.; Cretella, D.; Bonelli, M.; Fumarola, C.; Cavazzoni, A.; Digiacomo, G.; Flammini, L.; Barocelli, E.; Minari, R.; Naldi, N.; et al. Trastuzumab emtansine delays and overcomes resistance to the third-generation EGFR-TKI osimertinib in NSCLC EGFR mutated cell lines. *J. Exp. Clin. Cancer Res.* **2017**, *36*, 174. [CrossRef]
32. La Monica, S.; Minari, R.; Cretella, D.; Flammini, L.; Fumarola, C.; Bonelli, M.; Cavazzoni, A.; Digiacomo, G.; Galetti, M.; Madeddu, D.; et al. Third generation EGFR inhibitor osimertinib combined with pemetrexed or cisplatin exerts long-lasting anti-tumor effect in EGFR-mutated pre-clinical models of NSCLC. *J. Exp. Clin. Cancer Res.* **2019**, *38*, 222. [CrossRef]

33. Christodoulou, I.; Goulielmaki, M.; Devetzi, M.; Panagiotidis, M.; Koliakos, G.; Zoumpourlis, V. Mesenchymal stem cells in preclinical cancer cytotherapy: A systematic review. *Stem Cell Res. Ther.* **2018**, *9*, 336. [CrossRef]
34. Park, C.W.; Kim, K.S.; Bae, S.; Son, H.K.; Myung, P.K.; Hong, H.J.; Kim, H. Cytokine secretion profiling of human mesenchymal stem cells by antibody array. *Int. J. Stem Cells* **2009**, *2*, 59–68. [CrossRef] [PubMed]
35. Wu, Y.; Hoogduijn, M.J.; Baan, C.C.; Korevaar, S.S.; de Kuiper, R.; Yan, L.; Wang, L.; van Besouw, N.M. Adipose Tissue-Derived Mesenchymal Stem Cells Have a Heterogenic Cytokine Secretion Profile. *Stem Cells Int.* **2017**, *2017*, 4960831. [CrossRef] [PubMed]
36. Abdul Rahim, S.N.; Ho, G.Y.; Coward, J.I. The role of interleukin-6 in malignant mesothelioma. *Transl. Lung Cancer Res.* **2015**, *4*, 55–66.
37. Assoni, A.; Coatti, G.; Valadares, M.C.; Beccari, M.; Gomes, J.; Pelatti, M.; Mitne-Neto, M.; Carvalho, V.M.; Zatz, M. Different Donors Mesenchymal Stromal Cells Secretomes Reveal Heterogeneous Profile of Relevance for Therapeutic Use. *Stem Cells Dev.* **2017**, *26*, 206–214. [CrossRef] [PubMed]
38. Shrihari, T.G. Dual role of inflammatory mediators in cancer. *Ecancermedicalscience* **2017**, *11*, 721. [CrossRef]
39. Lee, M.W.; Ryu, S.; Kim, D.S.; Lee, J.W.; Sung, K.W.; Koo, H.H.; Yoo, K.H. Mesenchymal stem cells in suppression or progression of hematologic malignancy: Current status and challenges. *Leukemia* **2019**, *33*, 597–611. [CrossRef]
40. Pokrovskaya, L.A.; Zubareva, E.V.; Nadezhdin, S.V.; Lysenko, A.S.; Litovkina, T.L. Biological activity of mesenchymal stem cells secretome as a basis for cell-free therapeutic approach. *Res. Results Pharmacol.* **2020**, *6*, 57–68. [CrossRef]
41. Mirabdollahi, M.; Haghjooyjavanmard, S.; Sadeghi-Aliabadi, H. An anticancer effect of umbilical cord-derived mesenchymal stem cell secretome on the breast cancer cell line. *Cell Tissue Bank* **2019**, *20*, 423–434. [CrossRef]
42. Leuning, D.G.; Beijer, N.R.M.; du Fossé, N.A.; Vermeulen, S.; Lievers, E.; van Kooten, C.; Rabelink, T.J.; Boer, J. The cytokine secretion profile of mesenchymal stromal cells is determined by surface structure of the microenvironment. *Sci. Rep.* **2018**, *8*, 7716. [CrossRef] [PubMed]
43. Otsu, K.; Das, S.; Houser, S.D.; Quadri, S.K.; Bhattacharya, S.; Bhattacharya, J. Concentration-dependent inhibition of angiogenesis by mesenchymal stem cells. *Blood* **2009**, *113*, 4197–4205. [CrossRef]
44. Pessina, A.; Bonomi, A.; Coccè, V.; Invernici, G.; Navone, S.; Cavicchini, L.; Sisto, F.; Ferrari, M.; Viganò, L.; Locatelli, A.; et al. Mesenchymal stromal cells primed with paclitaxel provide a new approach for cancer therapy. *PLoS ONE* **2011**, *6*, e28321. [CrossRef] [PubMed]
45. Bonomi, A.; Sordi, V.; Dugnani, E.; Ceserani, V.; Dossena, M.; Coccè, V.; Cavicchini, L.; Ciusani, E.; Bondiolotti, G.; Piovani, G.; et al. Gemcitabine-releasing mesenchymal stromal cells inhibit in vitro proliferation of human pancreatic carcinoma cells. *Cytotherapy* **2015**, *17*, 1687–1695. [CrossRef] [PubMed]

 *cells*

Article

Short-Term Autophagy Preconditioning Upregulates the Expression of COX2 and PGE2 and Alters the Immune Phenotype of Human Adipose-Derived Stem Cells In Vitro

Rachel M. Wise [1,2], Sara Al-Ghadban [2,3], Mark A. A. Harrison [1,2], Brianne N. Sullivan [1,2], Emily R. Monaco [1], Sarah J. Aleman [1], Umberto M. Donato [1] and Bruce A. Bunnell [2,3,4,*]

[1] Neuroscience Program, Tulane Brain Institute, Tulane University School of Science & Engineering, New Orleans, LA 70118, USA; rachel.wise@med.uni-muenchen.de (R.M.W.); mharri26@tulane.edu (M.A.A.H.); bsulliv7@tulane.edu (B.N.S.); emilyrosemonaco@gmail.com (E.R.M.); saleman@tulane.edu (S.J.A.); udonato1@tulane.edu (U.M.D.)

[2] Center for Stem Cell Research & Regenerative Medicine, Tulane University School of Medicine, New Orleans, LA 70112, USA; sara.al-ghadban@unthsc.edu

[3] Department of Microbiology, Immunology and Genetics, University of North Texas Health Science Center, Fort Worth, TX 76107, USA

[4] Department of Pharmacology, Tulane University School of Medicine, New Orleans, LA 70112, USA

* Correspondence: bruce.bunnell@unthsc.edu

Citation: Wise, R.M.; Al-Ghadban, S.; Harrison, M.A.A.; Sullivan, B.N.; Monaco, E.R.; Aleman, S.J.; Donato, U.M.; Bunnell, B.A. Short-Term Autophagy Preconditioning Upregulates the Expression of COX2 and PGE2 and Alters the Immune Phenotype of Human Adipose-Derived Stem Cells In Vitro. *Cells* **2022**, *11*, 1376. https://doi.org/10.3390/cells11091376

Academic Editors: Joni H. Ylostalo, Nisha C. Durand and Christian Dani

Received: 25 January 2022
Accepted: 7 April 2022
Published: 19 April 2022

Abstract: Human adipose-derived stem cells (hASCs) are potent modulators of inflammation and promising candidates for the treatment of inflammatory and autoimmune diseases. Strategies to improve hASC survival and immunoregulation are active areas of investigation. Autophagy, a homeostatic and stress-induced degradative pathway, plays a crucial role in hASC paracrine signaling—a primary mechanism of therapeutic action. Therefore, induction of autophagy with rapamycin (Rapa), or inhibition with 3-methyladenine (3-MA), was examined as a preconditioning strategy to enhance therapeutic efficacy. Following preconditioning, both Rapa and 3-MA-treated hASCs demonstrated preservation of stemness, as well as upregulated transcription of cyclooxygenase-2 (COX2) and interleukin-6 (IL-6). Rapa-ASCs further upregulated TNFα-stimulated gene-6 (TSG-6) and interleukin-1 beta (IL-1β), indicating additional enhancement of immunomodulatory potential. Preconditioned cells were then stimulated with the inflammatory cytokine interferon-gamma (IFNγ) and assessed for immunomodulatory factor production. Rapa-pretreated cells, but not 3-MA-pretreated cells, further amplified COX2 and IL-6 transcripts following IFNγ exposure, and both groups upregulated secretion of prostaglandin-E2 (PGE2), the enzymatic product of COX2. These findings suggest that a 4-h Rapa preconditioning strategy may bestow the greatest improvement to hASC expression of cytokines known to promote tissue repair and regeneration and may hold promise for augmenting the therapeutic potential of hASCs for inflammation-driven pathological conditions.

Keywords: adipose tissue-derived stem cells (ASCs); autophagy; rapamycin; 3-methyladenine; immunosuppression; inflammation

1. Introduction

Adipose-derived mesenchymal stem cells (ASCs) have emerged as a promising cell-based therapeutic agent in regenerative medicine and tissue engineering. They possess significant advantages over their bone marrow-derived (BMSC) counterparts due to their ease of harvest, higher stem cell yield, enhanced secretion of immunomodulatory factors and reduced immunogenicity [1–3]. As reviewed by Ceccarelli and colleagues, ASCs' ability to regulate the immune microenvironment gives them immense translational potential in autoimmune, inflammatory, ischemic, and neurodegenerative disease states [4]. However, the lack of standardized cellular processing, donor-to-donor variability, and low ASC survival in the transplant environment remain major obstacles to translational success. ASCs exert their therapeutic benefit predominantly via the production and secretion of immune-modifying proteins and extracellular vesicles (EVs) [3,5–7]. In vitro and in vivo studies demonstrated that ASCs amplify their secretory activity following exposure to hypoxic and inflammatory conditions, which are biomimetic of many disease states. Specifically, these conditions serve to enhance ASC production of molecules like transforming growth factor-beta (*TGF-β*), indoleamine-pyrrole 2,3-dioxygenase (*IDO*), prostaglandin E2 (PGE2), and interleukin-10 (*IL-10*) which suppress T cell proliferation and shift macrophages from pro-inflammatory to anti-inflammatory/pro-regenerative phenotypes [5,8–13]. Of particular interest is the eicosanoid PGE2, which is synthesized from arachidonic acid by cyclooxygenase-2 (*COX2*), and can modify T cell and macrophage populations in several inflammatory [14,15] and autoimmune [16–21] diseases. Although preclinical research supports the efficacy of ASC-based therapeutics, the transition from preclinical to clinical testing has had limited success. The focus has therefore shifted toward optimizing strategies that enhance ASC post-transplant survival and immunosuppression with hopes of bridging this gap [22,23].

Autophagy, derived from the Greek words for "self-eating", is a multifunctional pathway with key roles in stem cell development, homeostatic maintenance, metabolism, secretory pathways, and stress response [24,25]. Under basal conditions, autophagy serves a housekeeping function by degrading damaged organelles and long-lived proteins via trafficking to the lysosome for enzymatic breakdown. However, stress conditions such as nutrient deprivation, hypoxia, and inflammation trigger signaling cascades that activate autophagic flux to maintain protein synthesis and promote survival [26]. Numerous pharmacological agents can be used to manipulate the autophagy pathway. Two of the most common are rapamycin (Rapa), an immunosuppressive drug that induces autophagy by silencing its negative regulator, mammalian target of rapamycin (mTOR); and 3-methyladenine (3-MA), which inhibits the initiation of autophagy through its blockade of vacuolar protein sorting 34 (VPS34) [27,28]. To date, few investigations have explored the correlation between human ASCs (hASCs) autophagy and immunomodulatory capacity [29–31]. In hASCs, Li and colleagues demonstrated that autophagy induction with short-term Rapa exposure suppressed activation of caspase 3 and enhanced overall survival in response to oxygen–glucose deprivation (OGD). In contrast, the inhibition of autophagy proteins with an shRNA approach exacerbated cell death [31]. The authors suggested that Rapa may protect hASCs by enhancing their survival rate after therapeutic administration into hypoxic tissue niches. Another study conducted by Kim and collaborators showed that Rapa preconditioning of hASCs increased both mRNA and protein expression of the immunosuppressive cytokines *IDO*, *IL-10*, and *TGF-β*, suppressed T helper 17 (T_h17) cells thereby promoting T regulatory cells (T_{regs}), and prolonged survival in a mouse model of graft-versus-host-disease (GVHD) [29]. Conversely, Javorka et al. demonstrated that Rapa did not enhance the expression of anti-inflammatory markers like *TSG-6*, *IDO*, and *COX2* in hASCs either at rest or following interferon gamma (IFNγ) stimulation [30]. These contradictory findings suggest that autophagy may play a complicated role in regulating hASC immunomodulatory capacity and may be dependent on undefined variables which must be investigated prior to clinical applications.

In this study, the impact of pharmacological induction and inhibition of autophagy on hASC stemness and immunomodulatory behavior was investigated by using Rapa and 3-MA, respectively. The effects of short-term (4 h) versus long-term (24 h) autophagy preconditioning were also compared to determine the optimal strategy for enhancing production of immune-modifying factors. The data demonstrate that hASCs maintained their stemness regardless of pharmacological agent or duration of exposure. However, expression of pro- and anti-inflammatory mediators differed between autophagy-induced and autophagy-inhibited cells, and also between exposure times, showing that transient versus prolonged manipulation of autophagy has different impacts on hASC physiology.

Interestingly, both Rapa and 3-MA-preconditioned hASCs exhibited a time-dependent increase of *COX2* gene expression and secretion of PGE2. Overall, our findings demonstrate that short-term preconditioning of hASCs with Rapa represents a novel strategy for enhancement of their immunomodulatory potential. This may prove beneficial in enhancing the translation of ASC-based therapies from animal models to human patients.

2. Materials and Methods

2.1. Cells and Cell Culture

Primary human ASCs (hASCs) were purchased from LaCell LLC (New Orleans, LA, USA). Each hASC cell line underwent full characterization individually prior to being pooled together [32–36]. Pooled hASCs from 5 healthy donors were thawed at passage 3 and maintained in complete culture medium (CCM) consisting of minimum essential medium alpha (Cat #: 12561; Gibco, Grand Island, NY, USA) supplemented with 10% heat-inactivated fetal bovine serum (FBS, Cat #: SH30396.03; Thermo Fisher, Waltham, MA, USA), and 1% penicillin-streptomycin (Cat #: 15140122; 10,000 U/mL, Thermo Fisher) in a humidified 5% CO_2 incubator. Media was changed every 2–3 days until cells achieved 80% confluence then harvested with 0.25% trypsin/1 mM EDTA (Cat #: 25200056; Thermo Fisher) and passaged. For all experiments in this study, cells were used at passage 5. Complete donor information is listed in Table 1.

Table 1. Donor Demographics.

Donor	Age	BMI
1	34	20.34
2	40	21.18
3	39	23.4
4	25	22.0
5	40	21.19
Average ± SD	36.5 ± 2.87	21.62 ± 0.52

2.2. Western Blot Analysis

hASCs were seeded at a density of 4×10^3 cells/cm^2. After 72 h, cells were preconditioned with either Rapa (500 nM in DMSO, Cat #: 553211; Millipore Sigma, Burlington, MA, USA), or 3-methyladenine (3-MA; 5 mM in CCM, Cat #: M9281; Millipore Sigma) for 1, 4, 12, 24, and 48 h. After autophagy preconditioning, hASCs were washed once with ice-cold 1X phosphate-buffered saline (PBS) and lysed with RIPA lysis buffer (Cat #: 89900; Thermo Fisher) supplemented with 1X protease inhibitor (Cat #: 1862209; Thermo Fisher) and 1X phosphatase inhibitor (Cat #: 1862495; Thermo Fisher). Protein samples were quantified by using the bicinchoninic acid assay (BCA, Cat #: 23225; Thermo Fisher), and 10 μg of protein lysate was resolved with SDS-PAGE on 4–12% Bis-Tris gels (Cat #: NW04122BOX; Thermo Fisher) in NuPAGE LDS sample buffer (Cat #: NP0007; Thermo Fisher) by using the XCell SureLock Mini-Cell Electrophoresis system (Thermo Fisher). Separated proteins were then transferred onto a PVDF membrane (Cat #: IB401032; Thermo Fisher) by using the iBLOT semi-dry transfer system (Thermo Fisher). The membrane was immediately blocked with 5% non-fat milk in TBS-T (150 mM NaCl, 0.1% Tween 20, 25 mM Tris-HCl, pH 7.6) for 1 h at room temperature (RT). The membrane was then probed with primary antibodies for LC3 (1:1000; Cat #: 27755; Cell Signaling Technologies, Danvers, MA, USA), p62/SQSTM1 (1:2000; Cat #: ab56416; Abcam, Cambridge, UK) and β-*actin* (1:2000; Cat #: 8457S; Cell Signaling Technologies) according to the manufacturer's protocol. After incubating with HRP-conjugated goat anti-mouse or goat anti-rabbit secondary antibodies (Cat #: 7076P2 or 70745; Cell Signaling Technologies), the membrane was incubated with Clarity Western ECL substrate (Cat #: 1705061; Biorad, Hercules, CA, USA) for 5 min and immediately imaged by using the ImageQuant 4000 Imaging System (GE Healthare, Chicago, IL, USA). Captured images were analyzed by using the ImageJ software (U. S. National Institutes of Health, Bethesda, MD, USA). Expression levels of all proteins were normalized to β-*actin*.

2.3. Flow Cytometry

For phenotypic analysis of hASC surface marker expression, cells were blocked with 1% CD16/CD32 in 1X PBS supplemented with 1% bovine serum albumin (BSA) and stained with the following fluorochrome-conjugated primary antibodies at 4 °C for 15 min: CD3 (Cat #: 562406, BD Biosciences, San Jose, CA, USA), CD14 (Cat #: IM2640U, Beckman-Coulter, Brea, CA, USA), CD31 (Cat #: 563651, BD Biosciences), CD45 (Cat #: A71117, Beckman-Coulter), CD73 (Cat #: 550257, BD Biosciences), CD90 (Cat #: 11-0909-42, Thermo Fisher, Waltham, MA, USA), and CD105 (Cat #: 17-1057-42, Thermo Fisher). Stained cells were then fixed with 1% paraformaldehyde (PFA) for 5 min at RT and at least 5000 events were captured by using a Gallios flow cytometer (Beckman Coulter) and analyzed by using Kaluza Analysis 2.1 software (Beckman Coulter).

2.4. Colony Forming Unit-Fibroblast (CFU-F) Assay

hASCs were seeded at a density of 250 cells per 10 cm^2 and allowed to adhere for 24 h. Cells were autophagy preconditioned for 4 h, washed, then cultured for 14 days to allow for colony formation. The medium was changed on day 7 and on day 14 the cells were washed twice with 1X PBS and stained with 3% crystal violet (MilliporeSigma, St. Louis, MO, USA) in methanol for 30 min at RT. The plates were then washed with deionized water until clear and the number of colonies larger than 2 mm in diameter were manually recorded.

2.5. RNA Isolation and Quantitative Reverse-Transcription PCR (qRT-PCR)

Total RNA was collected from lysed hASCs, and RNA extraction was performed by using the Qiagen RNeasy Plus mini kit (Cat #: 74136, Qiagen, Germantown, MD, USA). A total of 1 µg of mRNA was then used for cDNA synthesis by using the Applied Bioscience High-Capacity cDNA Reverse Transcription kit (Cat #: 4368814, Thermo Fisher). qRT-PCR was performed by using the SsoAdvanced Universal SYBR Green Supermix (Cat #: 1725271, Bio-Rad, Hercules, CA, USA). Exon-spanning human-specific primers were designed by using the Primer-BLAST online tool. Primer sequences used for qRT-PCR are listed in Table 2. All reactions were performed in duplicate. Analysis was performed by using the $2^{-\Delta\Delta Ct}$ method to calculate the relative fold-change in transcript expression after normalization to the reference gene, *β-actin*.

Table 2. Primer Sequences.

Gene	Forward (5′-3′)	Reverse (5′-3′)
Beta-actin	ACGTTGCTATCCAGGCTGTGCTAT	TTAATGTCACGCACGATTTCCCGC
ATG7	ATGATCCCTGTAACTTAGCCCA	CACGGAAGCAAACAACTTCAAC
LC3B	AAGGCGCTTACAGCTCAATG	CTGGGAGGCATAGACCATGT
P62	GCACCCCAATGTGATCTGC	CGCTACACAAGTCGTAGTCTGG
TGF-β	CAGTCACCATAGCAACACTC	CCTGGCCTGAACTACTATCT
IDO	TCTCATTTCGTGATGGAGACTGC	GTGTCCCGTTCTTGCATTTGC
TSG-6	AGAATTTGTGAGCAGCCCCT	GGCTGCTCGTTCAAGCCATA
IL-1β	CATGGGATAACGAGGCTTATG	CCACTTGTTGCTCCATATCC
IL-6	CCTTCCAAAGATGGCTGAAA	TGGCTTGTTCCTCACTACT
COX2	TTGCTGGCAGGGTTGCTGGTGGTA	CATCTGCCTGCTCTGGTCAATCGAA

2.6. Enzyme-Linked Immunosorbent Assay (ELISA)

To determine the concentration of secreted PGE2, hASCs were preconditioned with either control, Rapa, or 3-MA-containing media for 4 or 24 h. Following a wash to remove residual autophagy compounds, cells were exposed to human IFNγ (hIFNγ) (5 ng/mL in sterile water; Cat #: PHC4031; Thermo Fisher). After 24 h, the conditioned medium (CM) was collected, centrifuged to remove cellular debris, and stored at −80°C until use. Levels of PGE2 were then measured by using the Prostaglandin E2 Parameter Assay Kit (Cat #: KGE004B; R&D Systems, Minneapolis, MN, USA) according to the manufacturer's protocol.

Briefly, CM was thawed at RT and equal volumes from three independent experiments were pooled together, diluted three-fold, then incubated in the wells of a 96-well pre-coated plate at room temperature. Wells were incubated with a capture antibody, an HRP-conjugated PGE2 competitor, a substrate solution, and finally a stop solution. Absorbance was then read at 450 nm on a Synergy HTX plate reader (BioTek, Winooski, VT, USA). Each sample was standardized to the appropriate negative controls, and the PGE2 concentration was extrapolated from the standard curve.

2.7. Statistical Analysis

All data are presented as mean ±SEM of at least three independent experiments and GraphPad PRISM 8 was used to perform all statistical analyses. Results were compared by using one-way analysis of variance (ANOVA) followed by a Tukey's post-hoc test to analyze the differences between multiple groups. Asterisks (*) denote statistical significance: * $p < 0.05$; ** $p < 0.01$; and *** $p < 0.001$.

3. Results

3.1. The mTOR Inhibitor Rapamycin Induces, While the PI3K Inhibitor 3-MA Suppresses, Autophagy in hASCs

In this study, hASCs were treated with rapamycin (Rapa-ASCs) for 1, 4, 12, 24 and 48 h, then examined for the expression of key autophagy genes and proteins. Transcriptional analysis of Rapa-ASCs demonstrated significant upregulation of the autophagy genes *ATG7* and *LC3B* following prolonged, but not short-term exposure to Rapa (Supplementary Figure S1C), suggesting induction of autophagy. A Western blot analysis revealed increased protein levels of LC3-II/β-Actin and decreased protein levels of p62/β-Actin, however this failed to reach significance (Supplementary Figure S1A,B). To examine inhibition of autophagy, hASCs were treated with 3-MA (3MA-ASCs) for 1, 4, 12, and 24 h and then probed for changes in autophagy gene and protein levels. Due to previous reports of autophagy induction with prolonged exposure to 3-MA [37], the longest timepoint (48 h) was not included for this group. The 3MA-ASCs exhibited no change to LC3-II or p62 transcript or protein levels, indicating no induction of autophagosome formation (Supplementary Figure S2A,B).

3.2. Autophagy Preconditioning Does Not Alter hASCs Stem Cell Properties

To characterize the impact of autophagy preconditioning on stem cell characteristics, RapaASCs and 3MA-ASCs were examined for morphology, clonogenicity, and surface marker expression after either short-term (4 h) or long-term (24 h) treatment. Results demonstrated that hASCs maintained their spindle-like morphology with both short-term and long-term exposure to either Rapa or 3-MA (Figure 1A). Autophagy preconditioning did not significantly alter hASCs immunophenotype as measured by flow cytometric analysis of surface proteins (Figure 1B,C). In both Rapa-ASCs and 3MA-ASCs, positive expression of canonical MSC markers CD90 and CD105 ranged from 90–100% across all time points, and expression of CD73 ranged from 29.63% to 45.02% with no significant difference between groups (Figure 1B). Immunophenotype of hASCs was further confirmed by the absence of negative MSC markers including the pan T-cell marker CD3, the monocyte/macrophage marker CD14, the endothelial cell marker CD31, and the broad lymphohematopoietic lineage marker CD45. In both 3MA-ASCs and Rapa-ASCs, expression of these negative markers remained low and was not significantly different from untreated hASCs. Additionally, self-renewal ability of treated ASCs was measured by performing a colony-forming unit-fibroblast assay. Analysis of colony-forming units (CFUs) demonstrated no alteration in self-renewal capacity of Rapa-ASCs (Figure 1D). However, the 3MA-ASCs showed significantly reduced self-renewal capacity (Figure 1E).

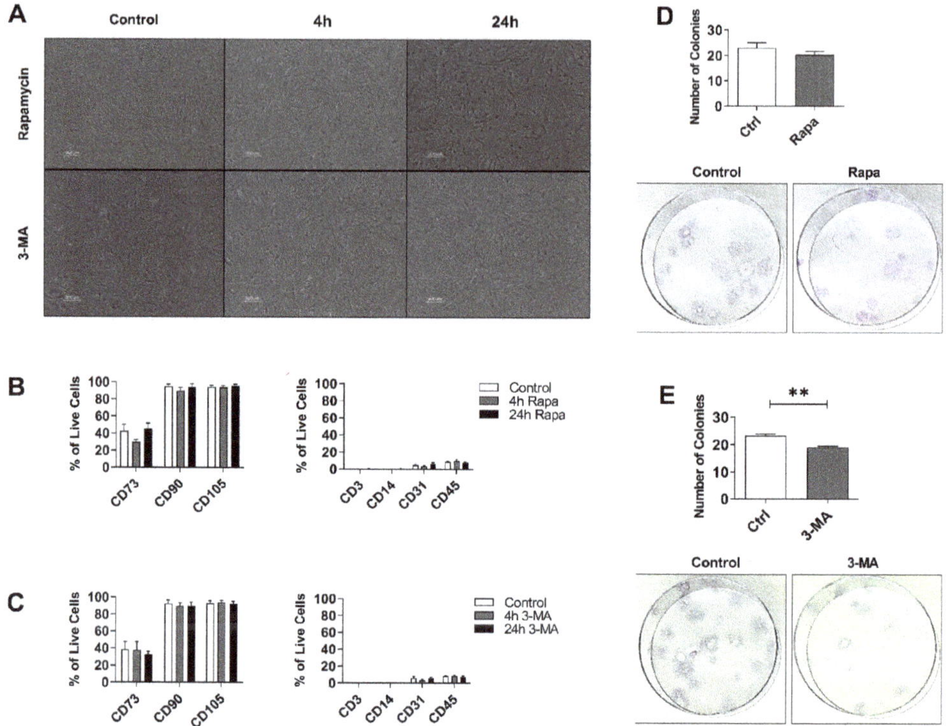

Figure 1. Autophagy preconditioning in hASCs does not alter self-renewal capacity or surface expression of MSC markers. (**A**) Representative images of hASCs morphology in culture after both short= and long-term exposure to autophagy preconditioning agents. Scale bar = 300 μm. Immunophenotype of Rapa-ASCs (**B**) and 3MA-ASCs (**C**) by flow cytometric analysis of surface proteins. Data are presented as mean ± SEM of 3 independent experiments. Quantifaction and images of colony-forming units by Rapa-ASCs (**D**) and 3MA-ASCs (**E**). Data are presented as mean ± SEM of 4 independent experiments. All data comparing 3 groups are analyzed by using one-way analysis of variance (ANOVA) with Tukey's post-hoc multiple comparisons, and CFU-F data are analyzed with unpaired student's *t*-test. Abbreviations: Rapa, Rapamycin; 3-MA, 3-methyladenine; CFU-F, colony forming units-fibroblasts. ** $p < 0.01$.

3.3. Autophagy Preconditioning of hASCs Alters Expression of Both Anti-Inflammatory and Pro-Inflammatory Mediators

To determine whether pharmacologically targeting the autophagy pathway alters the immune-modifying abilities of hASCs, cells were treated for either 4 or 24 h with Rapa or 3-MA and then analyzed with RT-qPCR for expression of common immune mediators. In Rapa-ASCs, 4 h, but not 24 h, treatment produced robust enhancement of *TSG-6* mRNA (Figure 2A). The opposite result was seen in 3MA-ASCs, with only the 24-h group demonstrating enhanced expression of *TGF-β* and, to a lesser extent, *TSG-6* ($p = 0.073$). Further analysis revealed similar time-dependent differences in expression of pro-inflammatory mediators (Figure 2B). In both Rapa-ASCs and 3MA-ASCs, *IL-6* was upregulated after 4 h, but not 24 h. Moreover, Rapa-ASCs, but not 3MA-ASCs, upregulated *IL-1β* after 4-h treatment. Interestingly, both 4-h Rapa-ASCs and 3MA-ASCs robustly upregulate *COX2* transcription relative to control cells. This effect was not mirrored in secreted protein, as ELISA analysis of conditioned medium revealed elevation of the *COX2* metabolite, PGE2, in 24-h treated 3MA-ASCs only (Figure 2C).

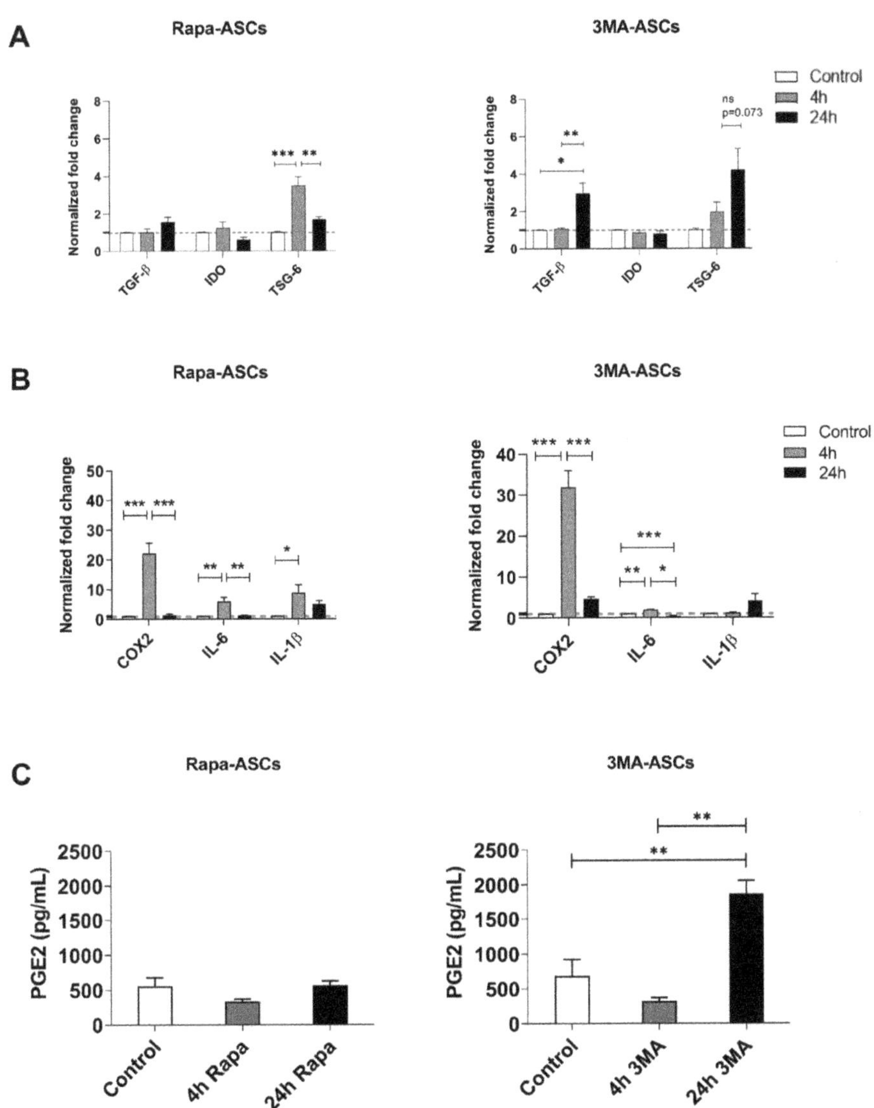

Figure 2. Autophagy preconditioning in hASCs alters expression of both anti-inflammatory and pro-inflammatory mediators. Rapa-ASCs (left) and 3MA-ASCs (right) were treated for 4 or 24 h, then relative expression levels of anti-inflammatory (**A**) and pro-inflammatory (**B**) genes were measured via RT-qPCR by using the ΔΔCt method. Data are presented as mean relative fold-change ± SEM of 4 independent experiments. (**C**) Secreted PGE2 levels were measured via ELISA in 24 h CM from control and autophagy preconditioned hASCs. Data are presented as means ± SEM of 3 independent experiments. Statistical analysis was performed by using one-way analysis of variance (ANOVA), and differences between the means are indicated with * $p < 0.05$, ** $p < 0.01$, *** $p < 0.001$. Abbreviations: *TGF-β*, transforming growth factor-beta; *IDO*, indoleamine 2,3-dioxygenase; *TSG-6*, TNF Alpha Induced Protein 6, *COX2*, cyclooxygenase 2; *IL-6*, interleukin-6; *IL-1β*, interleukin-1 beta; PGE2, prostaglandin E2.

3.4. Autophagy Preconditioning of hASCs Alters Response to Pro-Inflammatory Stimulation with hIFNγ

To investigate the effects of autophagy preconditioning on hASC immunomodulatory response to signals they may encounter following administration into an inflammatory microenvironment, cells were pre-treated for either 4 or 24 h followed by 24 h stimulation with the classic pro-inflammatory cytokine hIFNγ. Our findings demonstrate that, in comparison to unstimulated and non-pretreated control ASCs, 4 h Rapa-preconditioning resulted in significant elevation of *TGF-β*, *COX2*, and *IL-6*, whereas 24 h Rapa-preconditioning resulted in upregulation of *TGF-β*, *IDO*, *TSG-6*, and *COX2*, as well as the inflammatory cytokines *IL-6* and *IL-1β*. We also found that 4 h 3MA-preconditioning resulted in upregulation of *IDO*, *COX2*, *IL-6*, and *IL-1β*, whereas24 h 3MA-preconditioning only increased expression of *COX2* transcripts (Figure 3A,B). This contrasted with our non-preconditioned control ASCs, which in response to hIFNγ stimulation only displayed significant upregulation of *IDO* and *COX2* transcripts when compared to non- IFNγ-stimulated ASCs. Perhaps most interesting, we demonstrated that secretion of PGE2 was upregulated by both 4 h Rapa and 4 h 3MA, which was significantly higher than both unstimulated control and IFNγ-stimulated control cells, as well as both 24 h preconditioned groups (Figure 3C). Taken together, these findings show that although transcriptional activity is altered based on unique temporal and compound-dependent patterns, the robust upregulation of PGE2 release in both acute Rapa and 3-MA preconditioning strategies suggest the involvement of an autophagy-independent mechanism that is currently unknown.

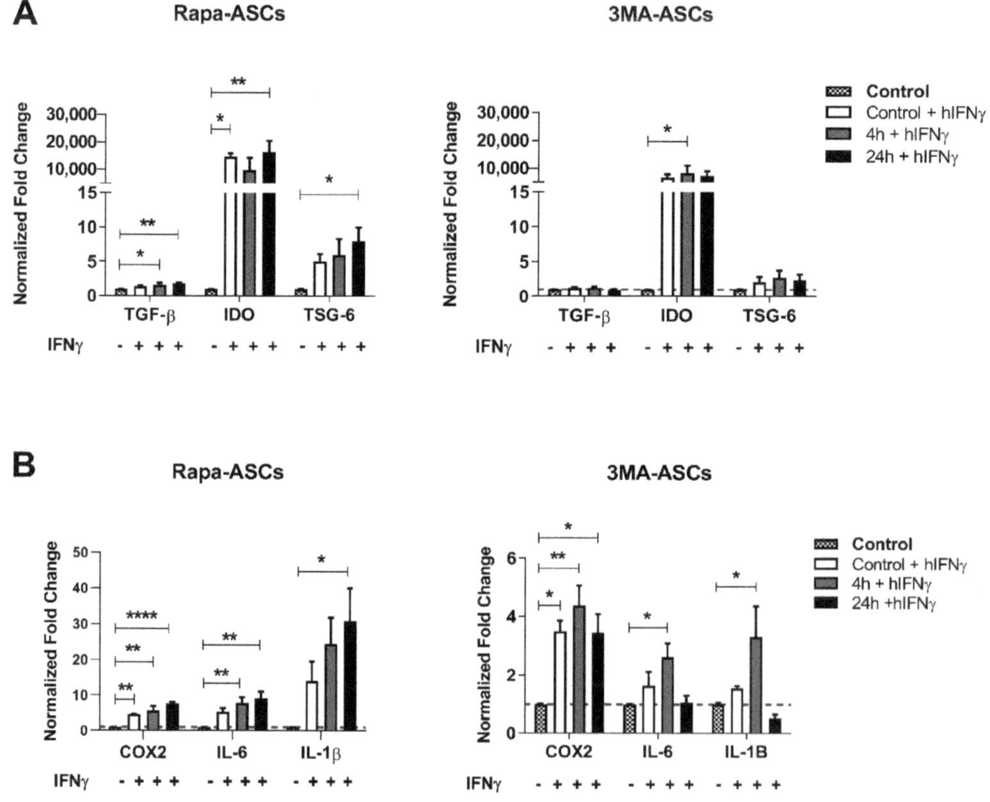

Figure 3. *Cont.*

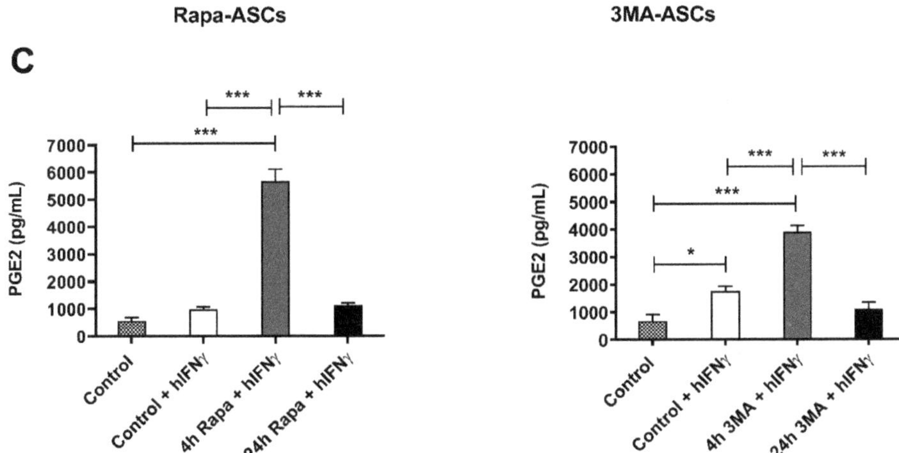

Figure 3. Autophagy preconditioning in hASCs alters their transcriptional and secretory response to pro-inflammatory stimulation with hIFNγ. Rapa-ASCs (left) and 3MA-ASCs (right) were preconditioned for 4 or 24 h with their respective autophagy compounds, then stimulated with hIFNγ (5 ng/mL) for 24 h. Following stimulation, anti-inflammatory (**A**) and pro-inflammatory (**B**) genes were measured by using RT-qPCR. Data are presented as means ± SEM of 4 independent experiments and normalized to controls that were not treated with hIFNγ. (**C**) Secreted PGE2 levels were measured in 24 h CM from autophagy-preconditioned and hIFNγ-stimulated hASCs. Data are presented as means ± SEM of 3 independent experiments. All statistical significance is determined by using one-way analysis of variance (ANOVA). Statistical differences between the means are indicated with * $p < 0.05$, ** $p < 0.01$, *** $p < 0.001$, **** $p < 0.0001$. Abbreviations: TGF-β, transforming growth factor-beta; IDO, indoleamine 2,3-dioxygenase; TSG-6, TNF Alpha Induced Protein 6, COX2, cyclooxygenase 2; IL-6, IL-1β, interleukin-6; interleukin-1 beta; PGE2, prostaglandin E2.

4. Discussion

ASCs have shown substantial preclinical promise as therapeutic tools in inflammatory, autoimmune, and neurodegenerative diseases [38]. However, the successful application of ASCs in human clinical trials has faced many challenges, including variable clinical efficacy outcomes, low post-transplant viability, and immunosuppressive potency [39]. Preconditioning strategies to improve immunomodulatory potency of MSCs for various anti-inflammatory and regenerative medicine applications is an active focus of the stem cell research field. Rapamycin has emerged as a promising candidate compound, and Rapa-treated MSCs from adipose [29,40], bone marrow [41], and umbilical cord [42] have been examined in preclinical models. In a 2016 report by Kim and colleagues, hASCs were treated with Rapa for 48 h prior to administration in a GvHD mouse model. Authors demonstrate elevated expression of IL-10, IDO, and TGF-β, and correlated this with enhanced production of anti-inflammatory cytokines, modulation of the T cell repertoire, and prevention of GvHD development [29]. However, recent reports have revealed temporally distinct actions of Rapa on mTORC1 and mTORC2 [43], and this has been correlated to dynamic effects on MSC immunomodulatory capacity. Indeed, a wide range of dosage and exposure time has yielded varying success and highlights the importance of defining the unique immunoregulatory consequences of these novel preconditioning strategies depending on experimental conditions (i.e., dose and exposure time), cell type, and therapeutic application. For example, short-term (2 h) treatment of rat BMSCs improved survival and repair of damaged myocardium following transplant into an ischemia/reperfusion model [41]. However, in an animal model of multiple sclerosis, short-term (4 h) preconditioning of human ASCs did not yield any therapeutic benefit and was in fact correlated with worsened disease measures [40]. Thus, to improve translational potential, it is critically important to define the dynamic immunomodulatory response of MSCs and fine-tune these strategies to each pathological situation.

In the present study, the impacts of compounds known to either induce or inhibit autophagy on hASC immunomodulatory potential was assessed. Our data showed that 24 h Rapa treatment increased the expression of the autophagy genes *ATG7* and *LC3B*, whereas the expression of LC3-II protein was upregulated, albeit slightly, suggestive of autophagic induction. Conversely, 3-MA, a class I and class III PI3K inhibitor, obstructed autophagy initiation in hASCs. 3MA-ASCs displayed no significant change to autophagy transcripts or protein levels. Both Rapa-ASCs and 3MA-ASCs maintain characteristic stem cell properties, including plastic adherence, fibroblast-like morphology, and immunophenotypic profile, as determined by flow cytometry. However, because the immunomodulatory strength of ASCs is due to their paracrine activity rather than proliferation and engraftment, this likely has little impact on therapeutic potential. These results denote the retention of basic MSC identifiers in autophagy preconditioned hASCs.

To determine the role of autophagy preconditioning in hASC immunosuppressive potential, both the transcription and secretion of known ASC-derived pro- and anti-inflammatory mediators was examined. Among these, the pleiotropic signaling molecule PGE2 was of particular interest due to its established role in MSC-based therapies for inflammatory and autoimmune diseases including MS [18,19], sepsis [14], inflammatory bowel disease (IBD) [20,21], ischemia-reperfusion injury [15], and arthritis [16,17]. hASC-derived PGE2 exerts its immunomodulatory functions through the inhibition of T cell activation, proliferation and production of pro-inflammatory cytokines [17,44–47], the generation of IL-10-producing T regulatory cells (Tregs) [17], and the promotion of M2 macrophage polarization [21,48]. PGE2 is synthesized from arachidonic acid by the enzyme COX2 and has been proposed as an important mechanism contributing to the immunoregulatory actions of MSCs. In a mouse model of sepsis, concurrent IV administration of BMSCs pre-stimulated with LPS resulted in prolonged survival, which was correlated with elevated PGE2 secretion and modulation of host macrophage populations [14]. PGE2 has also been shown to be critical for MSCs' ability to inhibit the proliferation of natural killer (NK) cells [49], the maturation of dendritic cells (DC) from monocytes [45,50], and the proliferation of PHA-stimulated T cells [45].

In this study, 4 h preconditioned Rapa-ASCs and 3MA-ASCs both exhibited robust upregulation of *COX2* gene expression. This induction was not seen in secreted PGE2, the protein substrate of COX2, with autophagy preconditioning alone in either 4 h treatment group; however, PEG2 was significantly increased in ASCs treated with 3-MA for 24 h. Following exposure to hIFNγ, elevated *COX2* transcription and PGE2 secretion was seen in both Rapa-ASCs and 3MA-ASCs with 4 h, but not 24 h, preconditioning. This indicates that short-term autophagy preconditioning may "prime" hASCs to respond more rapidly and more robustly to an environment mimicking the inflammatory state of autoimmune or neurodegenerative diseases [51–54].

These unique temporal dynamics of COX2 and PGE2 have previously been shown in BMSCs, and have been suggested to result from activation of the Akt/glycogen synthase kinase 3 Beta (GSK-3β) pathway rather than autophagy [55,56]. mTOR, a serine/threonine kinase which forms the catalytic subunit of the two distinct complexes mTORC1 and mTORC2, regulates a spectrum of cellular processes including cell growth, autophagy, cytoskeletal remodeling, proteostasis, and metabolism (elegantly reviewed here [57]). Rapamycin, an FDA-approved immunosuppressant, rapidly inhibits mTORC1 activity through interaction with the FRB (FKBP12/rapamycin-binding) domain [58]. Under physiological conditions mTORC1 substrates negatively regulate mTORC2 kinase activity [59]. With Rapa inhibition of mTORC1, mTORC2 kinase activity is disinhibited, leading to phosphorylation of Akt at Ser473 and subsequently disrupting the glycogen synthase kinase 3 beta (GSK3B)-dependent blockade of the COX2 promotor region, thereby activating COX2 transcription [37,55,60]. With prolonged Rapa exposure mTORC2 is also inhibited, possibly explaining the rapid yet transient COX2 upregulation [43]. The temporally defined actions of 3-MA may also explain the observed transcriptional changes. 3-MA persistently inhibits class I PI3K whereas its inhibition of the class III PI3K, VPS34, is transient [37]. This was correlated with transient inhibition of autophagy and upregulated COX2 levels in human BMSCs with short-term exposure, but induction of autophagy and suppression of COX2 with long-term treatment. Thus, it may be that the selective inhibition of mTORC1 kinase activity, independent of the effects on autophagy, is responsible for the altered immunophenotype of ASCs.

Wang and colleagues demonstrated that short-term, but not long-term, inhibition of mTORC1 with Rapa resulted in elevated *COX2* gene expression, PGE2 secretion, and inhibition of proliferation in PBMCs. However, ASCs represent a more abundant and readily available source of stem cells compared to BMSCs, and possess higher immunomodulatory potential [3]. Thus, the present study investigated whether ASCs demonstrate a similar temporally distinct response to Rapa treatment to determine if this therapeutic strategy may be extended to a novel cell type. Due to the observation that both Rapa and 3-MA elicited some shared immunoregulatory responses in hASCs, it may be that the effects of these preconditioning strategies are independent of autophagy, which was suggested by Chinnadurai and collaborators in 2015 [61]. However, in this study hBMSC were treated with 3-MA for more than 48 h, and based on the temporal dynamics of class I versus class III PI3K inhibition it is possible that the inhibitory effect of 3-MA had weakened or stopped. Our results, consistent with observations in BMSCs, highlight the importance of fully understanding the temporal aspects of autophagy preconditioning to optimize therapeutic potential.

Evidence from two studies indicates that PGE2 is only partially responsible for the immunosuppressive actions of hASC function as inhibition of PGE2 or its receptors does not fully abolish immunosuppressive capacity [46,62]. Other anti-inflammatory mediators, including *IDO*, *TSG-6*, and *TGF-β1*, significantly contribute to hASCs' suppression of innate and adaptive immune cells. IDO, an enzyme that breaks down the essential amino acid tryptophan, promotes polarization of M2b/c macrophages, induces proliferation and *IL-10* production of Tregs, inhibits proliferation of Th cells, and suppresses the cytolytic activity of natural killer (NK) cells [1,5,8,62–64]. *TSG-6* plays a pivotal role in M2 macrophage polarization and reduction of inflammation in both colitis and acute pancreatitis mouse models [65,66]. Moreover, TSG-6 has displayed autocrine activity which maintains stemness and downregulates IL-6 production in mouse BMSCs [67]. *TGF*-β1 is necessary for the suppression of dendritic cell maturation and subsequent promotion of FOXP3$^+$ Tregs [68,69]. Additionally, Rapa-treated hBMSCs showed upregulation of *TGF-β1*, which was indispensable for the inhibition of CD4$^+$ T cell proliferation [70]. The present study showed that *IDO* was expressed at low levels in resting hASCs and was unaffected by autophagy preconditioning itself. Upon hIFNγ stimulation, both 4 h 3MA-ASCs and 24 h Rapa-ASCs amplified *IDO* expression compared to control stimulated cells. *TSG-6* expression was not significantly changed in resting or hIFNγ-stimulated 3MA-ASCs but was highly upregulated in 4 h preconditioned Rapa-ASCs. Further upregulation of transcription was seen in 24 h preconditioned, hIFNγ-stimulated Rapa-ASCs. Finally, only 24 h preconditioned 3MA-ASCs exhibited upregulation of *TGF-β1* transcripts. After hIFNγ stimulation, both 4 h and 24 h Rapa-ASCs elevated *TGF-β1* gene expression. Overall, Rapa-ASCs demonstrate more upregulation of anti-inflammatory genes than 3MA-ASCs, suggesting Rapa preconditioning may be a more promising strategy for augmentation of hASC immunomodulatory ability.

The pleiotropic cytokine IL-6 can act as a pro-inflammatory or pro-regenerative agent dependent on its actions on its soluble and membrane-bound receptors, respectively [71,72]. In the context of hASCs, IL-6 production has been linked to repression of Th17 cells, induction of IL-10-producing Tregs, and enhanced recruitment and anti-inflammatory polarization of macrophages [73–76]. In our study, both 4 h Rapa-ASCs and 3MA-ASCs showed a significant increase in *IL-6* transcript levels. Following pro-inflammatory activation with hIFNγ, both 4- and 24 h Rapa-ASCs, and 4 h 3MA-ASCs showed enhanced *IL-6* transcription. Finally, we assessed the expression of *IL-1β*, the potent pro-inflammatory cytokine which negatively correlates with hASCs' ability to suppress CD4$^+$ T helper, CD8$^+$ T effector, and NK cell proliferation [77]. In our study, 24 h Rapa preconditioning and 4 h 3-MA preconditioning followed by stimulation with hIFNγ resulted in additional upregulation of *IL-1β*. Taken together, the upregulated *IL-6* and unaltered *IL-1β* expression levels suggest that 4 h Rapa-ASCs may be the most effective autophagy preconditioning strategy.

In summary, these data demonstrate that preconditioning with both the autophagy-inducer Rapa and the autophagy-inhibitor 3-MA had distinct effects on the expression of immunomodulatory factors in hASCs based on the duration of treatment and presence or absence of pro-inflammatory stimuli. The results suggest that the conflicting data in the autophagy literature may be due, at least in part, to varied treatment times, doses, and the unique inflammatory milieu. We also demonstrated that both compounds had similar impact on secretory activity of ASCs in response to inflammatory activation, which strongly suggests a shared molecular mechanism that is independent of their actions on autophagy. Based on our findings, we propose that a short-term preconditioning with Rapa may bestow the most robust enrichment of hASC immunomodulatory potential due to the preservation of stemness qualities, the enhancement of the *COX2*/PGE2 pathway, and the increase of the anti-inflammatory cytokines *TGF-β* and *IL-6* without concomitant elevation of pro-inflammatory *IL-1β*. This is consistent with previous investigations which suggest that short-term Rapa treatment in MSCs elicits greater improvement of immunomodulatory function, possibly resulting from its differential inhibition of MTORC1 and MTORC2 [41,55,78]. On the other hand, 3MA-ASCs and Rapa-ASCs both upregulate PGE2 secretion when activated with hIFNγ, indicating that autophagy inhibition with 3-MA may also have therapeutic benefit in certain disease contexts, as was proposed by Dang and colleagues in the experimental autoimmune encephalomyelitis (EAE) mouse model [19]. Future investigations involving co-cultures with innate and adaptive immune cells and administration to preclinical disease models will be needed to determine the influence of these novel preconditioning strategies on the immunosuppressive function of hASCs. These approaches may reveal whether our in vitro findings translate to enhanced immunomodulatory function and therapeutic efficacy in autoimmune, inflammatory, and neurodegenerative diseases in vivo.

5. Conclusions

Adipose-derived stem cells (ASCs) secrete a variety of anti-inflammatory molecules that shift immune cell activity away from driving inflammation and toward tissue regeneration and repair, showing immense therapeutic promise in several inflammatory, autoimmune, and neurodegenerative diseases. However, their efficacy is limited by poor survival and secretory activity after administration, representing a significant hurdle to success in clinical trials. To overcome these hurdles, a novel preconditioning strategy is proposed that targets a cellular pathway involved in stress response, survival, and secretory activity. Drawing on evidence from bone marrow-derived stem cells (BMSCs), we find that compounds commonly used to modify the autophagy pathway result in "primed" ASCs capable of producing higher levels of immunomodulatory genes and proteins. We demonstrate that PGE2, a key contributing factor to stem cell therapeutic efficacy in multiple sclerosis (MS), sepsis, inflammatory bowel disease (IBD), and arthritis, was highly upregulated in ASCs preconditioned with either short-term Rapa or 3-MA, indicating that the immunomodulatory ffects of these compounds may, in fact, derive from mechanisms of action beyond their impact on autophagy. We propose that this represents a promising strategy for enhancing the therapeutic potential and possibly the translational success of ASCs for inflammation-driven disease states that warrants further investigation and delineation of mechanisms involved.

Supplementary Materials: The following supporting information can be downloaded at: https://www.mdpi.com/article/10.3390/cells11091376/s1, Figure S1: Rapamycin induces autophagy in hASCs; Figure S2: 3-MA inhibits autophagy in hASCs.

Author Contributions: Conceptualization and design, R.M.W. and B.A.B.; Data collection, R.M.W., M.A.A.H., B.N.S., E.R.M., S.J.A., and U.M.D.; Data analysis and interpretation, R.M.W. and S.A.-G.; Funding and supervision, B.A.B.; Writing—original draft, R.M.W.; Writing—review & editing, R.M.W., S.A.-G., M.A.A.H., B.N.S., and B.A.B. All authors have read and agreed to the published version of the manuscript.

Funding: The funds used for these studies were private funds from Tulane University.

Institutional Review Board Statement: Not applicable.

Informed Consent Statement: Not applicable.

Data Availability Statement: The original contributions presented in the study are included in the article/supplementary material, further inquiries can be directed to the corresponding author/s.

Acknowledgments: The authors would like to thank Alan Tucker with the Flow Cytometry core for his expertise and assistance with these studies.

Conflicts of Interest: The authors declare that the research was conducted in the absence of any commercial or financial relationships that could be construed as a potential conflict of interest.

References

1. Zheng, G.; Qiu, G.; Ge, M.; He, J.; Huang, L.; Chen, P.; Wang, W.; Xu, Q.; Hu, Y.; Shu, Q.; et al. Human Adipose-Derived Mesenchymal Stem Cells Alleviate Obliterative Bronchiolitis in a Murine Model via IDO. *Respir. Res.* **2017**, *18*, 119. [CrossRef]
2. Gimble, J.M.; Katz, A.J.; Bunnell, B.A. Adipose-Derived Stem Cells for Regenerative Medicine. *Circ. Res.* **2007**, *100*, 1249–1260. [CrossRef]
3. Melief, S.M.; Zwaginga, J.J.; Fibbe, W.E.; Roelofs, H. Adipose Tissue-Derived Multipotent Stromal Cells Have a Higher Immunomodulatory Capacity Than Their Bone Marrow-Derived Counterparts. *Stem Cells Transl. Med.* **2013**, *2*, 455–463. [CrossRef]
4. Ceccarelli, S.; Pontecorvi, P.; Anastasiadou, E.; Napoli, C.; Marchese, C. Immunomodulatory Effect of Adipose-Derived Stem Cells: The Cutting Edge of Clinical Application. *Front. Cell Dev. Biol.* **2020**, 236. [CrossRef] [PubMed]
5. Serejo, T.R.T.; Silva-Carvalho, A.É.; Braga, L.D.d.C.F.; Neves, F.d.A.R.; Pereira, R.W.; Carvalho, J.L.d.; Saldanha-Araujo, F. Assessment of the Immunosuppressive Potential of INF-γ Licensed Adipose Mesenchymal Stem Cells, Their Secretome and Extracellular Vesicles. *Cells* **2019**, *8*, 22. [CrossRef] [PubMed]
6. Yang, D.; Wang, W.; Li, L.; Peng, Y.; Chen, P.; Huang, H.; Guo, Y.; Xia, X.; Wang, Y.; Wang, H.; et al. The Relative Contribution of Paracine Effect versus Direct Differentiation on Adipose-Derived Stem Cell Transplantation Mediated Cardiac Repair. *PLoS ONE* **2013**, *8*, e59020. [CrossRef]
7. Fontanilla, C.V.; Gu, H.; Liu, Q.; Zhu, T.Z.; Zhou, C.; Johnstone, B.H.; March, K.L.; Pascuzzi, R.M.; Farlow, M.R.; Du, Y. Adipose-Derived Stem Cell Conditioned Media Extends Survival Time of a Mouse Model of Amyotrophic Lateral Sclerosis. *Sci. Rep.* **2015**, *5*, 1–11. [CrossRef]
8. Crop, M.J.; Baan, C.C.; Korevaar, S.S.; IJzermans, J.N.M.; Pescatori, M.; Stubbs, A.P.; van IJcken, W.F.J.; Dahlke, M.H.; Eggenhofer, E.; Weimar, W.; et al. Inflammatory Conditions Affect Gene Expression and Function of Human Adipose Tissue-Derived Mesenchymal Stem Cells. *Clin. Exp. Immunol.* **2010**, *162*, 474–486. [CrossRef]
9. Domenis, R.; Cifù, A.; Quaglia, S.; Pistis, C.; Moretti, M.; Vicario, A.; Parodi, P.C.; Fabris, M.; Niazi, K.R.; Soon-Shiong, P.; et al. Pro Inflammatory Stimuli Enhance the Immunosuppressive Functions of Adipose Mesenchymal Stem Cells-Derived Exosomes. *Sci. Rep.* **2018**, *8*, 1–11. [CrossRef]
10. Wobma, H.M.; Tamargo, M.A.; Goeta, S.; Brown, L.M.; Duran-Struuck, R.; Vunjak-Novakovic, G. The Influence of Hypoxia and IFN-γ on the Proteome and Metabolome of Therapeutic Mesenchymal Stem Cells. *Biomaterials* **2018**, *167*, 226–234. [CrossRef] [PubMed]
11. Roemeling-Van Rhijn, M.; Mensah, F.K.F.; Korevaar, S.S.; Leijs, M.J.; Van Osch, G.J.V.M.; IJzermans, J.N.M.; Betjes, M.G.H.; Baan, C.C.; Weimar, W.; Hoogduijn, M.J. Effects of Hypoxia on the Immunomodulatory Properties of Adipose Tissue-Derived Mesenchymal Stem Cells. *Front. Immunol.* **2013**, *4*, 203. [CrossRef] [PubMed]
12. Seo, Y.; Shin, T.-H.; Kim, H.-S. Molecular Sciences Current Strategies to Enhance Adipose Stem Cell Function: An Update. *Int. J. Mol. Sci.* **2019**, *20*, 3827. [CrossRef]
13. Hsiao, S.T.; Lokmic, Z.; Peshavariya, H.; Abberton, K.M.; Dusting, G.J.; Lim, S.Y.; Dilley, R.J. Hypoxic Conditioning Enhances the Angiogenic Paracrine Activity of Human Adipose-Derived Stem Cells. *Stem Cells Dev.* **2013**, *22*, 1614–1623. [CrossRef]
14. Németh, K.; Leelahavanichkul, A.; Yuen, P.S.T.; Mayer, B.; Parmelee, A.; Doi, K.; Robey, P.G.; Leelahavanichkul, K.; Koller, B.H.; Brown, J.M.; et al. Bone Marrow Stromal Cells Attenuate Sepsis via Prostaglandin E 2-Dependent Reprogramming of Host Macrophages to Increase Their Interleukin-10 Production. *Nat. Med.* **2009**, *15*, 42–49. [CrossRef] [PubMed]

15. Liu, L.; He, Y.R.; Liu, S.J.; Hu, L.; Liang, L.C.; Liu, D.L.; Liu, L.; Zhu, Z.Q. Enhanced Effect of IL-1 β-Activated Adipose-Derived MSCs (ADMSCs) on Repair of Intestinal Ischemia-Reperfusion Injury via COX-2-PGE 2 Signaling. *Stem Cells Int.* **2020**, *2020*, 2803747. [CrossRef]
16. Bouffi, C.; Bony, C.; Courties, G.; Jorgensen, C.; Noël, D. IL-6-Dependent PGE2 Secretion by Mesenchymal Stem Cells Inhibits Local Inflammation in Experimental Arthritis. *PLoS ONE* **2010**, *5*, e14247. [CrossRef]
17. González, M.A.; González-Rey, E.; Rico, L.; Büscher, D.; Delgado, M. Treatment of Experimental Arthritis by Inducing Immune Tolerance with Human Adipose-Derived Mesenchymal Stem Cells. *Arthritis Rheum.* **2009**, *60*, 1006–1019. [CrossRef]
18. Matysiak, M.; Orlowski, W.; Fortak-Michalska, M.; Jurewicz, A.; Selmaj, K. Immunoregulatory Function of Bone Marrow Mesenchymal Stem Cells in EAE Depends on Their Differentiation State and Secretion of PGE2. *J. Neuroimmunol.* **2011**, *233*, 106–111. [CrossRef] [PubMed]
19. Dang, S.; Xu, H.; Xu, C.; Cai, W.; Li, Q.; Cheng, Y.; Jin, M.; Wang, R.-X.; Peng, Y.; Zhang, Y.; et al. Autophagy Regulates the Therapeutic Potential of Mesenchymal Stem Cells in Experimental Autoimmune Encephalomyelitis. *Autophagy* **2014**, *10*, 1301–1315. [CrossRef]
20. Yang, F.Y.; Chen, R.; Zhang, X.; Huang, B.; Tsang, L.L.; Li, X.; Jiang, X. Preconditioning Enhances the Therapeutic Effects of Mesenchymal Stem Cells on Colitis Through PGE2-Mediated T-Cell Modulation. *Cell Transplant.* **2018**, *27*, 1352–1367. [CrossRef]
21. Park, H.J.; Kim, J.; Saima, F.T.; Rhee, K.J.; Hwang, S.; Kim, M.Y.; Baik, S.K.; Eom, Y.W.; Kim, H.S. Adipose-Derived Stem Cells Ameliorate Colitis by Suppression of Inflammasome Formation and Regulation of M1-Macrophage Population through Prostaglandin E2. *Biochem. Biophys. Res. Commun.* **2018**, *498*, 988–995. [CrossRef]
22. Ceccariglia, S.; Cargnoni, A.; Silini, A.R.; Parolini, O. Autophagy: A Potential Key Contributor to the Therapeutic Action of Mesenchymal Stem Cells. *Autophagy* **2019**, *16*, 28–37. [CrossRef] [PubMed]
23. Jakovljevic, J.; Harrell, C.R.; Fellabaum, C.; Arsenijevic, A.; Jovicic, N.; Volarevic, V. Modulation of Autophagy as New Approach in Mesenchymal Stem Cell-Based Therapy. *Biomed. Pharmacother.* **2018**, *104*, 404–410. [CrossRef]
24. Glick, D.; Barth, S.; Macleod, K.F. Autophagy: Cellular and Molecular Mechanisms. *J. Pathol.* **2010**, *221*, 3–12. [CrossRef]
25. Chen, X.; He, Y.; Lu, F. Autophagy in Stem Cell Biology: A Perspective on Stem Cell Self-Renewal and Differentiation. *Stem Cells Int.* **2018**, *2018*, 9131397. [CrossRef]
26. Weiss, D.J.; English, K.; Krasnodembskaya, A.; Isaza-Correa, J.M.; Hawthorne, I.J.; Mahon, B.P. The Necrobiology of Mesenchymal Stromal Cells Affects Therapeutic Efficacy. *Front. Immunol.* **2019**, *10*, 1228. [CrossRef] [PubMed]
27. Li, J.; Kim, S.G.; Blenis, J. Rapamycin: One Drug, Many Effects. *Cell Metab.* **2014**, *19*, 373–379. [CrossRef] [PubMed]
28. Pyo, J.O.; Nah, J.; Jung, Y.K. Molecules and Their Functions in Autophagy. *Exp. Mol. Med.* **2012**, *44*, 73–80. [CrossRef]
29. Kim, K.-W.; Moon, S.-J.; Park, M.-J.; Kim, B.-M.; Kim, E.-K.; Lee, S.-H.; Lee, E.-J.; Chung, B.-H.; Yang, C.-W.; Cho, M.-L. Optimization of Adipose Tissue-Derived Mesenchymal Stem Cells by Rapamycin in a Murine Model of Acute Graft-versus-Host Disease. *Stem Cell Res. Ther.* **2015**, *6*, 202. [CrossRef]
30. Javorkova, E.; Vackova, J.; Hajkova, M.; Hermankova, B.; Zajicova, A.; Holan, V.; Krulova, M. The Effect of Clinically Relevant Doses of Immunosuppressive Drugs on Human Mesenchymal Stem Cells. *Biomed. Pharmacother.* **2018**, *97*, 402–411. [CrossRef]
31. Li, C.; Ye, L.; Yang, L.; Yu, X.; He, Y.; Chen, Z.; Li, L.; Zhang, D. Rapamycin Promotes the Survival and Adipogenesis of Ischemia-Challenged Adipose Derived Stem Cells by Improving Autophagy. *Cell. Physiol. Biochem.* **2017**, *44*, 1762–1774. [CrossRef]
32. Strong, A.L.; Bowles, A.C.; Wise, R.M.; Morand, J.P.; Dutreil, M.F.; Gimble, J.M.; Bunnell, B.A. Human Adipose Stromal/Stem Cells from Obese Donors Show Reduced Efficacy in Halting Disease Progression in the Experimental Autoimmune Encephalomyelitis Model of Multiple Sclerosis. *Stem Cells* **2016**, *34*, 614–626. [CrossRef] [PubMed]
33. Al-Ghadban, S.; Diaz, Z.T.; Singer, H.J.; Mert, K.B.; Bunnell, B.A. Increase in Leptin and PPAR-γ Gene Expression in Lipedema Adipocytes Differentiated in Vitro from Adipose-Derived Stem Cells. *Cells* **2020**, *9*, 430. [CrossRef]
34. Sabol, R.A.; Villela, V.A.; Denys, A.; Freeman, B.T.; Hartono, A.B.; Wise, R.M.; Harrison, M.A.A.; Sandler, M.B.; Hossain, F.; Miele, L.; et al. Obesity-Altered Adipose Stem Cells Promote Radiation Resistance of Estrogen Receptor Positive Breast Cancer through Paracrine Signaling. *Int. J. Mol. Sci.* **2020**, *21*, 2722. [CrossRef] [PubMed]
35. Strong, A.L.; Hunter, R.S.; Jones, R.B.; Bowles, A.C.; Dutreil, M.F.; Gaupp, D.; Hayes, D.J.; Gimble, J.M.; Levi, B.; McNulty, M.A.; et al. Obesity Inhibits the Osteogenic Differentiation of Human Adipose-Derived Stem Cells. *J. Transl. Med.* **2016**, *14*, 27. [CrossRef]
36. Scruggs, B.A.; Semon, J.A.; Zhang, X.; Zhang, S.; Bowles, A.C.; Pandey, A.C.; Imhof, K.M.P.; Kalueff, A.V.; Gimble, J.M.; Bunnell, B.A. Age of the Donor Reduces the Ability of Human Adipose-Derived Stem Cells to Alleviate Symptoms in the Experimental Autoimmune Encephalomyelitis Mo. *Stem Cells Transl. Med.* **2013**, *2*, 797–807. [CrossRef]
37. Wu, Y.-T.; Tan, H.-L.; Shui, G.; Bauvy, C.; Huang, Q.; Wenk, M.R.; Ong, C.-N.; Codogno, P.; Shen, H.-M. Dual Role of 3-Methyladenine in Modulation of Autophagy via Different Temporal Patterns of Inhibition on Class I and III Phosphoinositide 3-Kinase. *J. Biol. Chem.* **2010**, *285*, 10850–10861. [CrossRef] [PubMed]
38. Uccelli, A.; Moretta, L.; Pistoia, V. Mesenchymal Stem Cells in Health and Disease. *Nat. Rev. Immunol.* **2008**, *8*, 726–736. [CrossRef]
39. Patrikoski, M.; Mannerström, B.; Miettinen, S. Perspectives for Clinical Translation of Adipose Stromal/Stem Cells. *Stem Cells Int.* **2019**, *2019*, 5858247. [CrossRef] [PubMed]

40. Wise, R.M.; Harrison, M.A.A.; Sullivan, B.N.; Al-Ghadban, S.; Aleman, S.J.; Vinluan, A.T.; Monaco, E.R.; Donato, U.M.; Pursell, I.A.; Bunnell, B.A. Short-Term Rapamycin Preconditioning Diminishes Therapeutic Efficacy of Human Adipose-Derived Stem Cells in a Murine Model of Multiple Sclerosis. *Cells* **2020**, *9*, 2218. [CrossRef]
41. Li, Z.; Wang, Y.; Wang, H.; Wu, J.; Tan, Y. Rapamycin-Preactivated Autophagy Enhances Survival and Differentiation of Mesenchymal Stem Cells After Transplantation into Infarcted Myocardium. *Stem Cell Rev. Rep.* **2020**, *16*, 344–356. [CrossRef]
42. Zheng, J.; Li, H.; He, L.; Huang, Y.; Cai, J.; Chen, L.; Zhou, C.; Fu, H.; Lu, T.; Zhang, Y.; et al. Preconditioning of Umbilical Cord-derived Mesenchymal Stem Cells by Rapamycin Increases Cell Migration and Ameliorates Liver Ischaemia/Reperfusion Injury in Mice via the CXCR4/CXCL12 Axis. *Cell Prolif.* **2019**, *52*, 12546. [CrossRef]
43. Sarbassov, D.D.; Ali, S.M.; Sengupta, S.; Sheen, J.H.; Hsu, P.P.; Bagley, A.F.; Markhard, A.L.; Sabatini, D.M. Prolonged Rapamycin Treatment Inhibits MTORC2 Assembly and Akt/PKB. *Mol. Cell* **2006**, *22*, 159–168. [CrossRef]
44. Najar, M.; Raicevic, G.; Boufker, H.I.; Kazan, H.F.; De Bruyn, C.; Meuleman, N.; Bron, D.; Toungouz, M.; Lagneaux, L. Mesenchymal Stromal Cells Use PGE2 to Modulate Activation and Proliferation of Lymphocyte Subsets: Combined Comparison of Adipose Tissue, Wharton's Jelly and Bone Marrow Sources. *Cell. Immunol.* **2010**, *264*, 171–179. [CrossRef]
45. Yañez, R.; Oviedo, A.; Aldea, M.; Bueren, J.A.; Lamana, M.L. Prostaglandin E2 Plays a Key Role in the Immunosuppressive Properties of Adipose and Bone Marrow Tissue-Derived Mesenchymal Stromal Cells. *Exp. Cell Res.* **2010**, *316*, 3109–3123. [CrossRef] [PubMed]
46. Puissant, B.; Barreau, C.; Bourin, P.; Clavel, C.; Corre, J.; Bousquet, C.; Taureau, C.; Cousin, B.; Abbal, M.; Laharrague, P.; et al. Immunomodulatory Effect of Human Adipose Tissue-Derived Adult Stem Cells: Comparison with Bone Marrow Mesenchymal Stem Cells. *Br. J. Haematol.* **2005**, *129*, 118–129. [CrossRef]
47. Cui, L.; Shuo, Y.; Liu, W.; Li, N.; Zhang, W.; Cao, Y. Expanded Adipose-Derived Stem Cells Suppress Mixed Lymphocyte Reaction by Secretion of Prostaglandin E2. *Tissue Eng.* **2007**, *13*, 1185–1195. [CrossRef]
48. Manferdini, C.; Paolella, F.; Gabusi, E.; Gambari, L.; Piacentini, A.; Filardo, G.; Fleury-Cappellesso, S.; Barbero, A.; Murphy, M.; Lisignoli, G. Adipose Stromal Cells Mediated Switching of the Pro-Inflammatory Profile of M1-like Macrophages Is Facilitated by PGE2: In Vitro Evaluation. *Osteoarthr. Cartil.* **2017**, *25*, 1161–1171. [CrossRef]
49. Spaggiari, G.M.; Capobianco, A.; Abdelrazik, H.; Becchetti, F.; Mingari, M.C.; Moretta, L. Mesenchymal Stem Cells Inhibit Natural Killer-Cell Proliferation, Cytotoxicity, and Cytokine Production: Role of Indoleamine 2,3-Dioxygenase and Prostaglandin E2. *Blood* **2008**, *111*, 1327–1333. [CrossRef] [PubMed]
50. Spaggiari, G.M.; Abdelrazik, H.; Becchetti, F.; Moretta, L. MSCs Inhibit Monocyte-Derived DC Maturation and Function by Selectively Interfering with the Generation of Immature DCs: Central Role of MSC-Derived Prostaglandin E2. *Blood* **2009**, *113*, 6576–6583. [CrossRef] [PubMed]
51. Liu, J.; Gao, L.; Zang, D. Elevated Levels of IFN-γ in CSF and Serum of Patients with Amyotrophic Lateral Sclerosis. *PLoS ONE* **2015**, *10*, e0136937. [CrossRef] [PubMed]
52. Kallaur, A.P.; Oliveira, S.R.; Simao, A.N.C.; De Almeida, E.R.D.; Morimoto, H.K.; Lopes, J.; De Carvalho Jennings Pereira, W.L.; Andrade, R.M.; Pelegrino, L.M.; Borelli, S.D.; et al. Cytokine Profile in Relapsing-Remitting Multiple Sclerosis Patients and the Association between Progression and Activity of the Disease. *Mol. Med. Rep.* **2013**, *7*, 1010–1020. [CrossRef]
53. Oke, V.; Gunnarsson, I.; Dorschner, J.; Eketjäll, S.; Zickert, A.; Niewold, T.B.; Svenungsson, E. High Levels of Circulating Interferons Type I, Type II and Type III Associate with Distinct Clinical Features of Active Systemic Lupus Erythematosus. *Arthritis Res. Ther.* **2019**, *21*, 107. [CrossRef] [PubMed]
54. Belkhelfa, M.; Rafa, H.; Medjeber, O.; Arroul-Lammali, A.; Behairi, N.; Abada-Bendib, M.; Makrelouf, M.; Belarbi, S.; Masmoudi, A.N.; Tazir, M.; et al. IFN-γ and TNF-α Are Involved during Alzheimer Disease Progression and Correlate with Nitric Oxide Production: A Study in Algerian Patients. *J. Interf. Cytokine Res.* **2014**, *34*, 839–847. [CrossRef]
55. Wang, B.; Lin, Y.; Hu, Y.; Shan, W.; Liu, S.; Xu, Y.; Zhang, H.; Cai, S.; Yu, X.; Cai, Z.; et al. MTOR Inhibition Improves the Immunomodulatory Properties of Human Bone Marrow Mesenchymal Stem Cells by Inducing COX-2 and PGE2. *Stem Cell Res. Ther.* **2017**, *8*, 292. [CrossRef] [PubMed]
56. Lin, Y.-C.; Kuo, H.-C.; Wang, J.-S.; Lin, W.-W. Regulation of Inflammatory Response by 3-Methyladenine Involves the Coordinative Actions on Akt and Glycogen Synthase Kinase 3β Rather than Autophagy. *J. Immunol.* **2012**, *189*, 4154–4164. [CrossRef]
57. Saxton, R.A.; Sabatini, D.M. MTOR Signaling in Growth, Metabolism, and Disease. *Cell* **2017**, *168*, 960. [CrossRef]
58. Banaszynski, L.A.; Liu, C.W.; Wandless, T.J. Characterization of the FKBP-Rapamycin-FRB Ternary Complex. *J. Am. Chem. Soc.* **2005**, *127*, 4715–4721. [CrossRef]
59. Zhang, J.; Xu, K.; Liu, P.; Geng, Y.; Wang, B.; Gan, W.; Guo, J.; Wu, F.; Chin, Y.R.; Berrios, C.; et al. Inhibition of Rb Phosphorylation Leads to MTORC2-Mediated Activation of Akt. *Mol. Cell* **2016**, *62*, 929–942. [CrossRef]
60. Wang, Q.; Zhou, Y.; Wang, X.; Evers, B.M. Glycogen Synthase Kinase-3 Is a Negative Regulator of Extracellular Signal-Regulated Kinase. *Oncogene* **2005**, *25*, 43–50. [CrossRef]
61. Chinnadurai, R.; Copland, I.B.; Ng, S.; Garcia, M.; Prasad, M.; Arafat, D.; Gibson, G.; Kugathasan, S.; Galipeau, J. Mesenchymal Stromal Cells Derived from Crohn's Patients Deploy Indoleamine 2,3-Dioxygenase-Mediated Immune Suppression, Independent of Autophagy. *Mol. Ther.* **2015**, *23*, 1248–1261. [CrossRef] [PubMed]
62. Delarosa, O.; Lombardo, E.; Beraza, A.; Mancheño-Corvo, P.; Ramirez, C.; Menta, R.; Rico, L.; Camarillo, E.; García, L.; Abad, J.L.; et al. Requirement of IFN-γ-Mediated Indoleamine 2,3-Dioxygenase Expression in the Modulation of Lymphocyte Proliferation by Human Adipose-Derived Stem Cells. *Tissue Eng. Part A* **2009**, *15*, 2795–2806. [CrossRef] [PubMed]

63. Sun, M.; Sun, L.; Huang, C.; Chen, B.; Zhou, Z. Induction of Macrophage M2b/c Polarization by Adipose Tissue-Derived Mesenchymal Stem Cells. *J. Immunol. Res.* **2019**, *2019*, 7059680. [CrossRef] [PubMed]
64. DelaRosa, O.; Sánchez-Correa, B.; Morgado, S.; Ramírez, C.; Del Río, B.; Menta, R.; Lombardo, E.; Tarazona, R.; Casado, J.G. Human Adipose-Derived Stem Cells Impair Natural Killer Cell Function and Exhibit Low Susceptibility to Natural Killer-Mediated Lysis. *Stem Cells Dev.* **2012**, *21*, 1333–1343. [CrossRef] [PubMed]
65. Song, W.J.; Li, Q.; Ryu, M.O.; Ahn, J.O.; Ha Bhang, D.; Chan Jung, Y.; Youn, H.Y. TSG-6 Secreted by Human Adipose Tissue-Derived Mesenchymal Stem Cells Ameliorates DSS-Induced Colitis by Inducing M2 Macrophage Polarization in Mice. *Sci. Rep.* **2017**, *7*, 5187. [CrossRef] [PubMed]
66. Li, Q.; Song, W.J.; Ryu, M.O.; Nam, A.; An, J.H.; Ahn, J.O.; Bhang, D.H.; Jung, Y.C.; Youn, H.Y. TSG-6 Secreted by Human Adipose Tissue-Derived Mesenchymal Stem Cells Ameliorates Severe Acute Pancreatitis via ER Stress Downregulation in Mice. *Stem Cell Res. Ther.* **2018**, *9*, 255. [CrossRef]
67. Romano, B.; Elangovan, S.; Erreni, M.; Sala, E.; Petti, L.; Kunderfranco, P.; Massimino, L.; Restelli, S.; Sinha, S.; Lucchetti, D.; et al. TNF-Stimulated Gene-6 Is a Key Regulator in Switching Stemness and Biological Properties of Mesenchymal Stem Cells. *Stem Cells* **2019**, *37*, 973–987. [CrossRef]
68. Wang, Y.C.; Chen, R.F.; Brandacher, G.; Lee, W.P.A.; Kuo, Y.R. The Suppression Effect of Dendritic Cells Maturation by Adipose-Derived Stem Cells through TGF-B1 Related Pathway. *Exp. Cell Res.* **2018**, *370*, 708–717. [CrossRef]
69. English, K.; Ryan, J.M.; Tobin, L.; Murphy, M.J.; Barry, F.P.; Mahon, B.P. Cell Contact, Prostaglandin E2 and Transforming Growth Factor Beta 1 Play Non-Redundant Roles in Human Mesenchymal Stem Cell Induction of CD4+CD25Highforkhead Box P3+ Regulatory T Cells. *Clin. Exp. Immunol.* **2009**, *156*, 149–160. [CrossRef]
70. Gao, L.; Cen, S.; Wang, P.; Xie, Z.; Liu, Z.; Deng, W.; Su, H.; Wu, X.; Wang, S.; Li, J.; et al. Autophagy Improves the Immunosuppression of CD4+ T Cells by Mesenchymal Stem Cells Through Transforming Growth Factor-B1. *Stem Cells Transl. Med.* **2016**, *5*, 1496. [CrossRef]
71. Scheller, J.; Chalaris, A.; Schmidt-Arras, D.; Rose-John, S. The Pro- and Anti-Inflammatory Properties of the Cytokine Interleukin-6. *Biochim. Biophys. Acta Mol. Cell Res.* **2011**, *1813*, 878–888. [CrossRef] [PubMed]
72. Schett, G. Physiological Effects of Modulating the Interleukin-6 Axis. *Rheumatology* **2018**, *57* (Suppl. S2), ii43–ii50. [CrossRef] [PubMed]
73. Vasilev, G.; Ivanova, M.; Ivanova-Todorova, E.; Tumangelova-Yuzeir, K.; Krasimirova, E.; Stoilov, R.; Kyurkchiev, D. Secretory Factors Produced by Adipose Mesenchymal Stem Cells Downregulate Th17 and Increase Treg Cells in Peripheral Blood Mononuclear Cells from Rheumatoid Arthritis Patients. *Rheumatol. Int.* **2019**, *39*, 819–826. [CrossRef] [PubMed]
74. Ivanova-Todorova, E.; Bochev, I.; Dimitrov, R.; Belemezova, K.; Mourdjeva, M.; Kyurkchiev, S.; Kinov, P.; Altankova, I.; Kyurkchiev, D. Conditioned Medium from Adipose Tissue-Derived Mesenchymal Stem Cells Induces CD4+FOXP3+ Cells and Increases IL-10 Secretion. *J. Biomed. Biotechnol.* **2012**, *2012*, 295167. [CrossRef]
75. Pilny, E.; Smolarczyk, R.; Jarosz-Biej, M.; Hadyk, A.; Skorupa, A.; Ciszek, M.; Krakowczyk, Ł.; Kułach, N.; Gillner, D.; Sokół, M.; et al. Human ADSC Xenograft through IL-6 Secretion Activates M2 Macrophages Responsible for the Repair of Damaged Muscle Tissue. *Stem Cell Res. Ther.* **2019**, *10*, 93. [CrossRef]
76. Heo, S.C.; Jeon, E.S.; Lee, I.H.; Kim, H.S.; Kim, M.B.; Kim, J.H. Tumor Necrosis Factor-α-Activated Human Adipose Tissue-Derived Mesenchymal Stem Cells Accelerate Cutaneous Wound Healing through Paracrine Mechanisms. *J. Invest. Dermatol.* **2011**, *131*, 1559–1567. [CrossRef]
77. Sempere, J.M.; Martinez-Peinado, P.; Arribas, M.I.; Reig, J.A.; De La Sen, M.L.; Zubcoff, J.J.; Fraga, M.F.; Fernández, A.F.; Santana, A.; Roche, E. Single Cell-Derived Clones from Human Adipose Stem Cells Present Different Immunomodulatory Properties. *Clin. Exp. Immunol.* **2014**, *176*, 255–265. [CrossRef]
78. Girdlestone, J.; Pido-Lopez, J.; Srivastava, S.; Chai, J.; Leaver, N.; Galleu, A.; Lombardi, G.; Navarrete, C.V. Enhancement of the Immunoregulatory Potency of Mesenchymal Stromal Cells by Treatment with Immunosuppressive Drugs. *Cytotherapy* **2015**, *17*, 1188–1199. [CrossRef]

Review

Pre-Conditioning Methods and Novel Approaches with Mesenchymal Stem Cells Therapy in Cardiovascular Disease

Anthony Matta [1,2,3], Vanessa Nader [1,4], Marine Lebrin [1,5], Fabian Gross [1,5], Anne-Catherine Prats [6], Daniel Cussac [6], Michel Galinier [1] and Jerome Roncalli [1,5,6,*]

1 Department of Cardiology, Institute CARDIOMET, University Hospital of Toulouse, 31059 Toulouse, France; dr.anthonymatta@hotmail.com (A.M.); nader.e.vanessa@gmail.com (V.N.); lebrin.m@chu-toulouse.fr (M.L.); gross.f@chu-toulouse.fr (F.G.); galinier.m@chu-toulouse.fr (M.G.)
2 Faculty of Medicine, Holy Spirit University of Kaslik, Kaslik 446, Lebanon
3 Department of Cardiology, Intercommunal Hospital Centre Castres-Mazamet, 81100 Castres, France
4 Faculty of Pharmacy, Lebanese University, Beirut 6573/14, Lebanon
5 CIC-Biotherapies, University Hospital of Toulouse, 31059 Toulouse, France
6 INSERM I2MC—UMR1297, 31432 Toulouse, France; anne-catherine.prats@inserm.fr (A.-C.P.); daniel.cussac@inserm.fr (D.C.)
* Correspondence: roncalli.j@chu-toulouse.fr; Tel.: +33-56-132-3334; Fax: +33-56-132-2246

Abstract: Transplantation of mesenchymal stem cells (MSCs) in the setting of cardiovascular disease, such as heart failure, cardiomyopathy and ischemic heart disease, has been associated with good clinical outcomes in several trials. A reduction in left ventricular remodeling, myocardial fibrosis and scar size, an improvement in endothelial dysfunction and prolonged cardiomyocytes survival were reported. The regenerative capacity, in addition to the pro-angiogenic, anti-apoptotic and anti-inflammatory effects represent the main target properties of these cells. Herein, we review the different preconditioning methods of MSCs (hypoxia, chemical and pharmacological agents) and the novel approaches (genetically modified MSCs, MSC-derived exosomes and engineered cardiac patches) suggested to optimize the efficacy of MSC therapy.

Keywords: mesenchymal stem cells; preconditioning; exosome; engineered cardiac patches

Citation: Matta, A.; Nader, V.; Lebrin, M.; Gross, F.; Prats, A.-C.; Cussac, D.; Galinier, M.; Roncalli, J. Pre-Conditioning Methods and Novel Approaches with Mesenchymal Stem Cells Therapy in Cardiovascular Disease. *Cells* **2022**, *11*, 1620. https://doi.org/10.3390/cells11101620

Academic Editor: Joni H. Ylostalo

Received: 13 April 2022
Accepted: 10 May 2022
Published: 12 May 2022

Publisher's Note: MDPI stays neutral with regard to jurisdictional claims in published maps and institutional affiliations.

1. Introduction

Several clinical trials have established the safety of mesenchymal stem cell (MSC) therapy and have shown promising results in the setting of cardiovascular disease over the past decades [1,2]. In ischemic heart disease, the role of existing conventional therapy, including percutaneous coronary intervention, coronary artery bypass graft and medical treatment, is limited to prevent future ischemic events and further expansion of myocardial damage [3]. Unlike MSC transplantation, there are no effects on myocardial repair, lost myocardial tissue and cardiomyocytes regeneration. Data from the literature showed a reduction in scar burden, myocardial fibrosis and infarct size, a reversion of left ventricular remodeling and an improvement in cardiac function after MSC therapy [1,4,5].

MSCs are undifferentiated, multipotent and self-renewable cells recognized for their potential of differentiation [6,7] and paracrine activity [2,8–10]. MSCs secrete diverse biological active cytokines, growth factors, chemokines and miRNA, resulting in anti-fibrotic, anti-inflammatory, regenerative, proliferative, immunomodulatory and angiogenic effects [11–14]. Neovascularization, angiogenesis, cardiomyocytes apoptosis inhibition, myocardial repair enhancement and dead cardiomyocytes replacement are the major targets of MSC therapy within the context of myocardial infarction [2]. MSCs are present in different human organs, but usually isolated from the following three main sources: umbilical cord, adipose tissue and bone marrow [2]. The latter is commonly used, despite the fact that it provides a mixture of non-purified miscellaneous cells [15]. After injection, MSCs are able to home, accumulate and engraft with the adjacent cellular components of the injured tissue and, subsequently, recruit additional progenitor cells [15,16]. However, hypoxia and increased free radical concentration in the context of myocardial infarction generate a detrimental microenvironment for transplanted MSCs [17]. Thus, preconditioning of MSCs with hypoxia or pharmacological or chemical agents in addition to novel strategies, such as exosome-mediated MSCs, genetically modified MSCs and engineered cardiac patches, were performed for improving the overall efficacy of MSC transplantation (Figure 1). All these techniques promote MSC survival and their capacity to form a regenerative and proliferative environment. Herein, we review the different preconditioning methods and novel approaches with MSCs in the setting of ischemic cardiac disease.

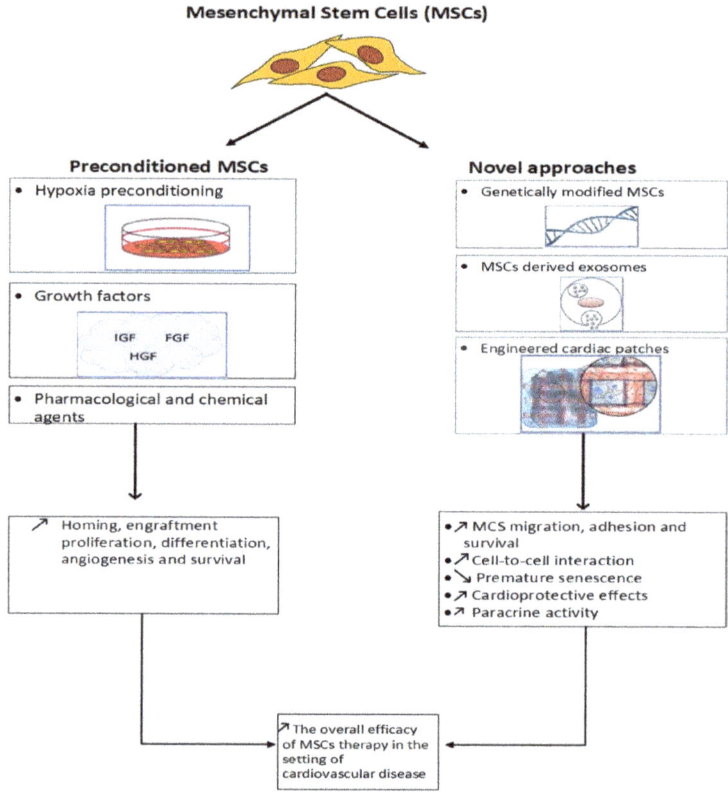

Figure 1. Figure illustrating mesenchymal stem cell (MSC) preconditioning methods, novel approaches and their main impacts.

2. Preconditioning Methods

2.1. Hypoxia-Preconditioned MSCs

The purpose of hypoxic preconditioning is to prolong the short survival time of grafted MSCs in the ischemic area, a major limitation of the therapeutic potential of stem cell therapy [18–20]. Indeed, hypoxic preconditioning increases the expression of protective factors against future hypoxic insult (hypoxia inducible factor-1 α (HIF-1α)), angiogenic factors (vascular epithelial growth factor, angiopoietin-1 and erythropoietin), pro-survival proteins (P65, P50 and P105) and anti-apoptotic proteins (Bcl-xl et Bcl-2) [21]. Previous study results showed that 24 h hypoxia exposure could dramatically amplify MSC proliferation and reduce their apoptosis by mainly activating the HIF-1α/Apelin/APJ axis [22]. First, HIF-1α modulates oxygen homeostasis and promotes cell function and tolerance in a hypoxic microenvironment [23]. It plays a crucial role in cardiomyocytes protection against ischemia-reperfusion injury by regulating mitochondrial reactive oxygen species [24] and heme oxgenase-1 [25]. Then, the inhibition of inflammatory reaction and apoptosis, up-regulation of collagen matrix and glycolysis, stimulation of angiogenesis and improvement of oxygen delivery are mediated by HIF-1α [26]. Second, the stimulation of Apelin/APJ enhances MSC survival and differentiation [27]. Moreover, hypoxia preconditioning activates other pathways, such as SDF-1α/CXCR4 axis implicated in MSC migration, detention and homing [28,29], PI3K/Akt signaling pathway that blocks cell death [30] and GRP78 that interferes in angiogenic cytokine secretion and cell migration [31]. A recent study revealed that extracellular vesicles from hypoxia-preconditioned MSCs may partly alleviate myocardial injury by targeting the thioredoxin-interacting protein-mediated HIF-1α pathway [32]. The evidence suggests that transplantation of hypoxia-preconditioned MSCs in the setting of myocardial infarction results in better cardiovascular outcomes by enhancing MSC engraftment, proliferation, differentiation, survival and paracrine activity [33,34]. Furthermore, it has shown that hypoxic preconditioning enhances survival and proangiogenic capacity of human first trimester chorionic villus-derived MSCs for fetal tissue engineering [35]. Lastly, we spotlight that different percentages of hypoxia have different outcomes. For example, 1% hypoxia extends MSC lifespan and maintains their proliferation rate [36,37]. In addition, 2% and 5 % hypoxia increased MSC number and viability [34]. Upregulation of stemness-related genes was observed with 3% hypoxia [38,39]. In other words, severe hypoxia (<1%) activates glycolytic metabolism and induces MSC quiescence, whereas moderate hypoxia (3–5%) stimulates MSC proliferation [40–42]. Although, short duration exposure to hypoxia (24 h) yields a better result than that of longer duration (72 h).

2.2. Preconditioning with Pharmacological and Chemical Agents

Numerous growth factors, drugs and pharmacological and chemical substances have been used for MSC preconditioning (Table 1). For example, the treatment of MSCs with IGF-1 showed a positive impact on survival, detrimental infarct consequences (infarct size, ventricular remodeling and fibrosis) and pro-inflammatory cytokines [43]. HGF promotes MSC differentiation into cardiomyocytes, whereas the effect of IGF-1 on MSC potential of differentiation remains uncertain [44,45]. On the other hand, pretreatment with bFGF improves stem cells' homing ability to the infarct zone and angiogenesis [46]. The pretreatment of MSCs with growth factor combinations (FGF-2, IGF-1 and BMP-2) leads to stronger engraftment, better viability in hypoxic situations, enhanced cell to cell communication and greater cytoprotective effects [47]. The results of a recent study showed superior cardiac function recovery and vasculogenesis in the infarcted myocardium 6 weeks after an injection of treated MSCs with SDF-1α in a rat model [48]. Beyond growth factors, variant biological active substances have been tested to improve the therapeutic efficacy of MSC therapy. Indeed, MSC pretreatment with angiotensin II potentiates the paracrine activity, angiogenesis, gap junction formation and global clinical outcome, by up-regulating the expression of VEGF, Cx43 with no effects on the differentiation mechanisms [49]. In addition, the left ventricular cardiac function and cardiomyogenic transdifferentiation have been significantly improved after transplantation of pioglitazone pretreated MSCs [50]. Thus, it seems a promising preconditioning method to predict cardiomyogenesis. Furthermore, pretreatment of MSCs with atorvastatin significantly improved cardiac function, reduced infarct size, decreased serum marker level of inflammation and fibrosis, inhibited apoptosis and enhanced survival of implanted MSCs, via activating the subtype eNOS of nitric oxide synthase [51]. Atorvastatin also improved the migration capacity of MSCs by increasing the expression of CXCR4 [52]. Benefits on MSC survival and differentiation have been observed with simvastatin pretreated MSCs [53]. Statin pretreatment positive outcomes have been also observed after transplantation of sevoflurane-preconditioned MSCs, which increase the expression of HIF-1α, HIF-2α, VEGF and p/Akt/Akt [54]. Transplantation of LPS-(lipopolysaccharide) preconditioned MSCs in the setting of myocardial infarction improves their biological and functional characteristics by up-regulating VEGF, phosphorylated Akt and TLR4 pathway [55]. Thereby, longer survival of transplanted cells, intense neovascularization and greater amelioration of left ventricular ejection fraction have been reported [55]. Vitamine E decreases oxidative stress and H_2O_2-related senescence by up-regulating the expression of VEGF, TGF-β and LDH [56]. The proliferation ability of MSCs has been promoted with astragaloside IV by inhibiting the translocation of NF-*k*Bp65 [57], apple ethanol extract by inducing the phosphorylation of eIF4E, p44, p70S6K, MAPK, eIF48, p44/42, mTOR and S6RP [58], oxytocin by activating the Akt/ERK1/2 axis [59], LL-37 by activating the MAPK pathway [60] and migration inhibitory factor by releasing VEGF, BFGF, HGF and IGF [61]. Although, the migration and homing abilities of MSCs have been improved with deferoxamine by expressing HIF-1α, CXCR4, CCR2, MMP-2 and MMP-9 [62], IL-1β by producing different cytokines, chemokines and adhesions molecules [63] and TGF-β1 by triggering the canonical SMADs [64]. In addition, the improvement of cardiovascular stem cell therapeutic outcomes has been associated with transplantation of 2,4-dinitrophenol [65], oxytocin [66] and dimethyloxalyglycine [67] pretreated MSCs. Finally, our group has shown that melatonin (pineal hormone to protect tissue from oxidative damage) pretreated MSCs modulate survival, differentiation and antifibrotic activity of cardiac fibroblasts [68]. Our results showed that MSCs significantly improved morphological and functional cardiac parameters two weeks after injection. However, the partial recovery of ventricular ejection fraction was maintained up to two months only when MSC survival was increased by melatonin treatment. These data indicate that the increased number of viable cells is critical for the amplification of the beneficial effects of MSCs on injured myocardium and ventricular function recovery. These properties of MSCs opened new perspective for understanding the mechanisms of action of MSCs and anticipated their potential therapeutic effects.

Table 1. MSC preconditioning with pharmacological and chemical agents.

Agents	Effects on	References
IGF-1	survival, infarct consequences, pro-inflammatory cytokines	[43]
HGF	differentiation into cardiomyocytes	[44,45]
bFGF	stem cells homing and angiogenesis	[46]
FGF-2, IGF-1 and BMP-2 combination	engraftment, viability, cell to cell communication, cytoprotective effect	[47]
SDF-1α	cardiac function recovery and vasculogenesis	[48]
Angiotensin II	paracrine activity, angiogenesis and gap junction formation	[49]
Pioglitazone	cardiac function and cardiomyogenic trans differentiation	[50]
Atorvastatin	cardiac function, infarct size, serum markers level of inflammation and fibrosis, apoptosis, migration capacity and survival of implanted MSCs	[51,52]
Simvastatin	MSC survival and differentiation	[53]
Sevoflurane	homing, survival and differentiation	[54]
LPS (lipopolysaccharide)	biological and functional characteristics of MSCs	[55]
Vitamine E	decreases oxidative stress and H2O2-related senescence	[56]
Astragaloside	proliferation ability of MSCs	[57]
Apple ethanol	proliferation ability of MSCs	[58]
Oxytocin	proliferation ability of MSCs	[59]
LL-37	proliferation ability of MSCs	[60]
Deferoxamine	migration and homing abilities of MSC	[62]

Table 1. *Cont.*

Agents	Effects on	References
IL-1β	migration and homing abilities of MSCs	[63]
TGF-β1	migration and homing abilities of MSCs	[64]
2,4-dinitrophenol	cardiovascular stem cell therapeutic outcomes	[65]
Oxytocin	cardiovascular stem cell therapeutic outcomes	[66]
Dimethyloxalyglycine	cardiovascular stem cell therapeutic outcomes	[67]
Melatonin	survival, differentiation and antifibrotic activity	[68]

3. Novel Approaches

3.1. Genetic Modification of MSCs

Genetic modification of MSCs up-regulates the expression of specific genes implicated in MSC migration, adhesion, survival and premature senescence (Table 2). To begin, the migratory ability of MSCs has been promoted by overexpressing nuclear receptors (Nur1, Nur77) [69,70], integrin subunit-α4 [71], aquaporin-1 [72] and CXCR4/CXCR7 that serve as receptors for major cellular migratory process chemokine (SDF-1) [73,74]. Then, the overexpression of α(1,3)fucosyltransferase [75], focal adhesion kinase [76], integrin-linked kinase [77] and miR-9-5p [78] have been linked to stronger MSC adhesion and engraftment. However prolonged survival of transplanted MSCs has been demonstrated with overexpression of integrin-linked kinase that activates AKT, mTOR, JAK2/STAT3 signaling pathways [79,80], protein kinase Cε [81], Trkβ [82] and Gremlin1 [83]. The up-regulation of Sox2 and Oct4 genes accelerates cell transition from phase G1 into phase S, enhancing MSC proliferation, differentiation and anti-inflammatory effect [84,85]. EphB2 overexpression reduced premature senescence by suppressing mitochondrial reactive oxygen species accumulation, which triggers MSC senescence [86]. Transplantation of Kallikrein-1 genetically modified MSCs attenuates cardiac inflammation, cardiomyocytes apoptosis and myocardial fibrosis via VEGF, GSK-3β and NO signaling pathways activation [87–91]. Thus, pleiotropic, angiogenic proteolytic and cardioprotective effects have been attributed to Kallikrein-1 [91]. In the context of acute myocardial infarction, several clinical trials have demonstrated the therapeutic benefits of transplantation of genetically modified MSCs in animal models. For example, the target outcomes of MSC therapy were maintained for longer durations with transplanted Akt or angiopoietin1-MSCs [92]. Although, an injection of Bcl-2 or SDF-1α-or TNFR gene modified MSCs or miR-377 depleted MSCs potentiates the required efficacy of vascular density, cardiac function, infarct size and myocardial fibrosis [93–98]. Genetic modification of MSCs is applied using viral vectors, such as adenoviral, lentiviral and retroviral vectors for nucleic acid delivery [99], non-viral delivery systems, such as plasmid DNA, polymers, nanoplasmids, liposomes and DNA minicircles [100–102] and the novel gene-editing technology, clustered regularly interspaced short palindromic repeats (CRISPR/Cas9) [103]. This last technique allows one to insert a new sequence in the genome via homology-directed repair, which could rectify an acquired gene mutation or provoke a knock-in or knock-out mutation or suppress a specific gene expression [103].

Table 2. Outcomes of genetic modifications of MSCs.

Function	Up-Regulating Genes	References
Improved MSC migration	Nur1, Nur7	[69,70]
	Integrin subunit- α4	[71]
	Aquaporin-1	[72]
	CXCR4/VXCR7	[73,74]
Improved MSC adhesion and engraftment	α(1,3)fucosyltransferase	[75]
	Focal adhesin kinase	76]
	Integrin-linked kinase	[77]
	miR-9-5-p	[78]
Prolonged MSC survival	Integrin-linked kinase	[79,80]
	Protein kinase Cε	[81]
	Trkβ	[82]
	Gremlin 1	[83]
Enhanced MSC proliferation and differentiation	Sox2 and Oct4	[84,85]
Reduced premature senescence	EphB2	[86]
Sustained therapeutic efficacy	AktAngiopoietin 1	[92]
Better outcomes in setting of acute myocardial infarction	Bcl-2	[93]
	SDF-1α	[95]
	TNFR	[97]
	miR-377	[98]

3.2. MSCs Derived-Exosomes

Exosomes are classified as extracellular vesicles that are continuously produced and released by various hematopoietic and non-hematopoietic cells [104–106]. Exosomes interfere in variant cell to cell interaction pathways that are implicated in different physiological and pathological patterns [107]. Endocytosis, membrane fusion and membrane receptors represent the three exosomal mechanisms to regulate cell to cell communication [108]. Exosomes are mainly isolated for therapeutic application, either by ultrafiltration or ultracentrifugation-based methods [107]. Preclinical experimental animal models have demonstrated the therapeutic benefits of MSC-derived exosomes in the setting of myocardial infarction. An injection of MSC- derived miRNA-enriched exosomes have showed remarkable outcomes, such as reduction in infarct size and myocardial fibrosis with miR-22 via acting on MECP2 [109], enhancement of anti-apoptotic and cardioprotective effects with miR-221 by inhibiting PUMA expression [110], promotion of cardiac function recovery with miR-19a by suppressing PTEN and activating ERK pathways, respectively [111], and improvement in angiogenesis with miR-210 [112]. Overall, the transplantation of exosomes-derived MSCs leads to stronger cardioprotective effects [113] and reduction in the risk of tumorigenicity [114] than MSC-based therapies.

3.3. Engineered Cardiac Patches

Cell sheets and cell containing scaffolds represent the two forms of engineered cardiac patches [115]. Multiple cell types, such as endothelial cells, cardiac fibroblasts, pluripotent stem cells, cardiomyocytes, progenitor cells and smooth muscle cells have been incorporated into engineered cardiac patches [116–119]. Consequently, the replacement of damaged cardiomyocytes with functional cardiac cells is the ultimate target of engineered cardiac tissue transplantation. Promising results with evidence of remuscularization of the fibrotic myocardium have been demonstrated in numerous pre-clinical studies [120–125]. Indeed, cardiac function recovery has been observed in rats and minipigs after 4 weeks of transplantation of cell-free patches in the setting of acute anterior myocardial infarction [126]. Furthermore, the implantation of a bioengineered cardiac patch has shown superior therapeutic efficacy compared to that of decellularized placenta and human-induced pluripotent stem cells for myocardial repair, mediated by growth and pro-angiogenic factors that promote engraftment, neovascularization and paracrine function [127]. However, the need for a huge quantity of exogenous cardiac cells to refill the injured myocardium and stable electromechanical coupling between the transplanted cardiac patches and host tissue for long-term engraftment are the main challenges for this novel cardiac approach [128]. Thereby, larger and thicker vascularized cardiac patches that are synchronized with the circulatory and electromechanical systems of the native myocardium are required to overcome these limitations. The safety concern of cardiac patch therapy was limited to arrhythmias, which were generally transient and non-fatal [129–131]. It is noteworthy that a recently published study has revealed the efficacy of upscaled engineered heart tissue to improve left ventricular function and reduce the infarct size in the context of ischemic myocardial disease without documenting a significant difference in arrhythmogenicity, compared to a cell-free patch group in a rabbit model [132].

Overall, the therapeutic benefits of MSCs have been demonstrated in the treatment of ischemic cardiomyopathies [133]; however, the limited engraftment and poor survival of MSCs injected into an ischemic heart hindered the efficacy of the treatment. The use of scaffolds and polymeric supports to provide transplanted cells anchorage, a straightforward approach to circumvent this limitation, has already been tested [134]. Indeed, a robust therapeutic benefit of ADSCs when transplanted with a collagen scaffold in a preclinical porcine model of myocardial infarction, compared with cells without a collagen scaffold, has been successfully demonstrated. The functional improvement in cardiac function and myocardial remodeling after ADSC-collagen scaffold transplantation was associated with increased cell engraftment [135]. The positive preclinical results obtained using different biomaterials and cell types invited researchers to test whether these experimental procedures could be translated into the clinical setting. Thus, the phase I MAGNUM clinical trial was designed with the purpose of comparing the effects exerted by bone-marrow mononucleated cells-seeded cellularized collagen matrices with those exerted by cells alone, in patients presenting left ventricular post-ischemic myocardial scars. The results were promising because no treatment-related serious adverse events were reported during the follow-up period and heart functionality and mechanical parameters improved significantly in patients who received the cellularized patches. In other words, clinically, this procedure seems to be safe, feasible, and effective [136]. We mention that one of the first clinical trials on engineered heart muscle in patients with terminal heart failure is ongoing, BioVAT-HF (ClinicalTrials.gov: NCT04396899). However, a recent report of in-human transplantation of an allogenic-induced pluripotent stem cell-derived cardiomyocytes patch into the epicardium of the anterior and lateral walls via the fourth intercostal space in a patient with ischemic cardiomyopathy has been currently published [137]. This report signals the safety and efficacy of these patches on NYHA class, left ventricular end systolic volume and Vo2 peak at the 1-year follow-up after transplantation [137]. Moreover, the ESCORT trial on six patients referred to cardiac surgery has also demonstrated the technical feasibility of producing clinical-grade human embryonic stem cell-derived cardiovascular progenitors delivered in a fibrin epicardial patch, and supported their short- and medium-term safety, thereby, setting the grounds for adequately powered efficacy studies [138]. Finally, the translation of preclinical findings to the first clinical results requires the creation of cardiac scaffolds following all the GMP regulatory and quality requirements in order to test their safety as potential therapeutic products. The CARDIOPATCH Interreg Sudoe program aims to create a 2.0 version patch (v2.0) with growth factors and genetically improved mesenchymal cells and iPS-derived cardiac cells that improve cell survival of both the implanted cells and the ischemic cardiac tissue, as well as their pro-angiogenic capacity.

4. MSCs Perspectives

As is known for most new therapies, the progression of MSC therapy has been hard, slow and punctuated by difficulties. The available evidence proves the safety of MSC transplantation, which represents a new, hopeful strategy for the management of cardiovascular disease, particularly ischemic and non-ischemic heart failure [139–141]. Up to date, numerous Phase I and Phase II trials have demonstrated promising results with regenerative medicine in the setting of heart failure and myocardial infarction [2]. The findings from these trials are divergent. However, several important points have not yet been defined, such as the preferred cell source, preparation method, appropriate dose and recommended manner of administration. Defining these parameters constitutes an important step towards establishing a standard approach with MSC therapy and ensuring result reproducibility. The results from pivotal phase III trials are required to support the clinical application of MSC therapy in the cardiovascular field. Recently, stem cell therapy was approved for the management of complex perianal fistulas in Crohn's disease [142]. We emphasize that pre-conditioning methods have contributed to overcome numerous hurdles, such as injected cell migration, engraftment, proliferation, differentiation and survival, resulting in stronger efficacy and better outcomes. Furthermore, recent studies have proved the benefits of mechanical stimulation on MSCs and the surrounding microenvironment and showed the interest of its application for bone regeneration therapy [143]. Lastly, engineered cardiac patch technology represents a revolution in stem cell therapy for cardiovascular disease, but manufacturing larger and thicker constructs that are suitably vascularized and incorporated with the electromechanical and circulatory systems of the vernacular myocardium is necessary for the clinical translation step.

5. Conclusions

To conclude, transplantation of pre-conditioned MSCs results in better therapeutic efficacy in the setting of cardiovascular disease, especially with moderate hypoxia preconditioning. In parallel, the available novel techniques are able to overcome the limitations (MSCs homing ability, engraftment and survival) of this regenerative medicine, promoting stronger cardiovascular outcomes. Starting translational engineered cardiac patch practice from pre-clinical trials in animal models to in-human trials may change our future management of heart failure.

Funding: This research was partly supported by CARDIOPATCH (SOE4/P1/E1063) project co-funded by the Interreg Sudoe program.

Institutional Review Board Statement: Not applicable.

Informed Consent Statement: Not applicable.

Data Availability Statement: Not applicable.

Conflicts of Interest: The authors declare no conflict of interest.

References

1. Guijarro, D.; Lebrin, M.; Lairez, O.; Bourin, P.; Piriou, N.; Pozzo, J.; Lande, G.; Berry, M.; Le Tourneau, T.; Cussac, D.; et al. Intramyocardial transplantation of mesenchymal stromal cells for chronic myocardial ischemia and impaired left ventricular function: Results of the MESAMI 1 pilot trial. *Int. J. Cardiol.* **2016**, *209*, 258–265. [CrossRef] [PubMed]
2. Razeghian-Jahromi, I.; Matta, A.G.; Canitrot, R.; Zibaeenezhad, M.J.; Razmkhah, M.; Safari, A.; Nader, V.; Roncalli, J. Surfing the clinical trials of mesenchymal stem cell therapy in ischemic cardiomyopathy. *Stem Cell Res.* **2021**, *12*, 361. [CrossRef] [PubMed]
3. Awada, H.K.; Hwang, M.P.; Wang, Y. Towards comprehensive cardiac repair and regeneration after myocardial infarction: Aspects to consider and proteins to deliver. *Biomaterials* **2016**, *82*, 94–112. [CrossRef] [PubMed]
4. Hare, J.M.; Fishman, J.E.; Gerstenblith, G.; DiFede Velazquez, D.L.; Zambrano, J.P.; Suncion, V.Y.; Tracy, M.; Ghersin, E.; Johnston, P.V.; Brinker, J.A.; et al. Comparison of allogeneic vs autologous bone marrow–derived mesenchymal stem cells delivered by transendocardial injection in patients with ischemic cardiomyopathy: The POSEIDON randomized trial. *JAMA* **2012**, *308*, 2369–2379. [CrossRef]

5. Butler, J.; Hamo, C.E.; Udelson, J.E.; O'Connor, C.; Sabbah, H.N.; Metra, M.; Shah, S.J.; Kitzman, D.W.; Teerlink, J.R.; Bernstein, H.S.; et al. Reassessing Phase II Heart Failure Clinical Trials: Consensus Recommendations. *Circ. Heart Fail.* **2017**, *10*, e003800. [CrossRef]
6. Pittenger, M.F. Multilineage potential of adult human mesenchymal stem cells. *Science* **1999**, *284*, 143–147. [CrossRef]
7. Chamberlain, G.; Fox, J.; Ashton, B.; Middleton, J. Concise review: Mesenchymal stem cells: Their phenotype, differentiation capacity, immunological features, and potential for homing. *Stem Cells* **2009**, *25*, 2739–2749. [CrossRef]
8. Caplan, A.I. Adult mesenchymal stem cells for tissue engineering versus regenerative medicine. *J. Cell Physiol.* **2007**, *213*, 341–347. [CrossRef]
9. Wei, X.; Yang, X.; Han, Z.P.; Qu, F.F.; Shao, L.; Shi, Y.F. Mesenchymal stem cells: A new trend for cell therapy. *Acta Pharm. Sin.* **2013**, *34*, 747–754. [CrossRef]
10. Castro-Manrreza, M.E.; Montesions, J.J. Immunoregulation by msenchymal stem cells: Biological aspects and clinical applications. *J. Imminol. Res.* **2015**, *2015*, 394917.
11. Kuraitis, D.; Giordano, C.; Ruel, M.; Musaro, A.; Suuronen, E.J. Exploiting extracellular matrix-stem cell interactions: A review of natural materials for therapeutic muscle regeneration. *Biomaterials* **2012**, *33*, 428–443. [CrossRef] [PubMed]
12. Khubutiya, M.S.; Vagabov, A.V.; Temmov, A.A.; Sklifas, A.N. Paracrine mechanisms of proliferative, anti-apoptotic and anti-inflammatory effects of mesenchymal stromal cells in models of acute organ injury. *Cryotherapy* **2014**, *16*, 579–585. [CrossRef] [PubMed]
13. Gnecchi, M.; Zhang, Z.; Ni, A.; Dzau, V.J. Paracrine mechanisms in adult stem cell signaling and therapy. *Circ. Res.* **2008**, *103*, 1204–1219. [CrossRef] [PubMed]
14. Williams, A.R.; Hare, J.M. Mesenchymal stem cells: Biology, pathophysiology, translational findings, and therapeutic implications for cardiac disease. *Circ. Res.* **2011**, *109*, 923–940. [CrossRef]
15. Quevedo, H.C.; Hatzistergos, K.E.; Oskouei, B.N.; Feigenbaum, G.S.; Rodriguez, J.E.; Valdes, D.; Pattany, P.M.; Zambrano, J.P.; Hu, Q.; McNiece, I.; et al. Allogeneic mesenchymal stem cells restore cardiac function in chronic ischemic cardiomyopathy via trilineage differentiating capacity. *Proc. Natl. Acad. Sci. USA* **2009**, *106*, 14022–14027. [CrossRef]
16. Barbash, I.M.; Chouraqui, P.; Baron, J.; Feinberg, M.S.; Etzion, S.; Tessone, A.; Miller, L.; Guetta, E.; Zipori, D.; Kedes, L.H.; et al. Systemic delivery of bone marrow-derived mesenchymal stem cells to the infarcted myocardium: Feasibility, cell migration, and body distribution. *Circulation* **2003**, *108*, 863–868. [CrossRef]
17. Brodarac, A.; Šarić, T.; Oberwallner, B.; Mahmoodzadeh, S.; Neef, K.; Albrecht, J. Susceptibility of murine induced pluripotent stem cell-derived cardiomyocytes to hypoxia and nutrient deprivation. *Stem Cell Res.* **2015**, *6*, 83. [CrossRef]
18. Zhang, M.; Methot, D.; Poppa, V.; Fujio, Y.; Walsh, K.; Murry, C.E. Cardiomyocyte grafting for cardiac repair: Graft cell death and anti-death strategies. *J. Mol. Cell Cardiol.* **2001**, *33*, 907–921. [CrossRef]
19. Reinecke, H.; Murry, C.E. Cell grafting for cardiac repair. *Methods Mol. Biol.* **2003**, *219*, 97–112.
20. Toma, C.; Pittenger, M.F.; Cahill, K.S.; Byrne, B.J.; Kessler, P.D. Human mesenchymal stem cells differentiate to a cardiomyocyte phenotype in the adult murine heart. *Circulation* **2002**, *105*, 93–98. [CrossRef]
21. Hu, X.; Yu, S.P.; Fraser, J.L.; Lu, Z.; Ogle, M.E.; Wang, J.A.; Wei, L. Transplantation of hypoxia-preconditioned of mesenchymal stem cells improves infarcted heart function via enhanced survival of implanted cells and angiogenis. *J. Thorac. Cardiovasc. Surg.* **2008**, *135*, 799–808. [CrossRef]
22. Hou, J.; Wang, L.; Long, H.; Wu, H.; Wu, Q.; Zhong, T.; Chen, X.; Zhou, C.; Guo, T.; Wang, T. Hypoxia preconditioning promotes cardiac stem cell survival and cardiogenic differentiation in vitro involving activation of the HIF-1α/Apelin/APJ axis. *Stem Cell Res.* **2017**, *8*, 215. [CrossRef] [PubMed]
23. Chen, J.; Kang, J.G.; Keyvanfar, K.; Youn, N.S.; Hwang, P.M. Long-term adaptation to hypoxia preserves hematopoietic stem cell function. *Exp. Hematol.* **2016**, *44*, 866–873. [CrossRef] [PubMed]
24. Cai, Z.; Zhong, H.; Bosch-Marce, M.; Fox-Talbot, K.; Wang, L.; Wei, C.; Trush, M.A.; Semenza, G.L. Complete loss of ischaemic preconditiong-induced cardioprotection in mice with partial deficiency of HIF-1 alpha. *Cardiovasc. Res.* **2008**, *77*, 463–470. [CrossRef] [PubMed]
25. Ockaili, R.; Natarajan, R.; Salloum, F.; Fisher, B.J.; Jones, D.; Fowler III, A.A.; Kukreja, R.C. HIF-1 activation attenuates postischemic myocardial injury: Role for heme oxygenase-1 in modulating microvascular chemokine generation. *Am. J. Physiol. Heart Circ. Physiol.* **2005**, *289*, H542–H548. [CrossRef]
26. Loor, G.; Schumacker, P.T. Role of hypoxia-inducible factor in cell survival during myocardial ischemia-reperfusion. *Cell Death Differ.* **2008**, *15*, 686–690. [CrossRef] [PubMed]
27. Zhang, N.K.; Cao, Y.; Zhu, Z.M.; Zheng, N.; Wang, L.; Xu, X.H.; Gao, L.R. Activation of endogenous cardiac stem cells by apelin-13 in infarcted rat heart. *Cell Transpl.* **2016**, *2*, 1645–1652. [CrossRef]
28. Tang, Y.L.; Zhu, W.; Cheng, M.; Chen, L.; Zhang, J.; Sun, T.; Kishore, R.; Phillips, M.I.; Losordo, D.W.; Qin, G. Hypoxic preconditioning enhances the benefit of cardiac progenitor cell therapy for treatment of myocardial infarction by inducing CXCR4 expression. *Circ. Res.* **2009**, *104*, 1209–1216. [CrossRef]
29. Kucia, M.; Reca, R.; Miekus, K.; Wanzeck, J.; Wojakowski, W.; Janowska-Wieczorek, A.; Ratajczak, J.; Ratajczak, M.Z. Trafficking of normal stem cells and metastasis of cancer stem cells involve similar mechanisms: Pivotal role of the SDF-1-CXCR4 axis. *Stem Cells* **2005**, *23*, 879–894. [CrossRef]

30. Alvarez-Tejado, M.; Naranjo-Suarez, S.; Jiménez, C.; Carrera, A.C.; Landázuri, M.O.; del Peso, L. Hypoxia induces the activation of the phosphatidylinositol 3-kinase/Akt cell survival pathway in PC12 cells: Protective role in apoptosis. *J. Biol. Chem.* **2001**, *276*, 22368–22374. [CrossRef]
31. Lee, J.H.; Yoon, Y.M.; Lee, S.H. Hypoxic preconditioning promotes the bioactivities of mesenchymal stem cells via the HIF-1α-GRP78-Akt axis. *Int. J. Mol. Sci.* **2017**, *18*, 1320. [CrossRef] [PubMed]
32. Mao, C.Y.; Zhang, T.T.; Li, D.J.; Zhou, E.; Fan, Y.Q.; He, Q.; Wang, C.Q.; Zhang, J.F. Extracellular vesicles from hypoxia-preconditioned mesenchymal stem cells alleviates myocardial injury by targeting thioredoxin-interacting protein-mediated hypoxia-inducible factor-1α pathway. *World J. Stem Cells* **2022**, *14*, 183–199. [CrossRef] [PubMed]
33. Raziyeva, K.; Smagulova, A.; Kim, Y.; Smagul, S.; Nurkesh, A.; Saparov, A. Preconditioned and genetically modified stem cells for myocardial infarction treatment. *Int. J. Mol. Sci.* **2020**, *21*, 7301. [CrossRef] [PubMed]
34. Bahir, B.; Choudhery, M.S.; Hussain, I. Hypoxic preconditioning as a strategy to maintain the regenerative potential of mesenchymal stem cells. In *Regenerative Medicine*; Choudhery, M., Ed.; IntechOpen: London, UK, 2020.
35. Hao, D.; Chuanchao, H.; Ma, B.; Lankford, L.; Reynaga, L.; Farmer, D.L.; Guo, F.; Wang, A. Hypoxic preconditioning enhances survival and proangiogenic capacity of human first trimester chorionic villus-derived mesenchymal stem cells for fetal tissue engineering. *Stem Cells Int.* **2019**, *2019*, 9695239. [CrossRef]
36. Lee, C.W.; Kang, D.; Kim, A.K.; Kim, D.Y.; Kim, D.I. Improvement of cell cycle lifespan and genetic damage susceptibility of human mesenchymal stem cells by hypoxic priming. *Int. J. Stem Cells* **2018**, *11*, 61. [CrossRef]
37. Suryawan, I.G.R.; Pikir, B.S.; Rantam, F.A.; Ratri, A.K.; Nugraha, R.A. Hypoxic preconditioning promotes survival of human adipocyte mesenchymal stem cell via expression of prosurvival and proangiogenic biomarkers. *F1000Research* **2021**, *10*, 843. [CrossRef]
38. Safwani, W.K.; Choi, J.R.; Yong, K.W.; Ting, I.; Adenan, N.A.; Pingguan-Murphy, B. Hypoxia enhances the viability, growth and chondrogenic potential of cryopreserved human adipose-derived stem cells. *Cryobiology* **2017**, *75*, 91–99. [CrossRef]
39. Werle, S.B.; Chagastelles, P.; Pranke, P.; Casagrande, L. Hypoxia upregulates the expression of the pluripotency markers in the stem cells from human deciduous teeth. *Clin. Oral Investig.* **2019**, *23*, 199–207. [CrossRef]
40. Obradovic, H.; Krstic, J.; Trivanovic, D.; Mojsilovic, S.; Okic, I.; Kukolj, T.; Ilic, V.; Jaukovic, A.; Terzic, M.; Bugarski, D. Improving Stemness and Functional Features of Mesenchymal Stem Cells from Wharton's Jelly of a Human Umbilical Cord by Mimicking the Native, Low Oxygen Stem Cell Niche. *Placenta* **2019**, *82*, 25–34. [CrossRef]
41. Park, S.E.; Kim, H.; Kwon, S.; Choi, S.; Oh, S.; Ryu, G.H.; Jeon, H.B.; Chang, J.W. Pressure Stimuli Improve the Proliferation of Wharton's Jelly-Derived Mesenchymal Stem Cells under Hypoxic Culture Conditions. *Int. J. Mol. Sci.* **2020**, *21*, 7092. [CrossRef]
42. Moniz, I.; Ramalho-Santos, J.; Branco, A.F. Differential oxygen exposure modulates mesenchymal stem cell metabolism and proliferation through mTOR signaling. *Int. J. Mol. Sci.* **2022**, *23*, 3749. [CrossRef] [PubMed]
43. Guo, J.; Zheng, D.; Li, W.F.; Li, H.R.; Zhang, A.D.; Li, Z.C. Insulin-like growth factor 1 treatment of MSCs attenuates inflammation and cardiac dysfunction following MI. *Inflammation* **2014**, *37*, 2156–2163. [CrossRef] [PubMed]
44. Zhang, G.W.; Gu, T.X.; Guan, X.Y.; Sun, X.J.; Qi, X.; Li, X.Y.; Wang, X.B.; Lv, F.; Yu, L.; Jiang, D.Q.; et al. HGF and IGF-1 promote protective effects of allogeneic BMSC transplantation in rabbit model of acute myocardial infarction. *Cell Prolif.* **2015**, *48*, 661–670. [CrossRef]
45. Farzaneh, M.; Rahimi, F.; Alishahi, M.; Khoshnam, S.E. Paracrine Mechanisms Involved in Mesenchymal Stem Cell Differentiation into Cardiomyocytes. *Curr. Stem Cell Res.* **2019**, *14*, 9–13. [CrossRef]
46. Ling, L.; Gu, S.; Cheng, Y.; Ding, L. bFGF promotes Sca-1+ cardiac stem cell migration through activation of the PI3K/Akt pathway. *Mol. Med. Rep.* **2018**, *17*, 2349–2356. [CrossRef] [PubMed]
47. Hahn, J.Y.; Cho, H.J.; Kang, H.J.; Kim, T.S.; Kim, M.H.; Chung, J.H.; Bae, J.W.; Oh, B.H.; Park, Y.B.; Kim, H.S. Pre-treatment of mesenchymal stem cells with a combination of growth factors enhances gap junction formation, cytoprotective effect on cardiomyocytes, and therapeutic efficacy for myocardial infarction. *J. Am. Coll Cardiol.* **2008**, *51*, 933–943. [CrossRef]
48. Esmaeili, R.; Darbandi-Azar, A.; Sadeghpour, A.; Majidzadeh, K.; Eini, L.; Jafarbeik-Iravani, N.; Hoseinpour, P.; Vajhi, A.; Oghabi Bakhshaiesh, T.; Masoudkabir, F.; et al. Mesenchymal stem cells pretreatment with stromal-derived factor-1 alpha augments cardiac function and angiogenesis in infarcted myocardium. *Am. J. Med. Sci.* **2021**, *361*, 765–775. [CrossRef]
49. Liu, C.; Fan, Y.; Zhou, L.; Zhu, H.Y.; Song, Y.C.; Hu, L.; Wang, Y.; Li, Q.P. Pretreatment of mesenchymal stem cells with angiotensin II enhances paracrine effects, angiogenesis, gap junction formation and therapeutic efficacy for myocardial infarction. *Int. J. Cardiol.* **2015**, *188*, 22–32. [CrossRef]
50. Shinmura, D.; Togashi, I.; Miyoshi, S.; Nishiyama, N.; Hida, N.; Tsuji, H.; Tsuruta, H.; Segawa, K.; Tsukada, Y.; Ogawa, S.; et al. Pretreatment of human mesenchymal stem cells with pioglitazone improved efficiency of cardiomyogenic transdifferentiation and cardiac function. *Stem Cells* **2011**, *29*, 357–366. [CrossRef]
51. Song, L.; Yang, Y.J.; Dong, Q.T.; Qian, H.Y.; Gao, R.L.; Qiao, S.B.; Shen, R.; He, Z.X.; Lu, M.J.; Zhao, S.H.; et al. Atorvastatin enhance efficacy of mesenchymal stem cells treatment for swine myocardial infarction via activation of nitric oxide synthase. *PLoS ONE* **2013**, *8*, e65702. [CrossRef]
52. Li, N.; Yang, Y.J.; Qian, H.Y.; Li, Q.; Zhang, Q.; Li, X.D.; Dong, Q.T.; Xu, H.; Song, L.; Zhang, H. Intravenous administration of atorvastatin-pretreated mesenchymal stem cells improves cardiac performance after acute myocardial infarction: Role of CXCR4. *Am. J. Transl. Res.* **2015**, *7*, 1058–1070. [PubMed]

53. Yang, Y.J.; Qian, H.Y.; Huang, J.; Li, J.J.; Gao, R.L.; Dou, K.F.; Yang, G.S.; Willerson, J.T.; Geng, Y.J. Combined therapy with simvastatin and bone-marrow derived mesenchymal stem cells increases benefits in infarcted swine hearts. *Arter. Thromb Vasc. Biol.* **2009**, *29*, 2076–2082. [CrossRef] [PubMed]
54. Sun, X.; Fang, B.; Zhao, X.; Zhang, G.; Ma, H. Preconditioning of mesenchymal stem cells by sevoflurane to improve their therapeutic potential. *PLoS ONE* **2014**, *9*, e90667.
55. Yao, Y.; Zhang, F.; Wang, L.; Zhang, G.; Wang, Z.; Chen, J.; Gao, X. Lipopolysaccharide preconditioning enhances the efficacy of mesenchymal stem cells transplantation in a rat model of acute myocardial infarction. *J. Biomed. Sci.* **2009**, *16*, 74. [CrossRef]
56. Bhatti, F.U.; Mehmood, A.; Latief, N.; Zahra, S.; Cho, H.; Khan, S.N.; Riazuddin, S. Vitamin E protects rat mesenchymal stem cells against hydrogen peroxide-induced oxidative stress *in vitro* and improves their therapeutic potential in surgically-induced rat model of osteoarthritis. *Osteoarthr. Cartil.* **2017**, *25*, 321–331. [CrossRef]
57. Li, M.; Yu, L.; She, T.; Gan, Y.; Liu, F.; Hu, Z.; Chen, Y.; Li, S.; Xia, H.; Xia, H. Astragaloside IV attenuates Toll-like receptor 4 expression via NF-kappaB pathway under high glucose condition in mesenchymal stem cells. *Eur. J. Pharm.* **2012**, *696*, 203–209. [CrossRef]
58. Lee, J.; Shin, M.S.; Kim, M.O.; Jang, S.; Oh, S.W.; Kang, M.; Jung, K.; Park, Y.S.; Lee, J. Apple ethanol extract promotes proliferation of human adult stem cells, which involves the regenerative potential of stem cells. *Nutr. Res.* **2016**, *36*, 925–936. [CrossRef]
59. Noiseux, N.; Borie, M.; Desnoyers, A.; Menaouar, A.; Stevens, L.M.; Mansour, S.; Danalache, B.A.; Roy, D.C.; Jankowski, M.; Gutkowska, J. Preconditioning of stem cells by oxytocin to improve their therapeutic potential. *Endocrinology* **2012**, *153*, 5361–5372. [CrossRef]
60. Yang, Y.; Choi, H.; Seon, M.; Cho, D.; Bang, S.I. LL-37 stimulates the functions of adipose-derived stromal/stem cells via early growth response 1 and the MAPK pathway. *Stem Cell Res.* **2016**, *7*, 58. [CrossRef]
61. Xia, W.; Zhang, F.; Xie, C.; Jiang, M.; Hou, M. Macrophage migration inhibitory factor confers resistance to senescence through CD74-dependent AMPK-FOXO3a signaling in mesenchymal stem cells. *Stem Cell Res.* **2015**, *6*, 82. [CrossRef]
62. Najafi, R.; Sharifi, A.M. Deferoxamine preconditioning potentiates mesenchymal stem cell homing *in vitro* and in streptozotocin-diabetic rats. *Expert Opin. Biol.* **2013**, *13*, 959–972. [CrossRef] [PubMed]
63. Carrero, R.; Cerrada, I.; Lledo, E.; Dopazo, J.; García-García, F.; Rubio, M.P.; Trigueros, C.; Dorronsoro, A.; Ruiz-Sauri, A.; Montero, J.A.; et al. IL1beta induces mesenchymal stem cells migration and leucocyte chemotaxis through NF-kappaB. *Stem Cell Rev.* **2012**, *8*, 905–916. [CrossRef] [PubMed]
64. Dubon, M.J.; Yu, J.; Choi, S.; Park, K.S. Transforming growth factor beta induces bone marrow mesenchymal stem cell migration via noncanonical signals and N-cadherin. *J. Cell Physiol.* **2017**, *233*, 201–213. [CrossRef] [PubMed]
65. Khan, I.; Ali, A.; Akhter, M.A.; Naeem, N.; Chotani, M.A.; Mustafa, T.; Salim, A. Preconditioning of mesenchymal stem cells with 2,4-dinitrophenol improves cardiac function in infarcted rats. *Life Sci.* **2016**, *162*, 60–69. [CrossRef]
66. Kim, Y.S.; Kwon, J.S.; Hong, M.H.; Kang, W.S.; Jeong, H.Y.; Kang, H.J.; Jeong, M.h.; Ahn, Y. Restoration of angiogenic capacity of diabetes-insulted mesenchymal stem cells by oxytocin. *BMC Cell Biol.* **2013**, *14*, 38. [CrossRef]
67. Liu, X.B.; Wang, J.A.; Ji, X.Y.; Yu, S.P.; Wei, L. Preconditioning of bone marrow mesenchymal stem cells by prolyl hydroxylase inhibition enhances cell survival and angiogenesis *in vitro* and after transplantation into the ischemic heart of rats. *Stem Cell Res.* **2014**, *5*, 111. [CrossRef]
68. Mias, C.; Lairez, O.; Trouche, E.; Roncalli, J.; Calise, D.; Seguelas, M.H.; Ordener, C.; Piercecchi-Marti, M.D.; Auge, N.; Salvayre, A.N.; et al. Mesenchymal stem cells promote matrix metalloproteinase secretion by cardiac fibroblasts and reduce cardiac ventricular fibrosis after myocardial infarction. *Stem Cells* **2009**, *27*, 2734–2743. [CrossRef]
69. Cornelissen, A.S.; Maijenburg, M.W.; Nolte, M.A.; Voermans, C. Organ-specific migration of mesenchymal stromal cells: Who, when, where and why? *Immunol. Lett.* **2015**, *168*, 159–169. [CrossRef]
70. Maijenburg, M.W.; Gilissen, C.; Melief, S.M.; Kleijer, M.; Weijer, K.; Ten Brinke, A.; Roelofs, H.; Van Tiel, C.M.; Veltman, J.A.; de Vries, C.J.; et al. Nuclear receptors Nur77 and Nurr1 modulate mesenchymal stromal cell migration. *Stem Cells Dev.* **2012**, *21*, 228–238. [CrossRef]
71. Andrzejewska, A.; Nowakowski, A.; Janowski, M.; Koniusz, S.; Walczak, P.; Grygorowicz, T.; Jablonska, A.; Lukomska, B. In vitro and in vivo functional studies on itga4 overexpressing human bone marrow mesenchymal stem cells. *J Regen Med.* **2015**, *4*, 4.
72. Pelagalli, A.; Nardelli, A.; Lucarelli, E.; Zannetti, A.; Brunetti, A. Autocrine signals increase ovine mesenchymal stem cells migration through aquaporin-1 and CXCR4 overexpression. *J. Cell Physiol.* **2018**, *233*, 6241–6249. [CrossRef] [PubMed]
73. De Becker, A.; Van Riet, I. Homing and migration of mesenchymal stromal cells: How to improve the efficacy of cell therapy? *World J. Stem Cells* **2016**, *8*, 73–87. [CrossRef] [PubMed]
74. Ullah, M.; Liu, D.D.; Thakor, A.S. Mesenchymal stromal cell homing: Mechanisms and strategies for improvement. *iScience* **2019**, *15*, 421–438. [CrossRef] [PubMed]
75. Lo, C.Y.; Weil, B.R.; Palka, B.A.; Momeni, A.; Canty, J.M.; Neelamegham, S. Cell surface glycoengineering improves selectin-mediated adhesion of mesenchymal stem cells (MSCs) and cardiosphere-derived cells (CDCs): Pilot validation in porcine ischemia-reperfusion model. *Biomaterials* **2016**, *74*, 19–30. [CrossRef] [PubMed]
76. Wang, H.; Wang, X.; Qu, J.; Yue, Q.; Hu, Y.; Zhang, H. VEGF enhances the migration of MSCs in neural differentiation by regulating focal adhesion turnover. *J. Cell Physiol.* **2015**, *230*, 2728–2742. [CrossRef] [PubMed]

77. Song, S.W.; Chang, W.; Song, B.W.; Song, H.; Lim, S.; Kim, H.J.; Cha, M.J.; Choi, E.; Im, S.H.; Chang, B.C.; et al. Integrin-linked kinase is required in hypoxic mesenchymal stem cells for strengthening cell adhesion to ischemic myocardium. *Stem Cells* **2009**, *27*, 1358–1365. [CrossRef]
78. Li, X.; He, L.; Yue, Q.; Lu, J.; Kang, N.; Xu, X.; Wang, H.; Zhang, H. MiR-9-5p promotes MSC migration by activating β-catenin signaling pathway. *Am. J. Physiol. Cell Physiol.* **2017**, *313*, C80–C93. [CrossRef]
79. Zeng, B.; Liu, L.; Wang, S.; Dai, Z. ILK regulates MSCs survival and angiogenesis partially through AKT and mTOR signaling pathways. *Acta HistoChem.* **2017**, *119*, 400–406. [CrossRef]
80. Mao, Q.; Liang, X.L.; Wu, Y.F.; Pang, Y.H.; Zhao, X.J.; Lu, Y.X. ILK promotes survival and self-renewal of hypoxic MSCs via the activation of lncTCF7-Wnt pathway induced by IL-6/STAT3 signaling. *Gene* **2019**, *26*, 165–176. [CrossRef]
81. He, H.; Zhao, Z.H.; Han, F.S.; Liu, X.H.; Wang, R.; Zeng, Y.J. Overexpression of protein kinase C ε improves retention and survival of transplanted mesenchymal stem cells in rat acute myocardial infarction. *Cell Death Dis.* **2016**, *7*, e2056. [CrossRef]
82. Heo, H.; Yoo, M.; Han, D.; Cho, Y.; Joung, I.; Kwon, Y.K. Upregulation of TrkB by forskolin facilitated survival of MSC and functional recovery of memory deficient model rats. *BioChem. Biophys. Res. Commun.* **2013**, *431*, 796–801. [CrossRef] [PubMed]
83. Xiang, Q.; Hong, D.; Liao, Y.; Cao, Y.; Liu, M.; Pang, J.; Zhou, J.; Wang, G.; Yang, R.; Wang, M.; et al. Overexpression of gremlin1 in mesenchymal stem cells improves hindlimb ischemia in mice by enhancing cell survival. *J. Cell Physiol.* **2017**, *232*, 996–1007. [CrossRef] [PubMed]
84. Han, S.M.; Han, S.H.; Coh, Y.R.; Jang, G.; Chan Ra, J.; Kang, S.K.; Lee, H.W.; Youn, H.Y. Enhanced proliferation and differentiation of Oct4- and Sox2-overexpressing human adipose tissue mesenchymal stem cells. *Exp. Mol. Med.* **2014**, *46*, e101. [CrossRef]
85. Li, Q.; Han, S.M.; Song, W.J.; Park, S.C.; Ryu, M.O.; Youn, H.Y. Anti-inflammatory effects of Oct4/Sox2-overexpressing human adipose tissue-derived mesenchymal stem cells. *Vivo* **2017**, *31*, 349–356. [CrossRef] [PubMed]
86. Jung, Y.H.; Lee, H.J.; Kim, J.S.; Lee, S.J.; Han, H.J. EphB2 signaling-mediated Sirt3 expression reduces MSC senescence by maintaining mitochondrial ROS homeostasis. *Free Radic. Biol. Med.* **2017**, *110*, 368–380. [CrossRef]
87. Agata, J.; Chao, L.; Chao, J. Kallikrein gene delivery improves cardiac reserve and attenuates remodeling after myocardial infarction. *Hypertension* **2002**, *40*, 653–659. [CrossRef]
88. Yao, Y.Y.; Yin, H.; Shen, B.; Smith, R.S.; Liu, Y.; Gao, L.; Chao, L.; Chao, J. Tissue kallikrein promotes neovascularization and improves cardiac function by the Akt-glycogen synthase kinase-3beta pathway. *Cardiovasc. Res.* **2008**, *80*, 354–364. [CrossRef]
89. Bledsoe, G.; Chao, L.; Chao, J. Kallikrein gene delivery attenuates cardiac remodeling and promotes neovascularization in spontaneously hypertensive rats. *Am. J. Physiol. Heart Circ. Physiol.* **2003**, *285*, H1479–H1488. [CrossRef]
90. Yao, Y.Y.; Yin, H.; Shen, B.; Chao, L.; Chao, J. Tissue kallikrein infusion prevents cardiomyocyte apoptosis, inflammation and ventricular remodeling after myocardial infarction. *Regul. Pept.* **2007**, *140*, 12–20. [CrossRef]
91. Devetzi, M.; Goulielmaki, M.; Khoury, N.; Spandidos, D.A.; Sotiropoulo, G.; Christodoulo, I.; Zoumpourlis, V. Genetically-modified stem cells in treatment of human diseases: Tissue kallikrein (KLK1)-based targeted therapy. *Int. J. Mol. Med.* **2018**, *41*, 1177–1186. [CrossRef]
92. Shujia, J.; Haider, H.K.; Idris, N.M.; Lu, G.; Ashraf, M. Stable therapeutic effects of mesenchymal stem cell-based multiple gene delivery for cardiac repair. *Cardiovasc. Res.* **2008**, *77*, 525–533. [CrossRef] [PubMed]
93. Li, W.; Ma, N.; Ong, L.L.; Nesselmann, C.; Klopsch, C.; Ladilov, Y.; Furlani, D.; Piechaczek, C.; Moebius, J.M.; Lützow, K.; et al. Bcl-2 engineered MSCs inhibited apoptosis and improved heart function. *Stem Cells* **2007**, *25*, 2118–2127. [CrossRef] [PubMed]
94. Grunewald, M.; Avraham, I.; Dor, Y.; Bachar-Lustig, E.; Itin, A.; Jung, S.; Chimenti, S.; Landsman, L.; Abramovitch, R.; Keshet, E. VEGF-induced adult neovascularization: Recruitment, retention, and role of accessory cells. *Cell* **2006**, *124*, 175–189. [CrossRef] [PubMed]
95. Tang, J.; Wang, J.; Yang, J.; Kong, X.; Zheng, F.; Guo, L.; Zhang, L.; Huang, Y. Mesenchymal stem cells over-expressing SDF-1 promote angiogenesis and improve heart function in experimental myocardial infarction in rats. *Eur. J. CardioThorac. Surg.* **2009**, *36*, 644–650. [CrossRef] [PubMed]
96. Tang, J.; Wang, J.; Zheng, F.; Kong, X.; Guo, L.; Yang, J.; Zhang, L.; Huang, Y. Combination of chemokine and angiogenic factor genes and mesenchymal stem cells could enhance angiogenesis and improve cardiac function after acute myocardial infarction in rats. *Mol. Cell BioChem.* **2010**, *339*, 107–118. [CrossRef] [PubMed]
97. Bao, C.; Guo, J.; Lin, G.; Hu, M.; Hu, Z. TNFR gene-modified mesenchymal stem cells attenuate inflammation and cardiac dysfunction following MI. *Scand. Cardiovasc. J.* **2008**, *42*, 56–62. [CrossRef]
98. Wen, Z.; Huang, W.; Feng, Y.; Cai, W.; Wang, Y.; Wang, X.; Liang, J.; Wani, M.; Chen, J.; Zhu, P.; et al. MicroRNA-377 regulates mesenchymal stem cell-induced angiogenesis in ischemic hearts by targeting VEGF. *PLoS ONE* **2014**, *9*, e104666.
99. Lundstrom, K. Viral vectors in gene therapy. *Diseases* **2018**, *6*, 42. [CrossRef]
100. Yin, H.; Kanasty, R.L.; Eltoukhy, A.A.; Vegas, A.J.; Dorkin, J.R.; Anderson, D.G. Non-viral vectors for gene-based therapy. *Nat. Rev. Genet.* **2014**, *15*, 541–555. [CrossRef]
101. Hardee, C.L.; Arévalo-Soliz, L.M.; Hornstein, B.D.; Zechiedrich, L. Advances in non-viral DNA vectors for gene therapy. *Genes* **2017**, *8*, 65. [CrossRef]
102. Suschak, J.J.; Williams, J.A.; Schmaljohn, C.S. Advancements in DNA vaccine vectors, non-mechanical delivery methods, and molecular adjuvants to increase immunogenicity. *Hum. Vaccin. Immunother.* **2017**, *13*, 2837–2848. [CrossRef] [PubMed]
103. Zhang, J.H.; Adikaram, P.; Pandey, M.; Genis, A.; Simonds, W.F. Optimization of genome editing through CRISPR-Cas9 engineering. *Bioengineered* **2016**, *7*, 166–174. [CrossRef] [PubMed]

104. Trams, E.G.; Lauter, C.J.; Salem, N.; Heine, U. Exfoliation of membrane ecto-enzymes in the form of micro-vesicles. *Biochim. Biophys. Acta* **1981**, *645*, 63–70. [CrossRef]
105. Potolicchio, I.; Carven, G.J.; Xu, X.; Stipp, C.; Riese, R.J.; Stern, L.J.; Santambrogio, L. Proteomic analysis of microglia-derived exosomes: Metabolic role of the aminopeptidase CD13 in neuropeptide catabolism. *J. Immunol.* **2005**, *175*, 2237–2243. [CrossRef]
106. Février, B.; Raposo, G. Exosomes: Endosomal-derived vesicles shipping extracellular messages. *Curr. Opin. Cell Biol.* **2004**, *16*, 415–421. [CrossRef]
107. Nikfarjam, S.; Rezaie, J.; Zolbanine, N.M.; Jafari, R. Mesenchymal stem cell derived-exosomes: A modern approach in translational medicine. *J. Transl. Med.* **2020**, *18*, 449. [CrossRef]
108. Loyer, X.; Vion, A.C.; Tedgui, A.; Boulanger, M. Microvesicles as cell-cell messengers in cardiovascular diseases. *Circ. Res.* **2014**, *114*, 345–353. [CrossRef]
109. Feng, Y.; Huang, W.; Wani, M.; Yu, X.; Ashraf, M. Ischemic preconditioning potentiates the protective effect of stem cells through secretion of exosomes by targeting Mecp2 via miR-22. *PLoS ONE* **2014**, *9*, e88685. [CrossRef]
110. Yu, B.; Gong, M.; Wang, Y.; Millard, R.W.; Pasha, Z.; Yang, Y.; Ashraf, M.; Xu, M. Cardiomyocyte protection by GATA-4 gene engineered mesenchymal stem cells is partially mediated by translocation of miR-221 in microvesicles. *PLoS ONE* **2013**, *8*, e73304. [CrossRef]
111. Yu, B.; Kim, H.W.; Gong, M.; Wang, J.; Millard, R.W.; Wang, Y.; Ashraf, M.; Xu, M. Exosomes secreted from GATA-4 overexpressing mesenchymal stem cells serve as a reservoir of anti-apoptotic microRNAs for cardioprotection. *Int. J. Cardiol.* **2015**, *182*, 349–360. [CrossRef]
112. Wang, N.; Chen, C.; Yang, D.; Liao, Q.; Luo, H.; Wang, X.; Zhou, F.; Yang, X.; Yang, J.; Zeng, C.; et al. Mesenchymal stem cells-derived extracellular vesicles, via miR-210, improve infarcted cardiac function by promotion of angiogenesis. *Biochim. Biophys. Acta Mol. Basis Dis.* **2017**, *1863*, 2085–2092. [CrossRef] [PubMed]
113. De Abreu, R.C.; Fernandes, H.; Da Costa Martins, P.A.; Sahoo, S.; Emanueli, C.; Ferreira, L. Native and bioengineered extracellular vesicles for cardiovascular therapeutics. *Nat. Rev. Cardiol.* **2020**, *17*, 685–697. [CrossRef] [PubMed]
114. Barkholt, L.; Flory, E.; Jekerle, V.; Lucas-Samuel, S.; Ahnert, P.; Bisset, L.; Büscher, D.; Fibbe, W.; Foussat, A.; Kwa, M.; et al. Risk of tumorigenicity in mesenchymal stromal-cell based therapies-bridging scientific observations and regulatory viewpoints. *Cytotherapy* **2013**, *15*, 753–759. [CrossRef] [PubMed]
115. Zhang, J. Engineered tissue patch for cardiac cell therapy. *Curr. Treat. Options Cardiovasc.* **2015**, *17*, 399. [CrossRef]
116. Zhang, J.; Zhu, W.; Radisic, M.; Vunjak-Novakovic, G. Can we engineer a human cardiac patch for therapy? *Circ. Res.* **2018**, *123*, 244–265. [CrossRef]
117. Zhou, P.; Pu, W.T. Recounting cardiac cellular composition. *Circ. Res.* **2016**, *118*, 368–370. [CrossRef]
118. Fleischer, S.; Vunjak-Novakovic, G. Cardiac tissue engineering: From repairing to modeling the human heart. In *Encyclopedia of Tissue Engineering and Regenerative Medicine*, 1st ed.; Academic Press: Oxford, UK, 2019; pp. 131–144.
119. Chong, J.J.; Yang, X.; Don, C.W.; Minami, E.; Liu, Y.W.; Weyers, J.J.; Mahoney, W.M.; Van Biber, B.; Cook, S.M.; Palpant, N.J.; et al. Human embryonic-stem-cell-derived cardiomyocytes regenerate non-human primate hearts. *Nature* **2014**, *510*, 273–277. [CrossRef]
120. Ye, L.; Chang, Y.H.; Xiong, Q.; Zhang, P.; Zhang, L.; Somasundaram, P.; Lepley, M.; Swingen, C.; Su, L.; Wendel, J.S.; et al. Cardiac repair in a porcine model of acute myocardial infarction with human induced pluripotent stem cell-derived cardiovascular cells. *Cell Stem Cell* **2014**, *15*, 750–761. [CrossRef]
121. Weinberger, F.; Breckwoldt, K.; Pecha, S.; Kelly, A.; Geertz, B.; Starbatty, J.; Yorgan, T.; Cheng, K.H.; Lessmann, K.; Stolen, T.; et al. Cardiac repair in guinea pigs with human engineered heart tissue from induced pluripotent stem cells. *Sci. Transl. Med.* **2016**, *8*, 363. [CrossRef]
122. Shiba, Y.; Fernandes, S.; Zhu, W.Z.; Filice, D.; Muskheli, V.; Kim, J.; Palpant, N.J.; Gantz, J.; Moyes, K.W.; Reinecke, H.; et al. Human ES-cell-derived cardiomyocytes electrically couple and suppress arrhythmias in injured hearts. *Nature* **2012**, *489*, 322–325. [CrossRef]
123. Wei, K.; Serpooshan, V.; Hurtado, C.; Diez-Cunado, M.; Zhao, M.; Maruyama, S.; Zhu, W.; Fajardo, G.; Noseda, M.; Nakamura, K.; et al. Epicardial FSTL1 reconstitution regenerates the adult mammalian heart. *Nature* **2015**, *525*, 479–485. [CrossRef] [PubMed]
124. Serpooshan, V.; Zhao, M.; Metzler, S.A.; Wei, K.; Shah, P.B.; Wang, A.; Mahmoudi, M.; Malkovskiy, A.V.; Rajadas, J.; Butte, M.J.; et al. The effect of bioengineered acellular collagen patch on cardiac remodeling and ventricular function post myocardial infarction. *Biomaterials* **2013**, *34*, 9048–9055. [CrossRef] [PubMed]
125. Masumoto, H.; Nakane, T.; Tinney, J.P.; Yuan, F.; Ye, F.; Kowalski, W.J.; Minakata, K.; Sakata, R.; Yamashita, J.K.; Keller, B.B. The myocardial regenerative potential of three-dimensional engineered cardiac tissues composed of multiple human iPS cell-derived cardiovascular cell lineages. *Sci. Rep.* **2016**, *6*, 29933. [CrossRef] [PubMed]
126. Wang, L.; Liu, Y.; Ye, G.; He, Y.; Li, B.; Guan, Y.; Gong, B.; Mequanint, K.; Xing, M.M.Q.; Qiu, X. Injectable and conductive cardiac patches repair infarcted myocardium in rats and minipigs. *Nat. Biomed. Eng.* **2021**, *5*, 1157–1173. [CrossRef]
127. Jiang, Y.; Sun, S.J.; Zhen, Z.; Wei, R.; Zhang, N.; Liao, S.Y.; Tse, H.F. Myocardial repair of bioengineered cardiac patches with decellularized placental scaffold and human-induced pluripotent stem cells in a rat model of myocardial infarction. *Stem Cell Res.* **2021**, *12*, 13. [CrossRef]
128. Wang, L.; Serpooshan, V.; Zhang, J. Engineering human cardiac muscle patch constructs for prevention of post-infarction LV remodeling. *Front. Cardiovasc. Med.* **2021**, *8*, 621781. [CrossRef]

129. Liu, Y.W.; Chen, B.; Yang, X.; Fugate, J.A.; Kalucki, F.A.; Futakuchi-Tsuchida, A.; Couture, L.; Vogel, K.W.; Astley, C.A.; Baldessari, A.; et al. Human embryonic stem cell-derived cardiomyocytes restore function in infarcted hearts of non-human primates. *Nat. Biotechnol.* **2018**, *36*, 597–605. [CrossRef]
130. Shiba, Y.; Gomibuchi, T.; Seto, T.; Wada, Y.; Ichimura, H.; Tanaka, Y.; Ogasawara, T.; Okada, K.; Shiba, N.; Sakamoto, K.; et al. Allogeneic transplantation of iPS cell-derived cardiomyocytes regenerates primate hearts. *Nature* **2016**, *538*, 388–391. [CrossRef]
131. Romagnuolo, R.; Masoudpour, H.; Porta-Sánchez, A.; Qiang, B.; Barry, J.; Laskary, A.; Qi, X.; Massé, S.; Magtibay, K.; Kawajiri, H.; et al. Human embryonic stem cell-derived cardiomyocytes regenerate the infarcted pig heart but induce ventricular tachyarrhythmias. *Stem Cell Rep.* **2019**, *12*, 967–981. [CrossRef]
132. Jabbour, R.J.; Owen, T.J.; Pandey, P.; Reinsch, M.; Wang, B.; King, O.; Couch, L.S.; Pantou, D.; Pitcher, D.S.; Chowdhury, R.A.; et al. In vivo grafting of large engineered heart tissue patches for cardiac repair. *JCI Insight* **2021**, *6*, e144068. [CrossRef]
133. Guo, Y.; Yu, Y.; Hu, S.; Chen, Y.; Shen, Z. The therapeutic potential of mesenchymal stem cells for cardiovascular diseases. *Cell Death Dis.* **2020**, *11*, 349. [CrossRef] [PubMed]
134. Perez-Estenaga, I.; Prosper, F.; Pelacho, B. Allogeneic Mesenchymal Stem Cells and Biomaterials: The Perfect Match for Cardiac Repair? *Int. J. Mol. Sci.* **2018**, *19*, 3236. [CrossRef] [PubMed]
135. Araña, M.; Gavira, J.J.; Peña, E.; Gonzalez, A.; Abizanda, G.; Cilla, M.; Pérez, M.M.; Albiasu, E.; Aguado, N.; Casado, M.; et al. Epicardial delivery of collagen patches with adipose-derived stem cells in rat and minipig models of chronic myocardial infarction. *Biomaterials* **2014**, *35*, 143–151. [CrossRef] [PubMed]
136. Chachques, J.C.; Trainini, J.C.; Lago, N.; Cortes-Morichetti, M.; Schussler, O.; Carpentier, A. Myocardial assistance by grafting a new bioartificial upgraded myocardium (MAGNUM trial): Clinical feasibility study. *Ann. Thorac. Sur.* **2008**, *85*, 901–908. [CrossRef]
137. Miyagawa, S.; Kainuma, S.; Kawamura, T.; Suzuki, K.; Ito, Y.; Iseoka, H.; Ito, E.; Takeda, M.; Sasai, M.; Mochizuki-Oda, N.; et al. Transplantation of IPSC-derived cardiomyocyte patches for ischemic cardiomyopathy. *BMJ* **2022**, *2021*, 12. [CrossRef]
138. Menasché, P.; Vanneaux, V.; Hagège, A.; Bel, A.; Cholley, B.; Parouchev, A.; Cacciapuoti, I.; Al-Daccak, R.; Benhamouda, N.; Blons, H.; et al. Transplantation of Human Embryonic Stem Cell-Derived Cardiovascular Progenitors for Severe Ischemic Left Ventricular Dysfunction. *J. Am. Coll Cardiol.* **2018**, *71*, 429–438. [CrossRef]
139. Bolli, R.; Solankhi, M.; Tang, X.L.; Kahlon, A. Cell therapy in patients with heart failure: A comprehensive review and emerging concepts. *Cardiovasc. Res.* **2022**, *118*, 951–976. [CrossRef]
140. Arjmand, B.; Abedi, M.; Arabi, M.; Alavi-Moghadam, S.; Rezaei-Tavirani, M.; Hadavandkhani, M.; Tayanloo-Beik, A.; Kordi, R.; Roudsari, P.P.; Larijani, B. Regenerative medicine for the treatment of ischemic heart disease; status and future perspectives. *Front. Cell Dev. Biol.* **2021**, *9*, 704903. [CrossRef]
141. Chang, D.; Fan, T.; Gao, S.; Jin, Y.; Zhang, M.; Ono, M. Application of mesenchymal stem cell sheet to treatment of ischemic heart disease. *Stem Cell Res.* **2021**, *12*, 384. [CrossRef]
142. Panès, J.; Garcia-Olmo, D.; Assche, G.V.; Colombel, J.F.; Reinisch, W.; Baumgart, D.C.; Dignass, A.; Nachury, M.; Ferrante, M.; Kazemi-Shirazi, L.; et al. ADMIRE CD Study Group Collaborators. Long-term efficacy and safety of stem cell therapy (Cx601) for complex perianal fistulas in patients with Crohn's disease. *Gastroenterology* **2018**, *154*, 1334–1342. [CrossRef]
143. Sun, Y.; Wan, B.; Wang, R.; Zhang, B.; Luo, P.; Wang, D.; Nie, J.J.; Chen, D.; Wu, X. Mechanical stimulation on mesenchymal stem cells and surrounding microenvironments in bone regeneration: Regulations and applications. *Front. Cell Dev. Biol.* **2022**, *10*, 808303. [CrossRef] [PubMed]